Student Study Art Notebook

to accompany

Human Anatomy

Kenneth S. Saladin

Georgia College and State University

Higher Education

Boston   Burr Ridge, IL   Dubuque, IA   Madison, WI   New York   San Francisco   St. Louis
Bangkok   Bogotá   Caracas   Kuala Lumpur   Lisbon   London   Madrid   Mexico City
Milan   Montreal   New Delhi   Santiago   Seoul   Singapore   Sydney   Taipei   Toronto

The McGraw·Hill Companies

Student Study Art Notebook to accompany
HUMAN ANATOMY
KENNETH S. SALADIN

Published by McGraw-Hill Higher Education, an imprint of The McGraw-Hill Companies, Inc.,
1221 Avenue of the Americas, New York, NY 10020. Copyright © 2005 by The McGraw-Hill
Companies, Inc. All rights reserved.

This book is printed on recycled, acid-free paper containing
10% postconsumer waste.

RECYCLED

2 3 4 5 6 7 8 9 0 QPD/QPD 0 9 8 7 6 5 4

ISBN 0-07-297809-0

www.mhhe.com

DIRECTORY OF NOTEBOOK FIGURES

TO ACCOMPANY
SALADIN, HUMAN ANATOMY, 1/E

Chapter 1

Variation in Human Anatomy Figure 1.7	1
The Body's Structural Hierarchy Figure 1.8	2
Early Microscope Figure 1.11	3
Primate Hands Figure 1.13	3
Primate Phylogeny Figure 1.15	4

Atlas A

Anatomical Position Figure A.1	5
Positions of the Forearm Figure A.2	6
Anatomical Planes of Reference Figure A.3	6
Views of the Body in the Three Primary Anatomical Planes Figure A.4	7
The Adult Male and Female Bodies Figure A.5	8
Four Quadrants and Nine Regions of the Abdomen Figure A.6	9
The Major Body Cavities Figure A.7	10
The Major Body Cavities Figure A.7 Unlabeled	10
Parietal and Visceral Layers of Double-Walled Membranes Figure A.8	11
Transverse Section Through the Abdominal Cavity Figure A.9	11
Transverse Section Through the Abdominal Cavity Figure A.9 Unlabeled	12
Serous Membranes of the Abdominal Cavity Figure A.10	12
Serous Membranes of the Abdominal Cavity Figure A.10 Unlabeled	13
The Human Organisms Figure A.11	14
Superficial Anatomy of the Trunk Figure A.12	17
Superficial Anatomy of the Trunk Figure A.12 Unlabeled	18
Anatomy at the Level of the Rib Cage and Greater Omentum (male) Figure A.13	19
Anatomy at the Level of the Rib Cage and Greater Omentum (male) Figure A.13 Unlabeled	20
Anatomy at the Level of the Lungs and Intestines (male) Figure A.14	21
Anatomy at the Level of the Lungs and Intestines (male) Figure A.14 Unlabeled	22
Anatomy at the Level of the Retroperitoneal Viscera (female) Figure A.15	23
Anatomy at the Level of the Retroperitoneal Viscera (female) Figure A.15 Unlabeled	24
Anatomy at the Level of the Dorsal Body Wall (female) Figure A.16	25
Anatomy at the Level of the Dorsal Body Wall (female) Figure A.16 Unlabeled	26
Frontal View of the Thoracic Cavity Figure A.18	27
Transverse Section of the Thoracic Cavity Figure A.19	27
Frontal View of the Abdominal Cavity Figure A.20	28
Transverse Section of the Abdominal Cavity Figure A.21	29
Sagittal Section of the Pelvic Cavity Figure A.22	30

Chapter 2

Common Cell Shapes Figure 2.3	31
The Relationship Between Cell Surface Area and Volume Figure 2.4	32
Structure of a Generalized Cell Figure 2.5	33
Structure of a Generalized Cell Figure 2.5 Unlabeled	34
The Plasma Membrane Figure 2.6	35
Phospholipid Structure and Symbol Figure 2.7	36
Transmembrane Proteins Figure 2.8	37
Some Functions of Plasma Membrane Proteins Figure 2.9	37
Some Modes of Membrane Transport Figure 2.10	38
Modes of Vesicular Transport Figure 2.11	39
The Structure of Cilia Figure 2.13	40
The Structure of Cilia Figure 2.13 Unlabeled	41
Types of Intercellular Junctions Figure 2.14	42
Types of Intercellular Junctions Figure 2.14 Unlabeled	42
The Cytoskeleton Figure 2.15	43
The Cytoskeleton Figure 2.15 Unlabeled	43
Microtubules Figure 2.16	44
Major Organelles Figure 2.17	45

The Functional Relationship of Ribosomes
Figure 2.18   46
The Cell Cycle   Figure 2.19   46
Mitosis   Figure 2.20   47
Chromosomes   Figure 2.21   48

Chapter 3
Three-Dimensional Interpretation of
Two-Dimensional Images   Figure 3.1   48
Longitudinal, Cross, and Oblique Sections
Figure 3.2   49
Comparison of Simple and Stratified
Epithelia   Figure 3.3   49
External Surface (serosa) of the Small
Intestine   Figure 3.4   50
Kidney Tubules   Figure 3.5   51
Internal Surface (mucosa) of the Small
Intestine   Figure 3.6   52
Mucosa of the Trachea   Figure 3.7   53
Skin from the Sole of the Foot   Figure 3.9   54
Mucosa of the Vagina   Figure 3.10   54
Wall of a Follicle in the Ovary   Figure 3.11   55
Allantoic Duct of the Umbilical Cord
Figure 3.12   56
Tendons of the Hand   Figure 3.13   57
Spread of the Mesentery   Figure 3.14   57
Lymph Node   Figure 3.15   58
Adipose Tissue   Figure 3.16   59
Tendon   Figure 3.17   60
Dermis of the Skin   Figure 3.18   61
Fetal Skeleton   Figure 3.19   62
External Ear   Figure 3.20   62
Intervertebral Disc   Figure 3.21   63
Compact Bone   Figure 3.22   63
Blood Smear   Figure 3.23   64
Spinal Cord Smear   Figure 3.24   64
Skeletal Muscle   Figure 3.25   65
Cardiac Muscle   Figure 3.26   65
Smooth Muscle, Wall of the Small
Intestine   Figure 3.27   66
General Structure of an Exocrine Gland
Figure 3.28   66
Some Types of Exocrine Glands
Figure 3.29   67
Histology of a Mucous Membrane
Figure 3.30   67
Histology of a Mucous Membrane
Figure 3.30   Unlabeled   68

Chapter 4
Fertilization   Figure 4.1   69
Fertilization   Figure 4.1   Unlabeled   69
Migration of the Conceptus   Figure 4.2   70
Migration of the Conceptus   Figure 4.2
Unlabeled   70
Implantation   Figure 4.3   71
Implantation   Figure 4.3   Unlabeled   72
The Implanted Conceptus at 16 Days
Figure 4.4   73

The Implanted Conceptus at 16 Days
Figure 4.4   Unlabeled   73
Gastrulation   Figure 4.5   74
Gastrulation   Figure 4.5   Unlabeled   74
Embryonic Folding   Figure 4.6   75
Neurulation   Figure 4.7   76
The Pharyngeal Pouches   Figure 4.8   76
Development of the Embryo from 37
to 56 Days   Figure 4.10   77
The Embryonic Membranes
Figure 4.11   78
The Embryonic Membranes
Figure 4.11   Unlabeled   78
The Phases of Intrauterine Nutrition
Figure 4.12   79
Development of the Placenta and Fetal
Membranes   Figure 4.13   80
Development of the Placenta and Fetal
Membranes   Figure 4.13   Unlabeled   81
Growth of the Fetus   Figure 4.16   82
Disjunction and Nondisjunction
Figure 4.17   83
Down Syndrome   Figure 4.18   84

Chapter 5
Structure of the Skin and Subcutaneous
Tissue   Figure 5.1   85
Structure of the Skin and Subcutaneous
Tissue   Figure 5.1   Unlabeled   85
The Epidermis   Figure 5.2   86
The Epidermis   Figure 5.2   Unlabeled   86
Importance of the Skin in Nonverbal
Expression   Figure 5.3   88
Layers of the Dermis   Figure 5.4   88
Distribution of Subcutaneous Fat in Men
and Women   Figure 5.5   89
Structure of a Hair and its Follicle
Figure 5.7   89
Structure of a Hair and Its Follicle
Figure 5.7   Unlabeled   90
Basis of Hair Color and Texture
Figure 5.8   90
Anatomy of a Fingernail   Figure 5.9   91
Cutaneous Glands   Figure 5.10   91
Prenatal Development of the Epidermis
and Dermis   Figure 5.11   92
Prenatal Development of a Hair Follicle
Figure 5.12   93
Prenatal Development of a Hair Follicle
Figure 5.12   Unlabeled   94
Three Degrees of Burns   Figure 5.14   95

Chapter 6
Classification of Bones by Shape
Figure 6.1   96
General Anatomy of Long and Flat Bones
Figure 6.2   96
General Anatomy of Long and Flat Bones
Figure 6.2   Unlabeled   97

Bone Cells and Their Development
Figure 6.3   98
The Histology of Osseous Tissue
Figure 6.4   99
The Histology of Osseous Tissue
Figure 6.4   Unlabeled   100
Spongy Bone Structure in Relation to
Mechanical Stress   Figure 6.5   101
Distribution of Red and Yellow Bone
Marrow   Figure 6.6   102
Stages of Endochondral Ossification
Figure 6.7   103
Stages of Endochondral Ossification
Figure 6.7   Unlabeled   103
Stages of Intramembranous Ossification
Figure 6.10   104
Achondroplastic Dwarfism   Figure 6.12   104
Some Types of Bone Fractures
Figure 6.13   105
The Healing of a Bone Fracture
Figure 6.15   105

Chapter 7
The Adult Skeleton   Figure 7.1   106
The Adult Skeleton   Figure 7.1
Unlabeled   107
Surface Features of Bones   Figure 7.2   108
Surface Features of Bones   Figure 7.2
Unlabeled   109
The Skull, Anterior View   Figure 7.3   110
The Skull, Anterior View   Figure 7.3
Unlabeled   110
The Skull   Figure 7.4   111
The Skull   Figure 7.4   Unlabeled   111
Base of the Skull   Figure 7.5   113
Base of the Skull   Figure 7.5   Unlabeled   114
The Calvaria (skull)   Figure 7.6   116
The Calvaria (skull)   Figure 7.6
Unlabeled   117
Major Cavities of the Skull   Figure 7.7   117
Major Cavities of the Skull
Figure 7.7   Unlabeled   118
The Paranasal Sinuses   Figure 7.8   118
Cranial Fossae   Figure 7.9   119
The Right Temporal Bone   Figure 7.10   119
The Right Temporal Bone
Figure 7.10   Unlabeled   120
The Sphenoid Bone   Figure 7.11   121
The Sphenoid Bone   Figure 7.11
Unlabeled   122
The Ethmoid Bone   Figure 7.12   122
The Right Nasal Cavity   Figure 7.13   123
The Right Nasal Cavity   Figure 7.13
Unlabeled   123
The Left Orbit   Figure 7.14   124
The Left Orbit   Figure 7.14   Unlabeled   124
The Mandible   Figure 7.15   125
The Mandible   Figure 7.15   Unlabeled   125
The Hyoid Bone   Figure 7.16   126

Adaptations of the Skull for Bipedalism
Figure 7.17   126
The Vertebral Column   Figure 7.18   127
Curvatures of the Adult Vertebral Column
Figure 7.19   128
Comparison of Chimpanzee and Human
Vertebral Columns   Figure 7.20   128
Abnormal Spinal Curvatures   Figure 7.21   129
A Representative Vertebra and
Intervertebral Disc   Figure 7.22   129
Articulated Vertebrae   Figure 7.23   130
The Atlas and Axis   Figure 7.24   130
Typical Cervical, Thoracic, and Lumbar
Vertebrae   Figure 7.25   131
The Sacrum and Coccyx   Figure 7.26   131
The Thoracic Cage and Pectoral Girdle
Figure 7.27   132
The Thoracic Cage and Pectoral Girdle
Figure 7.27   Unlabeled   132
Anatomy of the Ribs   Figure 7.28   133
Articulation of Rib 6 with Vertebrae T5
and T6   Figure 7.29   133
The Fetal Skull Near the Time of Birth
Figure 7.31   134
The Fetal Skull Near the Time of Birth
Figure 7.31   Unlabeled   135
Development of the Vertebrae and
Intervertebral Discs   Figure 7.32   136
Development of a Thoracic Vertebrae
Figure 7.33   137
Spinal Curvature of the Newborn Infant
Figure 7.34   137
Skull Fractures   Figure 7.35   138
Injuries to the Vertebral Column
Figure 7.36   139

Chapter 8
The Right Clavicle   Figure 8.1   139
The Right Scapula   Figure 8.2   140
The Right Scapula   Figure 8.2   Unlabeled   140
The Right Humerus   Figure 8.3   141
The Right Humerus   Figure 8.3
Unlabeled   141
The Right Radius and Ulna   Figure 8.4   142
The Right Radius and Ulna   Figure 8.4
Unlabeled   142
The Right Wrist and Hand   Figure 8.5   143
The Right Wrist and Hand   Figure 8.5
Unlabeled   143
The Pelvic Girdle   Figure 8.6   144
The Pelvic Girdle   Figure 8.6   Unlabeled   145
The Right Os Coxae   Figure 8.7   145
The Right Os Coxae   Figure 8.7
Unlabeled   146
Chimpanzee and Human Os Coxae
Figure 8.8   146
Comparison of the Male and Female
Pelvic Girdles   Figure 8.9   147
The Right Femur and Patella   Figure 8.10   147

Adaptation of the Lower Limb for
Bipedalism   Figure 8.11   148
Fractures of the Femur   Figure 8.12   149
The Right Tibia and Fibula   Figure 8.13   150
The Right Foot   Figure 8.14   150
The Right Foot   Figure 8.14   Unlabeled   151
Some Adaptations of the Foot for
Bipedalism   Figure 8.15   151
Arches of the Foot   Figure 8.16   152

Chapter 9
Types of Fibrous Joints   Figure 9.1   152
Types of Sutures   Figure 9.2   153
Cartilaginous Joints   Figure 9.3   154
Structure of a Simple Synovial Joint
Figure 9.4   154
Tendon Sheaths and Other Bursae in
the Hand and Wrist   Figure 9.5   155
Tendon Sheaths and Other Bursae in
the Hand and Wrist   Figure 9.5
Unlabeled   156
The Six Types of Synovial Joints
Figure 9.6   157
Joint Flexion and Extension   Figure 9.7   158
Joint Abduction and Adduction
Figure 9.8   158
Elevation and Depression   Figure 9.9   159
Some Horizontal Joint Movements
Figure 9.10   159
Circumduction and Rotation   Figure 9.11   160
Joint Movements of the Forearm and
Thumb   Figure 9.12   161
Joint Movement of the Foot   Figure 9.13   162
The Temporomandibular Joint (TMJ)
Figure 9.14   163
The Shoulder (humeroscapular) Joint
Figure 9.15   164
The Shoulder (humeroscapular) Joint
Figure 9.15   Unlabeled   164
The Elbow Joint   Figure 9.16   168
Pulled Elbow   Figure 9.17   169
The Coxal (hip) Joint   Figure 9.18   170
The Knee Joint   Figure 9.19   171
The Knee Joint   Figure 9.19   Unlabeled   172
Anterior Dissection of the Knee Joint
Figure 9.20   174
Knee Injuries   Figure 9.21   175
The Talocrural (ankle) Joint and Ligaments
of the Right Foot   Figure 9.22   176

Chapter 10
The Connective Tissues of a Skeletal
Muscle   Figure 10.2   177
Classification of Muscles According to
Fascicle Orientation   Figure 10.3   177
A Muscle Group Acting on the Elbow
Figure 10.4   178
Basic Components of a Lever   Figure 10.5   178
Mechanical Advantage (MA)   Figure 10.6   179

The Three Classes of Levers   Figure 10.7   180
Structure of a Skeletal Muscle Fiber
Figure 10.8   181
Structure of a Skeletal Muscle Fiber
Figure 10.8   Unlabeled   182
Molecular Structure of Thick and
Thin Filaments   Figure 10.9   183
Muscle Striations and Their Molecular
Basis   Figure 10.10   184
A Neuromuscular Junction   Figure 10.12   184
A Motor Unit   Figure 10.13   185
The Principal Events in Muscle Contraction
and Relaxation   Figure 10.15   186
Cardiac and Smooth Muscle   Figure 10.17   188
Smooth Muscle Types   Figure 10.18   189
Layers of Visceral Muscle in the Wall of
the Esophagus   Figure 10.19   189
Embryonic Development of Skeletal
Muscle Fibers   Figure 10.20   190

Chapter 11
The Muscular System   Figure 11.1   191
The Muscular System   Figure 11.1
Unlabeled   192
Some Muscles of Facial Expression in
the Cadaver   Figure 11.2   195
Muscles of Facial Expression
Figure 11.3   196
Muscles of Facial Expression
Figure 11.3   Unlabeled   197
Expressions Produced by Several of
the Facial Muscles   Figure 11.4   198
Muscles of the Tongue and Pharynx
Figure 11.5   199
Muscles of Chewing   Figure 11.6   200
Muscles of the Neck   Figure 11.7   201
Triangles of the Neck   Figure 11.8   202
Muscles of the Shoulder and
Nuchal Regions   Figure 11.9   202
Muscles of Respiration   Figure 11.10   203
Cross Section of the Anterior
Abdominal Wall   Figure 11.11   203
Thoracic and Abdominal Muscles
Figure 11.13   204
Thoracic and Abdominal Muscles
Figure 11.13   Unlabeled   204
Neck, Back, and Gluteal Muscles
Figure 11.14   206
Muscles Acting on the Vertebral Column
Figure 11.15   207
Some Deep Back Muscles of the Cadaver
Figure 11.16   208
Muscles of the Pelvic Floor   Figure 11.17   209

Chapter 12
Actions of Some Thoracic Muscles on
the Scapula   Figure 12.1   210
Pectoral and Brachial Muscles
Figure 12.2   211

Muscles of the Chest and Arm of the
Cadaver   Figure 12.3   212
The Rotator Cuff   Figure 12.4   213
Actions of the Rotator Muscles on the
Forearm   Figure 12.5   214
Muscles of the Forearm   Figure 12.6   215
Serial Cross Sections Through the
Upper Limb   Figure 12.7   216
The Carpal Tunnel   Figure 12.8   217
Intrinsic Muscles of the Hand   Figure 12.9   218
Intrinsic Muscles of the Hand
Figure 12.9   Unlabeled   218
Muscles That Act on the Hip and
Femur   Figure 12.10   220
Gluteal Muscles   Figure 12.11   221
Gluteal Muscles   Figure 12.11
Unlabeled   221
Anterior Muscles of the Thigh   Figure 12.12   222
Anterior Superficial Thigh Muscles of
the Cadaver   Figure 12.13   223
Gluteal and Thigh Muscles   Figure 12.14   224
Anterior Muscles of the Leg   Figure 12.15   225
Superficial Muscles of the Leg,
Posterior Compartment   Figure 12.16   226
Deep Muscles of the Leg, Posterior and
Lateral Compartments   Figure 12.17   227
Superficial Muscles of the Leg of the
Cadaver   Figure 12.18   228
Serial Cross Sections Through the
Lower Limb   Figure 12.19   229
Intrinsic Muscles of the Foot
Figure 12.20   230

Atlas B
The Head and Neck   Figure B.1   231
The Thorax and Abdomen,
Ventral Aspect   Figure B.2   232
The Back and Gluteal Region   Figure B.3   233
The Pelvic Region   Figure B.4   234
The Axillary Region   Figure B.5   235
The Upper Limb, Lateral aspect
Figure B.6   236
The Antebrachium (forearm)   Figure B.7   236
The Wrist and Hand   Figure B.8   237
The Thigh and Knee   Figure B.9   239
The Leg and Foot, Lateral aspect
Figure B.10   240
The Leg and Foot, Medial Aspect
Figure B.11   241
The Leg and Foot, Dorsal Aspect
Figure B.12   242
The Foot   Figure B.13   243
The Foot   Figure B.14   244
Muscle Self-Test   Figure B.15   246

Chapter 13
The Nervous System   Figure 13.1   247
Subdivisions of the Nervous System
Figure 13.2   248

Functional Classes of Neurons
Figure 13.3   248
A Representative Neuron   Figure 13.4   249
A Representative Neuron
Figure 13.4   Unlabeled   250
Variation in Neuronal Structure
Figure 13.5   252
Neuroglia of the Central Nervous System
Figure 13.6   253
Formation of the Myelin Sheath
Figure 13.7   253
Unmyelinated Nerve Fibers   Figure 13.8   254
Myelinated and Unmyelinated Axons (TEM)
Figure 13.9   254
Synaptic Relationships Between Neurons
Figure 13.11   255
Structure of a Chemical Synapse
Figure 13.12   255
Four Types of Neuronal Circuits
Figure 13.13   256
Formation of the Neural Tube
Figure 13.14   256
Primary and Secondary Vesicles
of the Embryonic Brain   Figure 13.15   257
The Spinal Cord, Dorsal Aspect
Figure 14.1   258

Chapter 14
Cross Section of the Thoracic
Spinal Cord   Figure 14.2   259
Cross Section of the Thoracic
Spinal Cord   Figure 14.2   Unlabeled   259
Tracts of the Spinal Cord   Figure 14.3   260
Two Ascending Pathways of the CNS
Figure 14.4   261
Two Descending Pathways of the CNS
Figure 14.5   262
Anatomy of a Nerve   Figure 14.7   263
Anatomy of a Ganglion   Figure 14.8   263
The Spinal Nerve Roots and Plexuses,
Dorsal View   Figure 14.9   264
The Spinal Nerve Roots and Plexuses,
Dorsal View   Figure 14.9   Unlabeled   265
Branches of a Spinal Nerve in Relation to
the Spinal Cord and Vertebra (cross section)
Figure 14.10   266
Branches of a Spinal Nerve in Relation to
the Spinal Cord and Vertebra (cross section)
Figure 14.10   Unlabeled   266
Rami of the Spinal Nerves   Figure 14.12   267
The Cervical Plexus   Figure 14.13   268
The Brachial Plexus   Figure 14.14   269
The Lumbar Plexus   Figure 14.16   270
The Sacral and Coccygeal Plexuses
Figure 14.17   271
A Dermatome Map of the Body
Figure 14.18   272
A Representative Reflex Arc
Figure 14.19   273

Chapter 15

Surface Anatomy of the Brain
Figure 15.1 — 273
Medial Aspect of the Brain   Figure 15.2 — 275
Medial Aspect of the Brain   Figure 15.2
Unlabeled — 276
Directional Terms in CNS Anatomy
Figure 15.3 — 277
The Meninges of the Brain   Figure 15.4 — 278
Ventricles of the Brain   Figure 15.5 — 279
The Flow of Cerebrospinal Fluid
Figure 15.6 — 280
The Brainstem   Figure 15.7 — 281
The Brainstem   Figure 15.7   Unlabeled — 281
The Cerebellum   Figure 15.8 — 282
Motor Pathways Involving the Cerebellum
Figure 15.9 — 283
Cross Section of the Midbrain
Figure 15.10 — 284
The Reticular Formation   Figure 15.11 — 284
The Diencephalon   Figure 15.12 — 285
Lobes of the Cerebrum   Figure 15.13 — 285
Tracts of Cerebral White Matter
Figure 15.14 — 286
Histology of the Neocortex   Figure 15.15 — 287
The Basal Nuclei   Figure 15.16 — 288
The Limbic System   Figure 15.17 — 288
Some Functional Regions of the
Cerebral Cortex   Figure 15.18 — 289
The Primary Somesthetic Area
(postcentral gyrus)   Figure 15.19 — 289
The Primary Motor Area (precentral gyrus)
Figure 15.20 — 290
Lateralization of Cerebral Functions
Figure 15.22 — 291
Lateralization of Cerebral Functions
Figure 15.22   Unlabeled — 291
The Cranial Nerves   Figure 15.23 — 292
The Olfactory Nerve   Figure 15.24 — 293
The Optic Nerve   Figure 15.25 — 293
The Oculomotor Nerve   Figure 15.26 — 294
The Trochlear Nerve   Figure 15.27 — 294
The Trigeminal Nerve   Figure 15.28 — 294
The Abducens Nerve   Figure 15.29 — 295
The Facial Nerve   Figure 15.30 — 295
The Vestibulocochlear Nerve
Figure 15.31 — 296
The Glossopharyngeal Nerve
Figure 15.32 — 296
The Vagus Nerve   Figure 15.33 — 297
The Accessory Nerve   Figure 15.34 — 298
The Hypoglossal Nerve   Figure 15.35 — 298

Chapter 16

An Autonomic Reflex Arc in the Regulation
of Blood Pressure   Figure 16.1 — 299
Comparison of Somatic and Autonomic
Efferent Pathways   Figure 16.2 — 300
Sympathetic Pathways   Figure 16.4 — 301
Sympathetic Pathways Compared to
Somatic Efferent Pathways   Figure 16.5 — 302
Abdominal Components of the Sympathetic
Nervous System   Figure 16.6 — 303
Parasympathetic Pathways   Figure 16.7 — 304
Parasympathetic Pathways   Figure 16.7 — 305
Neurotransmitters and Receptors of the
Autonomic Nervous System   Figure 16.8 — 306
Dual Innervation of the Iris   Figure 16.9 — 307
Sympathetic and Vasomotor Tone
Figure 16.10 — 307

Chapter 17

Receptors of the General (somesthetic)
Senses   Figure 17.1 — 308
Receptive Fields of Sensory Neurons
Figure 17.2 — 309
Projection Pathways for Pain   Figure 17.3 — 310
Referred Pain   Figure 17.4 — 310
Taste Receptors   Figure 17.5 — 311
Gustatory Projection Pathways to
the Cerebral Cortex   Figure 17.6 — 312
Olfactory Receptors   Figure 17.7 — 313
Olfactory Receptors   Figure 17.7
Unlabeled — 314
Olfactory Projection Pathways in the Brain
Figure 17.8 — 315
External Anatomy of the Ear   Figure 17.9 — 316
Internal Anatomy of the Ear   Figure 17.10 — 316
Anatomy of the Inner Ear   Figure 17.11 — 317
Anatomy of the Cochlea   Figure 17.12 — 318
Mechanical Model of Auditory Function
Figure 17.14 — 318
Auditory Pathways in the Brain
Figure 17.15 — 319
The Saccule and Utricle   Figure 17.16 — 320
Structure and Function of the Semicircular
Ducts   Figure 17.17 — 321
Vestibular Projection Pathways in the Brain
Figure 17.18 — 321
Accessory Structures of the Orbit
Figure 17.20 — 322
Extrinsic Muscles of the Eye
Figure 17.21 — 323
Anatomy of the Eye   Figure 17.22 — 324
Anatomy of the Eye   Figure 17.22 — 324
Production and Reabsorption of
Aqueous Humor   Figure 17.23 — 325
Two Common Visual Defects and the
Effects of Corrective Lenses
Figure 17.26 — 325
Histology of the Retina   Figure 17.27 — 326
Rod and Cone Cells   Figure 17.28 — 327
The Duplicity Theory of Vision
Figure 17.29 — 328
The Visual Projection Pathway
Figure 17.30 — 329
Development of the Ear   Figure 17.31 — 330
Development of the Eye   Figure 17.32 — 331

Chapter 18

Major Organs of the Endocrine System
Figure 18.1 · 332

The Chemical Classes of Hormones
Figure 18.2 · 333

Communication by the Nervous and
Endocrine Systems · Figure 18.3 · 334

Gross Anatomy of the Pituitary Gland
Figure 18.4 · 335

Hormones and Target Organs of the
Anterior Pituitary Gland · Figure 18.6 · 336

The Thymus · Figure 18.7 · 337

The Thyroid Gland · Figure 18.8 · 337

The Parathyroid Glands · Figure 18.9 · 338

The Adrenal Glands · Figure 18.10 · 338

The Pancreatic Islets · Figure 18.11 · 339

Embryonic Development of the Pituitary
Gland and Thyroid · Figure 18.13 · 340

Embryonic Development of the Adrenal
Gland · Figure 18.14 · 341

Chapter 19

The Formed Elements of Blood · Figure 19.1 · 342

Separating the Plasma and Formed
Elements of Blood · Figure 19.2 · 343

The Structure of Erythrocytes · Figure 19.3 · 343

The Structure of Hemoglobin · Figure 19.5 · 344

Erythropoiesis · Figure 19.6 · 345

Chemical Basis of the ABO Blood Types
Figure 19.7 · 345

Leukopoiesis · Figure 19.9 · 346

A Megakaryocyte Producing Platelets
Figure 19.11 · 347

Chapter 20

General Schematic of the Cardiovascular
System · Figure 20.1 · 348

Position of the Heart in the Thoracic Cavity
Figure 20.2 · 349

External Anatomy of the Heart
Figure 20.3 · 350

External Anatomy of the Heart
Figure 20.3 · Unlabeled · 350

The Pericardium and Heart Wall
Figure 20.4 · 352

The Human Heart · Figure 20.5 · 352

Twisted Orientation of Myocardial
Muscle · Figure 20.6 · 353

Internal Anatomy of the Heart · Figure 20.7 · 353

Internal Anatomy of the Heart · Figure 20.7
Unlabeled · 354

The Heart Valves · Figure 20.8 · 355

The Heart Valves · Figure 20.8 · Unlabeled · 355

Operation of the Heart Valves · Figure 20.9 · 357

The Pathway of Blood Flow Through
the Heart · Figure 20.10 · 358

The Coronary Blood Vessels · Figure 20.11 · 359

The Cardiac Conduction System
Figure 20.13 · 360

Cardiac Muscle · Figure 20.14 · 361

Embryonic Development of the Heart
Figure 20.15 · 362

The Fetal Heart · Figure 20.16 · 363

Chapter 21

Histological Structure of Blood Vessels
Figure 21.1 · 364

Histological Structure of Blood Vessels
Figure 21.1 · Unlabeled · 365

A Neurovascular Bundle · Figure 21.2 · 366

Baroreceptors and Chemoreceptors in the
Arteries Superior to the Heart · Figure 21.4 · 366

Structure of a Continuous Capillary
Figure 21.5 · 367

Structure of a Fenestrated Capillary
Figure 21.6 · 367

A Sinusoid of the Liver · Figure 21.7 · 368

Pathways of Capillary Fluid Exchange
Figure 21.8 · 368

Perfusion of a Capillary Bed · Figure 21.9 · 369

Typical Distribution of the Blood in a
Resting Adult · Figure 21.10 · 370

The Skeletal Muscle Pump · Figure 21.11 · 370

Variations in Circulatory Pathways
Figure 21.12 · 371

The Pulmonary Circulation · Figure 21.13 · 372

The Major Systematic Arteries
Figure 21.14 · 373

The Major Systematic Arteries
Figure 21.14 · Unlabeled · 374

The Thoracic Aorta · Figure 21.15 · 375

Arteries of the Head and Neck
Figure 21.16 · 375

The Cerebral Arterial Circle · Figure 21.17 · 376

Arteries of the Upper Limb · Figure 21.18 · 376

Arteries of the Thorax · Figure 21.19 · 377

The Abdominal Aorta and its Major
Branches · Figure 21.20 · 378

Branches of the Celiac Trunk
Figure 21.21 · 379

The Mesenteric Arteries · Figure 21.22 · 380

The Mesenteric Arteries · Figure 21.22
Unlabeled · 380

Arteries of the Lower Limb · Figure 21.23 · 381

Arterial Flowchart of the Lower Limb
Figure 21.24 · 382

Arterial Pressure Points · Figure 21.25 · 383

The Major Systematic Veins · Figure 21.26 · 384

The Major Systematic Veins · Figure 21.26
Unlabeled · 385

Veins of the Head and Neck · Figure 21.27 · 386

Veins of the Upper Limb · Figure 21.28 · 387

Flowchart of the Azygos System
Figure 21.29 · 388

The Inferior Vena Cava and its Tributaries
Figure 21.30 · 388

Flowchart of the Hepatic Portal System
Figure 21.31 · 389

Anatomy of the Hepatic Portal System
Figure 21.32 **389**
Veins of the Lower Limb   Figure 21.33 **390**
Flowchart of the Lower Limb Veins
Figure 21.34 **391**
Development of Blood Vessels and
Primitive Erythrocytes from Embryonic
Blood Islands   Figure 21.35 **392**
Development of Some Major Arteries
from the Embryonic Aortic Arches
Figure 21.36 **393**
Major Embryonic Blood Vessels at
26 Days   Figure 21.37 **393**
Some Circulatory Changes Occurring
at Birth   Figure 21.38 **394**

Chapter 22
The Lymphatic System   Figure 22.1 **395**
Lymphatic Capillaries   Figure 22.3 **396**
Valve in a Lymphatic Vessel   Figure 22.4 **396**
Fluid Exchange Between the Circulatory
and Lymphatic Systems   Figure 22.5 **397**
Lymphatics of the Thoracic Region
Figure 22.6 **398**
Histology of the Red Bone Marrow
Figure 22.9 **398**
The Thymus   Figure 22.10 **399**
Anatomy of a Lymph Node   Figure 22.12 **400**
Some Areas of Lymph Node Concentration
Figure 22.13 **401**
Some Areas of Lymph Node Concentration
Figure 22.13   Unlabeled **402**
The Tonsils   Figure 22.14 **403**
The Spleen   Figure 22.15 **404**
The Life History and Migrations of B and
T Cells   Figure 22.16 **405**
Embryonic Development of the Lymphatic
Vessels and Lymph Nodes   Figure 22.18 **406**

Chapter 23
The Respiratory System   Figure 23.1 **407**
Anatomy of the Upper Limb Respiratory
Tract   Figure 23.2 **408**
Anatomy of the Upper Limb Respiratory
Tract   Figure 23.2   Unlabeled **408**
Anatomy of the Nasal Region
Figure 23.3 **409**
Anatomy of the Larynx   Figure 23.4 **410**
Action of Some of the Intrinsic Laryngeal
Muscles on the Vocal Cords   Figure 23.6 **410**
Anatomy of the Lower Respiratory Tract
Figure 23.7 **411**
Gross Anatomy of the Lungs   Figure 23.9 **412**
Pulmonary Alveoli   Figure 23.11 **413**
Respiratory Control Centers   Figure 23.13 **414**
The Peripheral Chemoreceptors of
Respiration   Figure 23.14 **415**
Embryonic Development of the
Respiratory System   Figure 23.15 **416**

Chapter 24
The Digestive System   Figure 24.1 **417**
The Digestive System   Figure 24.1
Unlabeled **418**
Tissue Layers of the Digestive Tract
Figure 24.2 **419**
Tissue Layers of the Digestive Tract
Figure 24.2   Unlabeled **420**
Serous Membranes Associated with
the Digestive Tract   Figure 24.3 **420**
The Oral Cavity   Figure 24.4 **421**
The Tongue   Figure 24.5 **421**
The Dentition and Ages at Which the
Teeth Erupt   Figure 24.6 **422**
Median Section of a Canine Tooth
and Its Alveolus   Figure 24.7 **423**
Permanent and Deciduous Teeth
in a Child's Skull   Figure 24.8 **423**
The Extrinsic Salivary Glands
Figure 24.9 **424**
Anatomy of the Salivary Glands
Figure 24.10 **425**
The Stomach   Figure 24.11 **426**
The Stomach   Figure 24.11   Unlabeled **426**
Microscopic Anatomy of the Stomach
Wall   Figure 24.12 **428**
Gross Anatomy of the Small Intestine
Figure 24.14 **429**
Intestinal Villi   Figure 24.15 **429**
Structure of a Villus   Figure 24.16 **430**
The Large Intestine   Figure 24.17 **431**
The Large Intestine   Figure 24.17
Unlabeled **432**
Gross Anatomy of the Liver
Figure 24.18 **433**
Microscopic Anatomy of the Liver
Figure 24.19 **434**
Gross Anatomy of the Gallbladder,
Pancreas, and Bile Passages
Figure 24.20 **435**
Gross Anatomy of the Gallbladder,
Pancreas, and Bile Passages
Figure 24.20   Unlabeled **435**
Histology of the Pancreas
Figure 24.21 **436**
Embryonic Development of the
Digestive Tract   Figure 24.22 **437**
Lateral View of the 5-week Embryo
Figure 24.23 **437**

Chapter 25
The Urinary System   Figure 25.1 **438**
Location of the Kidney   Figure 25.2 **439**
Location of the Kidney   Figure 25.2
Unlabeled **439**
Gross Anatomy of the Kidney
Figure 25.3 **440**
Renal Circulation   Figure 25.4 **440**
Structure of the Nephron   Figure 25.5 **441**

Structure of the Nephron   Figure 25.5
Unlabeled   **442**
Basic Steps in the Formation of Urine
Figure 25.6   **443**
The Renal Corpuscle   Figure 25.7   **443**
Structure of the Glomerulus
Figure 25.8   **444**
The Juxtaglomerular Apparatus
Figure 25.9   **444**
Anatomy of the Urinary Bladder
and Urethra   Figure 25.10   **445**
Embryonic Development of the
Urinary Tract   Figure 25.11   **446**
Embryonic Development of the Nephron
Figure 25.12   **446**

Chapter 26
The Male Perineum   Figure 26.1   **447**
The Male Reproductive System
Figure 26.2   **448**
The Male Reproductive System
Figure 26.2   Unlabeled   **449**
The Testis and Associated Structures
Figure 26.3   **450**
Anatomy of the Male Inguinal Region
and External Genitalia   Figure 26.5   **450**
Meiosis   Figure 26.6   **451**
Spermatogenesis   Figure 26.7   **452**
Spermiogenesis   Figure 26.8   **453**
The Mature Spermatozoon   Figure 26.9   **454**
Anatomy of the Penis   Figure 26.10   **455**
The Female Reproductive System
Figure 26.11   **455**

The Female Reproductive System
Figure 26.11   Unlabeled   **456**
Structure of the Ovary and the
Developmental Sequence
of the Ovarian Follicles   Figure 26.12   **456**
The Female Reproductive Tract and
Supportive Ligaments   Figure 26.13   **457**
The Female Reproductive Tract and
Supportive Ligaments   Figure 26.13
Unlabeled   **457**
Oogenesis (*left*) and Corresponding
Development of the Follicle (*right*)
Figure 26.14   **458**
Dissection of the Female Reproductive
Tract   Figure 26.17   **459**
Blood Supply to the Female Reproductive
Tract   Figure 26.20   **459**
Endometrial Changes Through the
Menstrual Cycle   Figure 26.21   **460**
The Female Perineum   Figure 26.22   **461**
Anatomy of the Lactating Breast
Figure 26.23   **462**
Sagittal Section of the Breast
of a Cadaver   Figure 26.24   **462**
Embryonic Development of the Male
and Female Reproductive Tracts
Figure 26.26   **463**
Embryonic Development of the Male
and Female Reproductive Tracts
Figure 26.26   Unlabeled   **464**
Development of the External Genitalia
Figure 26.27   **465**
Descent of the Testis   Figure 26.28   **466**

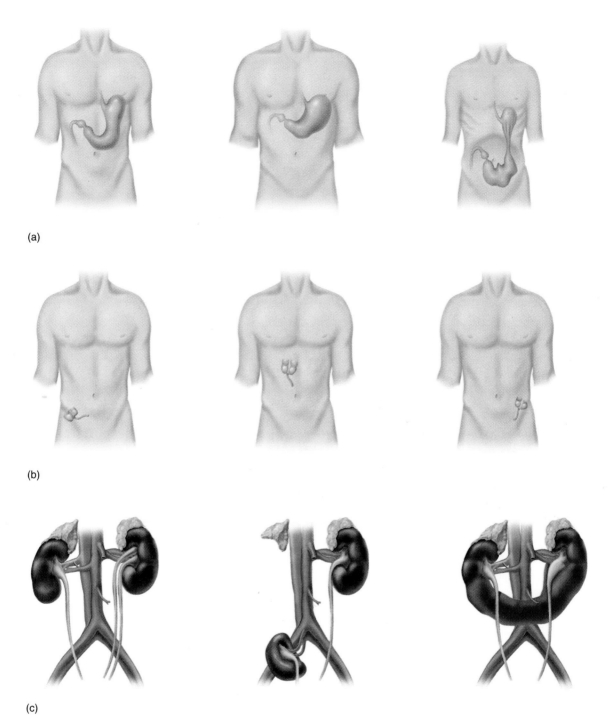

(a)

(b)

(c)

Variation in Human Anatomy
Figure 1.7

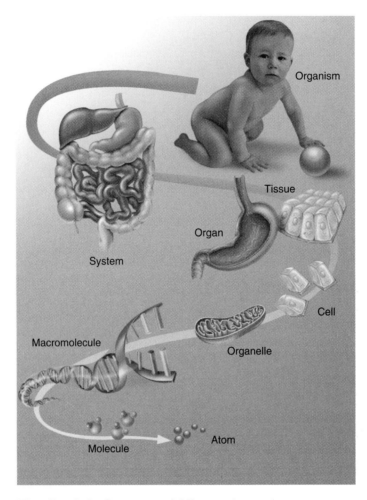

The Body's Structural Hierarchy
Figure 1.8

(b)

Early Microscope
Figure 1.11

b: Courtesy of the Armed
Forces Institute of Pathology

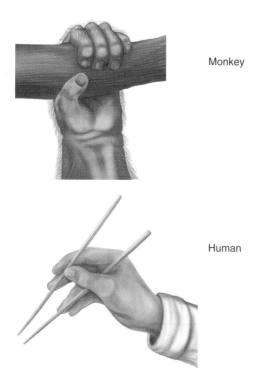

Monkey

Human

Primate Hands
Figure 1.13

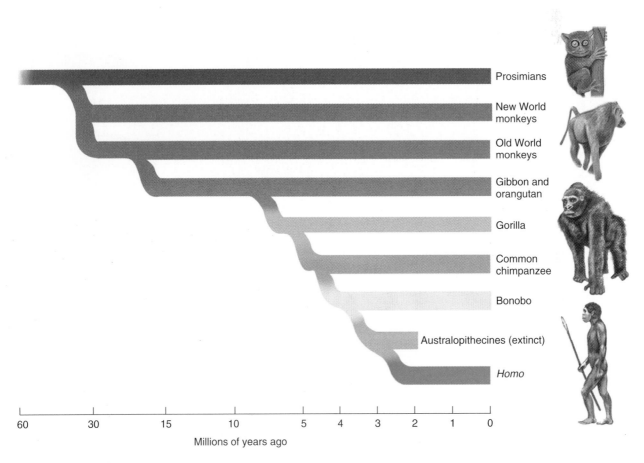

Primate Phylogeny
Figure 1.15

Anatomical Position
Figure A.1
© The McGraw-Hill Companies, Inc./
Joe DeGrandis, photographer

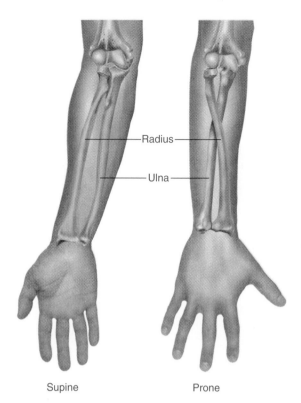

Radius

Ulna

Supine

Prone

Positions of the Forearm
Figure A.2

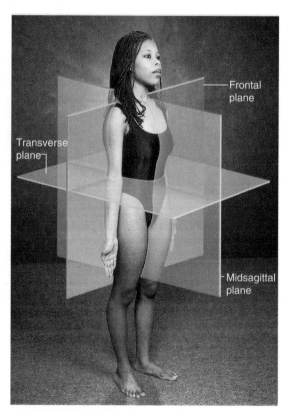

Frontal plane

Transverse plane

Midsagittal plane

Anatomical Planes of Reference
Figure A.3

© The McGraw-Hill Companies, Inc./
Joe DeGrandis, photographer

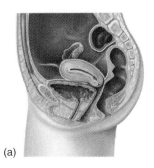

(a)

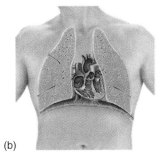

(b)

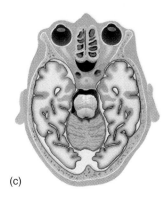

(c)

**Views of the Body in
the Three Primary
Anatomical Planes**
Figure A.4

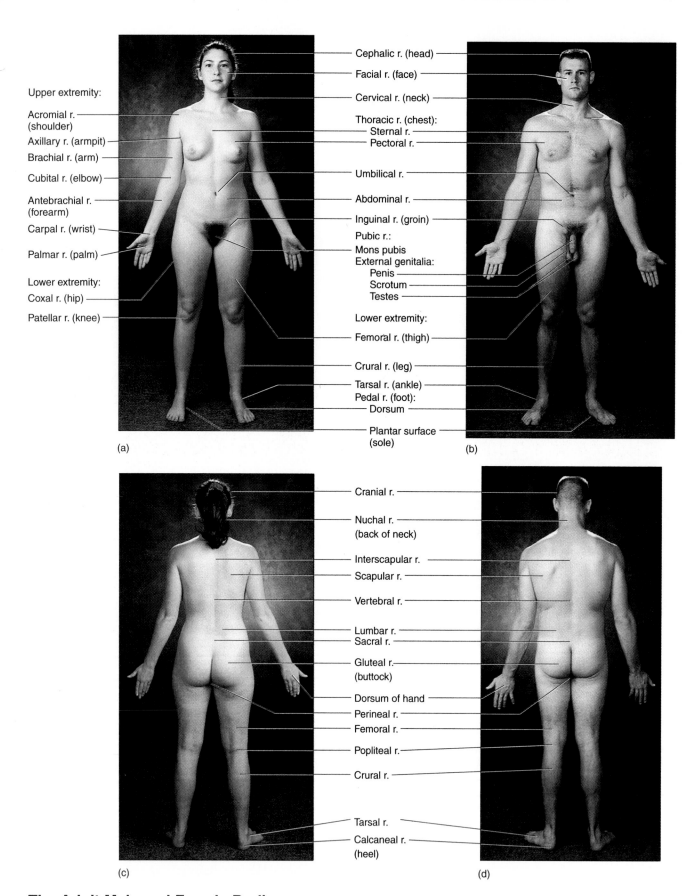

Upper extremity:
Acromial r. (shoulder)
Axillary r. (armpit)
Brachial r. (arm)
Cubital r. (elbow)
Antebrachial r. (forearm)
Carpal r. (wrist)
Palmar r. (palm)

Lower extremity:
Coxal r. (hip)
Patellar r. (knee)

(a)

Cephalic r. (head)
Facial r. (face)
Cervical r. (neck)
Thoracic r. (chest):
Sternal r.
Pectoral r.
Umbilical r.
Abdominal r.
Inguinal r. (groin)
Pubic r.:
Mons pubis
External genitalia:
Penis
Scrotum
Testes
Lower extremity:
Femoral r. (thigh)
Crural r. (leg)
Tarsal r. (ankle)
Pedal r. (foot):
Dorsum
Plantar surface (sole)

(b)

Cranial r.
Nuchal r. (back of neck)
Interscapular r.
Scapular r.
Vertebral r.
Lumbar r.
Sacral r.
Gluteal r. (buttock)
Dorsum of hand
Perineal r.
Femoral r.
Popliteal r.
Crural r.
Tarsal r.
Calcaneal r. (heel)

(c)

(d)

The Adult Male and Female Bodies
Figure A.5

a–d: © The McGraw-Hill Companies, Inc./Joe DeGrandis, photographer

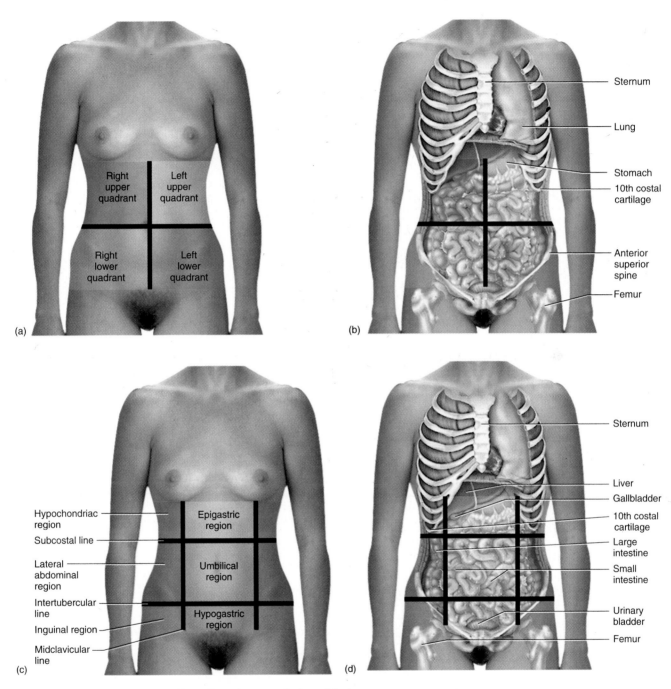

(a)

Right upper quadrant

Left upper quadrant

Right lower quadrant

Left lower quadrant

(b)

Sternum

Lung

Stomach

10th costal cartilage

Anterior superior spine

Femur

(c)

Hypochondriac region

Subcostal line

Lateral abdominal region

Intertubercular line

Inguinal region

Midclavicular line

Epigastric region

Umbilical region

Hypogastric region

(d)

Sternum

Liver

Gallbladder

10th costal cartilage

Large intestine

Small intestine

Urinary bladder

Femur

Four Quadrants and Nine Regions of the Abdomen
Figure A.6

9

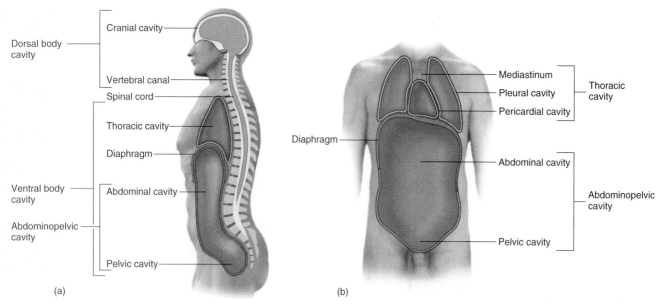

The Major Body Cavities
Figure A.7

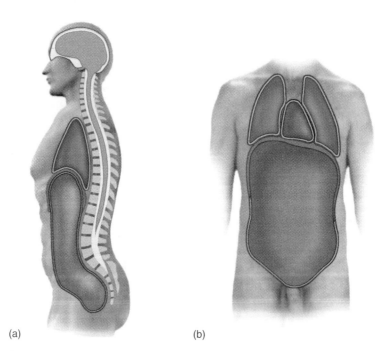

The Major Body Cavities
Figure A.7

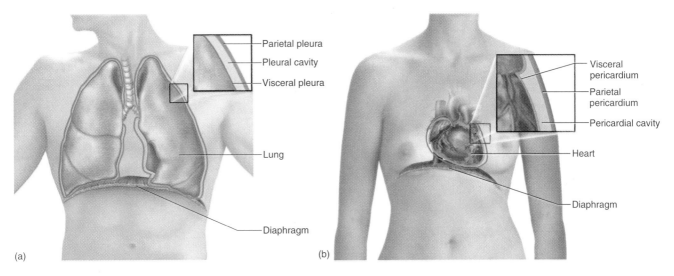

Parietal and Visceral Layers of Double-Walled Membranes
Figure A.8

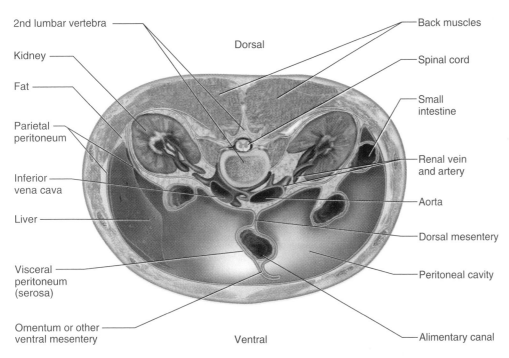

Transverse Section Through the Abdominal Cavity
Figure A.9

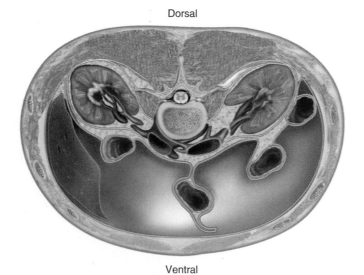

Dorsal

Ventral

Transverse Section Through the Abdominal Cavity
Figure A.9

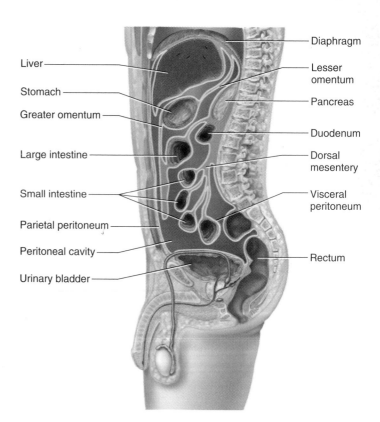

Liver

Stomach

Greater omentum

Large intestine

Small intestine

Parietal peritoneum

Peritoneal cavity

Urinary bladder

Diaphragm

Lesser omentum

Pancreas

Duodenum

Dorsal mesentery

Visceral peritoneum

Rectum

Serous Membranes of the Abdominal Cavity
Figure A.10

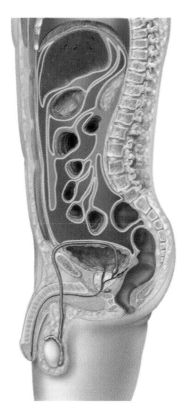

**Serous Membranes of
the Abdominal Cavity**
Figure A.10

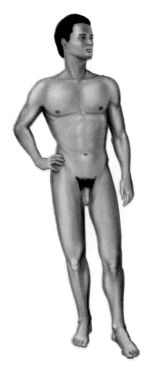

A.11a INTEGUMENTARY SYSTEM
Principal organs: Skin, hair, nails, cutaneous glands
Principal functions: Protection, water retention, thermoregulation, vitamin D synthesis, cutaneous sensation, nonverbal communication

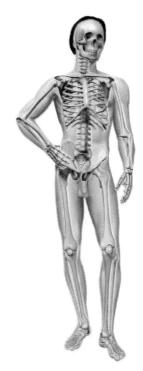

A.11b SKELETAL SYSTEM
Principal organs: Bones, cartilages, ligaments
Principal functions: Support, movement, protective enclosure of viscera, blood formation, electrolyte and acid-base balance

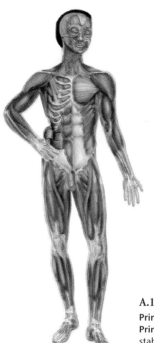

A.11c MUSCULAR SYSTEM
Principal organs: Skeletal muscles
Principal functions: Movement, stability, communication, control of body openings, heat production

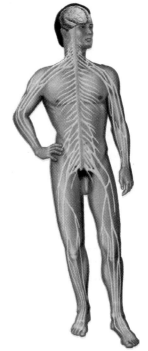

A.11d NERVOUS SYSTEM
Principal organs: Brain, spinal cord, nerves, ganglia
Principal functions: Rapid internal communication and coordination, sensation

The Human Organisms
Figure A.11

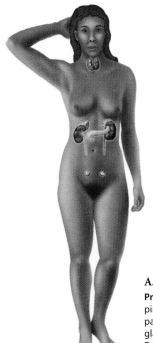

A.11*e* ENDOCRINE SYSTEM
Principal organs: Pituitary gland, pineal gland, thyroid gland, parathyroid glands, thymus, adrenal glands, pancreas, testes, ovaries
Principal functions: Internal chemical communication and coordination

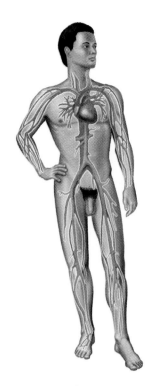

A.11*f* CIRCULATORY SYSTEM
Principal organs: Heart, blood vessels
Principal functions: Distribution of nutrients, oxygen, wastes, hormones, electrolytes, heat, immune cells, and antibodies; fluid, electrolyte, and acid-base balance

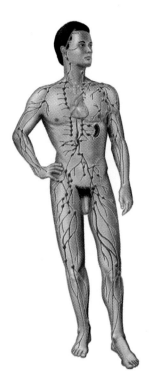

A.11*g* LYMPHATIC SYSTEM
Principal organs: Lymph nodes, lymphatic vessels, thymus, spleen, tonsils
Principal functions: Recovery of excess tissue fluid, detection of pathogens, production of immune cells, defense

A.11*h* RESPIRATORY SYSTEM
Principal organs: Nose, pharynx, larynx, trachea, bronchi, lungs
Principal functions: Absorption of oxygen, discharge of carbon dioxide, acid-base balance, speech

The Human Organisms (*Continued*)
Figure A.11

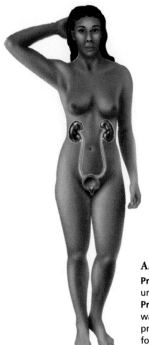

A.11*i* URINARY SYSTEM

Principal organs: Kidneys, ureters, urinary bladder, urethra
Principal functions: Elimination of wastes; regulation of blood volume and pressure; stimulation of red blood cell formation; control of fluid, electrolyte, and acid-base balance; detoxification

A.11*j* DIGESTIVE SYSTEM

Principal organs: Teeth, tongue, salivary glands, esophagus, stomach, small and large intestines, liver, pancreas
Principal functions: Nutrient breakdown and absorption; liver functions including metabolism of carbohydrates, lipids, proteins, vitamins, and minerals, synthesis of plasma proteins, disposal of drugs, toxins, and hormones, and cleansing of blood

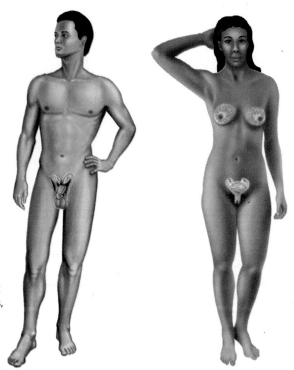

A.11*k* MALE REPRODUCTIVE SYSTEM

Principal organs: Testes, epididymides, spermatic ducts, seminal vesicles, prostate gland, bulbourethral glands, penis
Principal functions: Production and delivery of sperm

A.11*l* FEMALE REPRODUCTIVE SYSTEM

Principal organs: Ovaries, uterine tubes, uterus, vagina, vulva, mammary glands
Principal functions: Production of eggs, site of fertilization and fetal development, fetal nourishment, birth, lactation

The Human Organisms (*Continued*)
Figure A.11

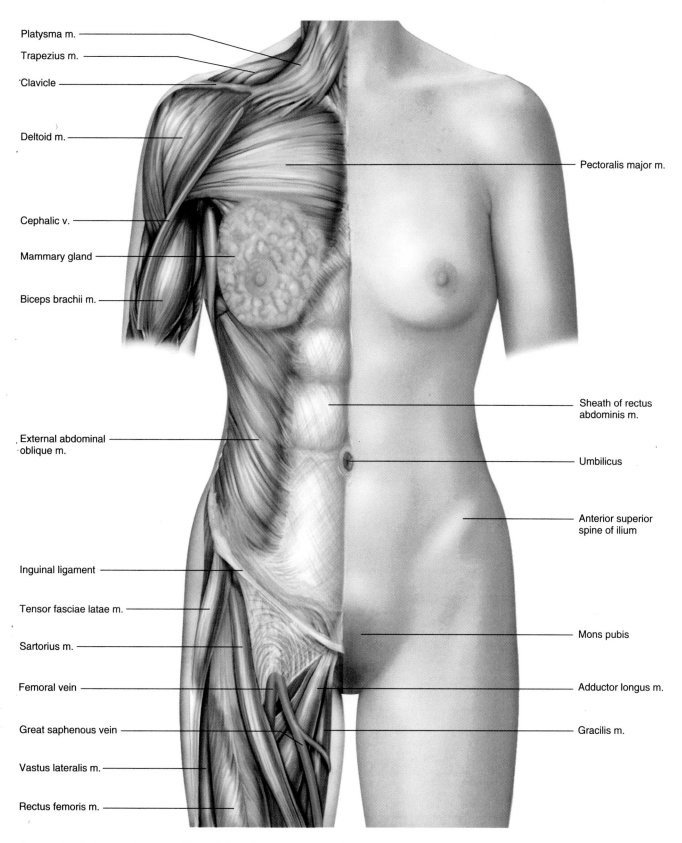

Platysma m.
Trapezius m.
Clavicle
Deltoid m.
Cephalic v.
Mammary gland
Biceps brachii m.
External abdominal oblique m.
Inguinal ligament
Tensor fasciae latae m.
Sartorius m.
Femoral vein
Great saphenous vein
Vastus lateralis m.
Rectus femoris m.

Pectoralis major m.
Sheath of rectus abdominis m.
Umbilicus
Anterior superior spine of ilium
Mons pubis
Adductor longus m.
Gracilis m.

Superficial Anatomy of the Trunk
Figure A.12

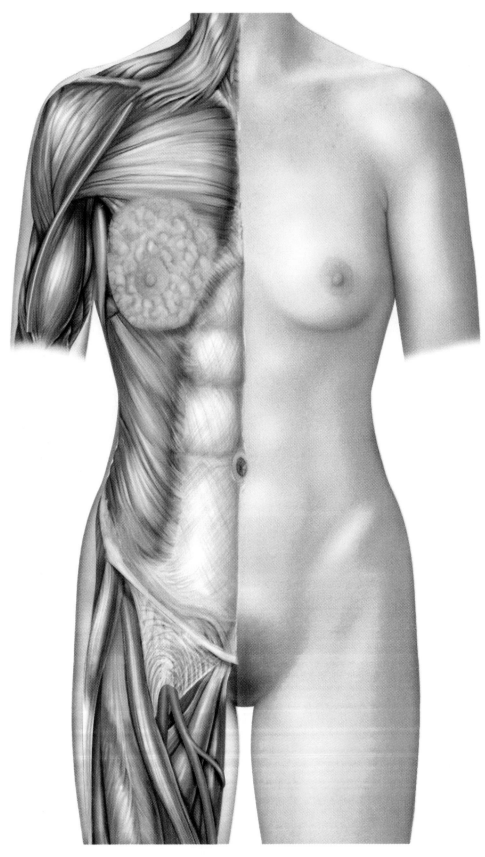

Superficial Anatomy of the Trunk
Figure A.12

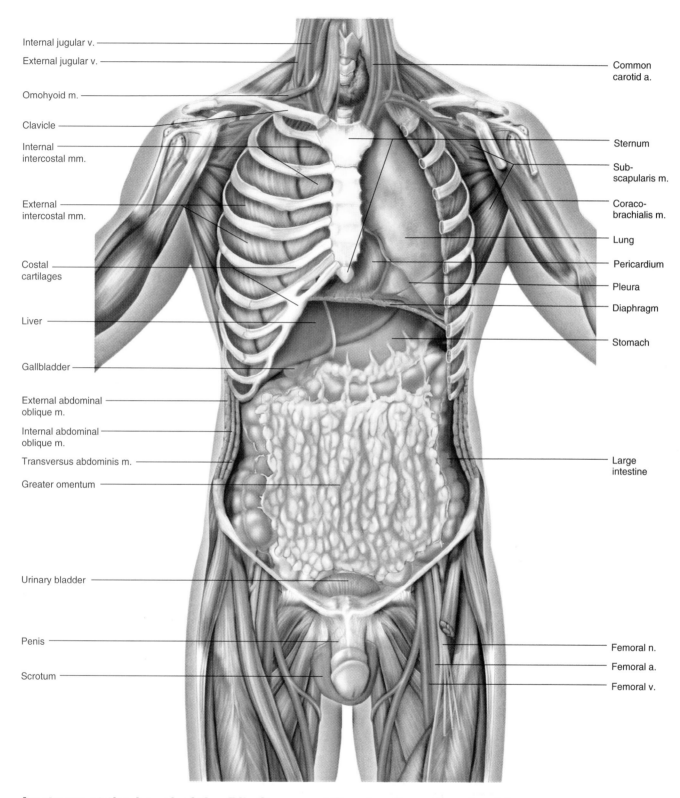

Internal jugular v.
External jugular v.
Omohyoid m.
Clavicle
Internal intercostal mm.
External intercostal mm.
Costal cartilages
Liver
Gallbladder
External abdominal oblique m.
Internal abdominal oblique m.
Transversus abdominis m.
Greater omentum
Urinary bladder
Penis
Scrotum

Common carotid a.
Sternum
Sub-scapularis m.
Coraco-brachialis m.
Lung
Pericardium
Pleura
Diaphragm
Stomach
Large intestine
Femoral n.
Femoral a.
Femoral v.

Anatomy at the Level of the Rib Cage and Greater Omentum (male)
Figure A.13

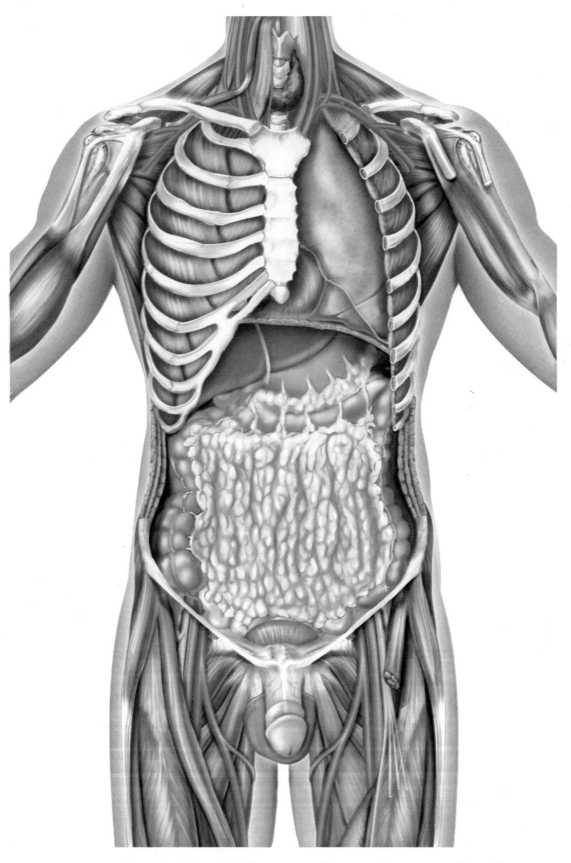

Anatomy at the Level of the Rib Cage and Greater Omentum (male)
Figure A.13

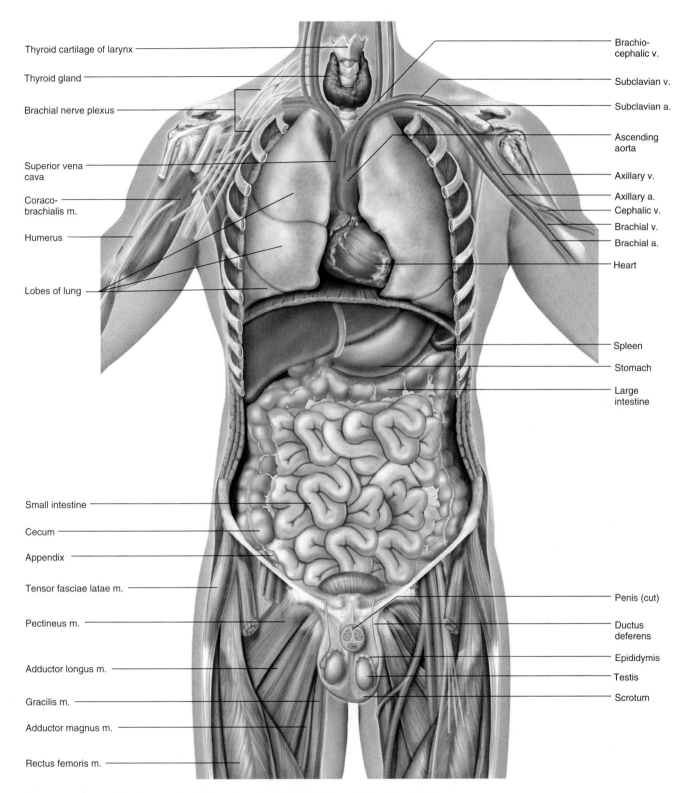

Thyroid cartilage of larynx

Thyroid gland

Brachial nerve plexus

Superior vena cava

Coraco-brachialis m.

Humerus

Lobes of lung

Small intestine

Cecum

Appendix

Tensor fasciae latae m.

Pectineus m.

Adductor longus m.

Gracilis m.

Adductor magnus m.

Rectus femoris m.

Brachio-cephalic v.

Subclavian v.

Subclavian a.

Ascending aorta

Axillary v.

Axillary a.

Cephalic v.

Brachial v.

Brachial a.

Heart

Spleen

Stomach

Large intestine

Penis (cut)

Ductus deferens

Epididymis

Testis

Scrotum

Anatomy at the Level of the Lungs and Intestines (male)
Figure A.14

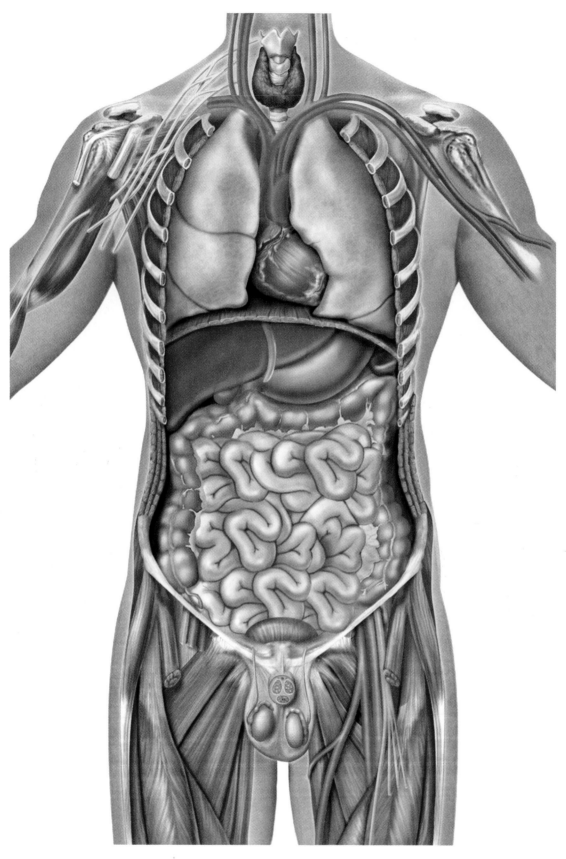

Anatomy at the Level of the Lungs and Intestines (male)
Figure A.14

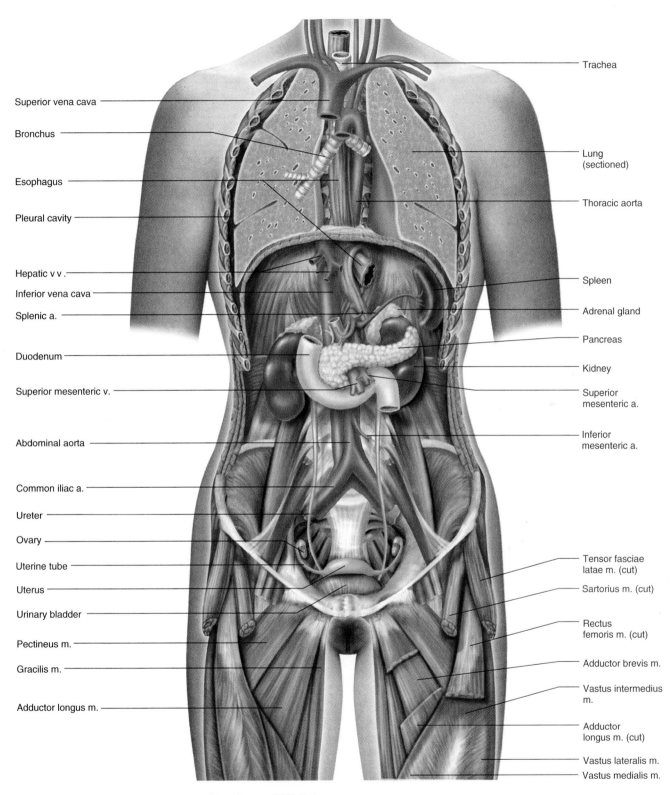

Trachea

Superior vena cava

Bronchus

Esophagus

Pleural cavity

Hepatic v v .

Inferior vena cava

Splenic a.

Duodenum

Superior mesenteric v.

Abdominal aorta

Common iliac a.

Ureter

Ovary

Uterine tube

Uterus

Urinary bladder

Pectineus m.

Gracilis m.

Adductor longus m.

Lung
(sectioned)

Thoracic aorta

Spleen

Adrenal gland

Pancreas

Kidney

Superior
mesenteric a.

Inferior
mesenteric a.

Tensor fasciae
latae m. (cut)

Sartorius m. (cut)

Rectus
femoris m. (cut)

Adductor brevis m.

Vastus intermedius
m.

Adductor
longus m. (cut)

Vastus lateralis m.

Vastus medialis m.

Anatomy at the Level of the Retroperitoneal Viscera (female)
Figure A.15

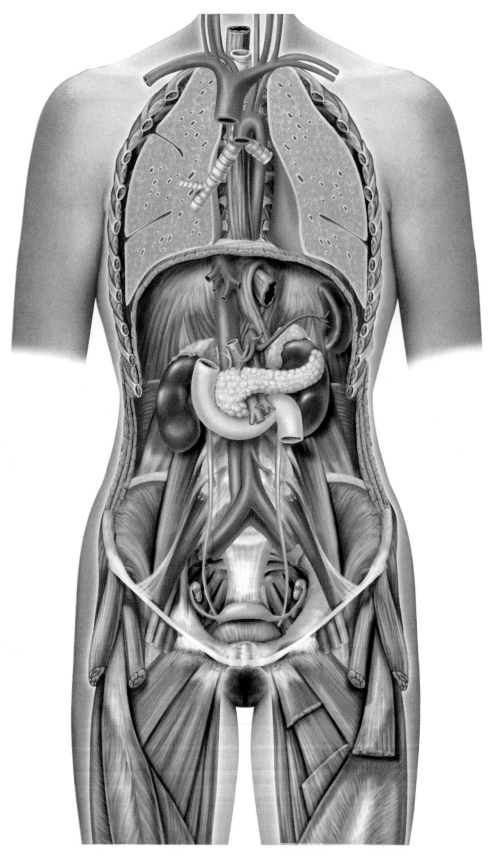

Anatomy at the Level of the Retroperitoneal Viscera (female)
Figure A.15

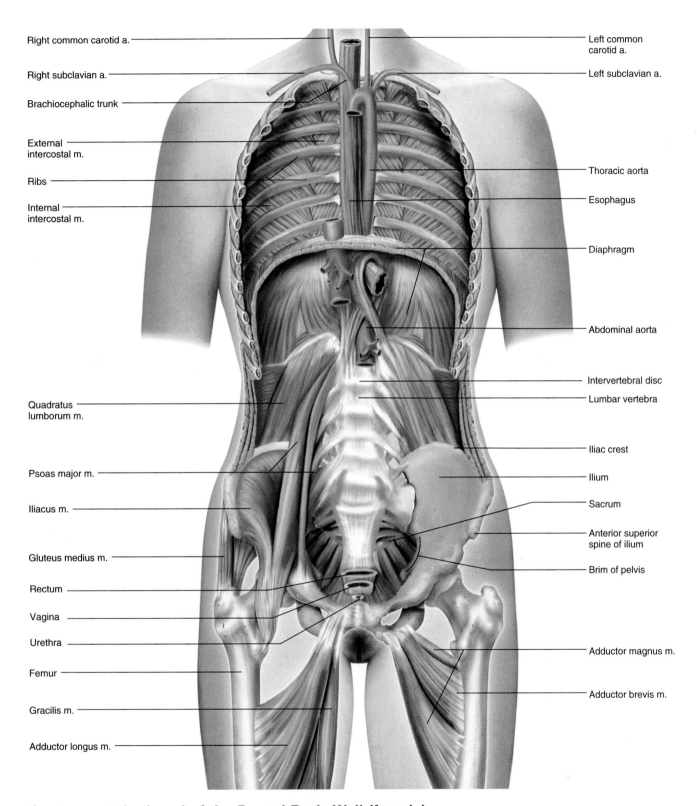

Right common carotid a.

Right subclavian a.

Brachiocephalic trunk

External intercostal m.

Ribs

Internal intercostal m.

Quadratus lumborum m.

Psoas major m.

Iliacus m.

Gluteus medius m.

Rectum

Vagina

Urethra

Femur

Gracilis m.

Adductor longus m.

Left common carotid a.

Left subclavian a.

Thoracic aorta

Esophagus

Diaphragm

Abdominal aorta

Intervertebral disc

Lumbar vertebra

Iliac crest

Ilium

Sacrum

Anterior superior spine of ilium

Brim of pelvis

Adductor magnus m.

Adductor brevis m.

Anatomy at the Level of the Dorsal Body Wall (female)
Figure A.16

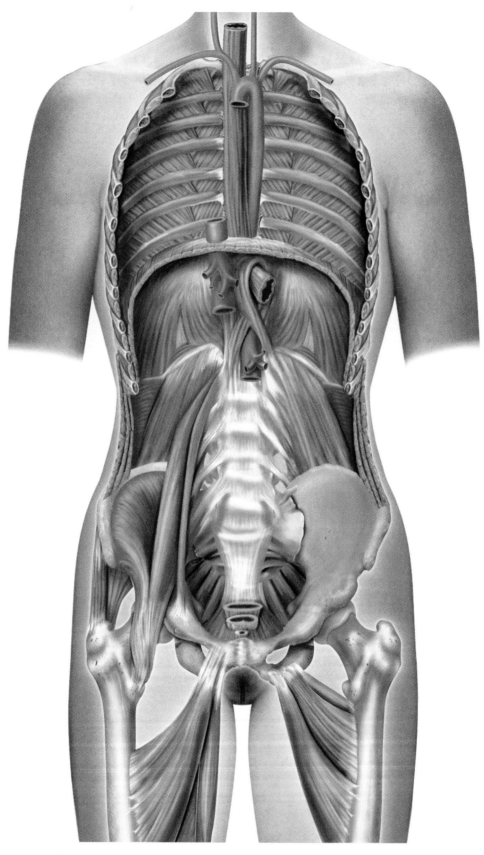

Anatomy at the Level of the Dorsal Body Wall (female)
Figure A.16

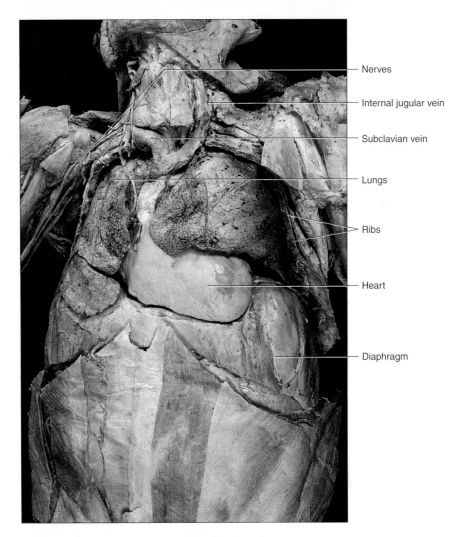

Nerves

Internal jugular vein

Subclavian vein

Lungs

Ribs

Heart

Diaphragm

Frontal View of the Thoracic Cavity
Figure A.18
© Tony Stone Images-Getty Images

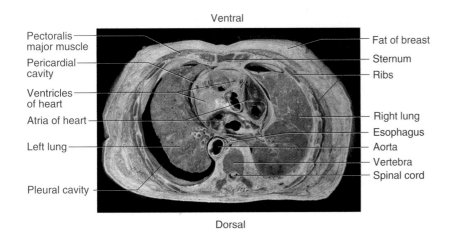

Ventral

Pectoralis major muscle

Pericardial cavity

Ventricles of heart

Atria of heart

Left lung

Pleural cavity

Fat of breast

Sternum

Ribs

Right lung

Esophagus

Aorta

Vertebra

Spinal cord

Dorsal

Transverse Section of the Thoracic Cavity
Figure A.19
© The McGraw-Hill Companies, Inc./Rebecca Gray, photographer/
Don Kincaid, dissections

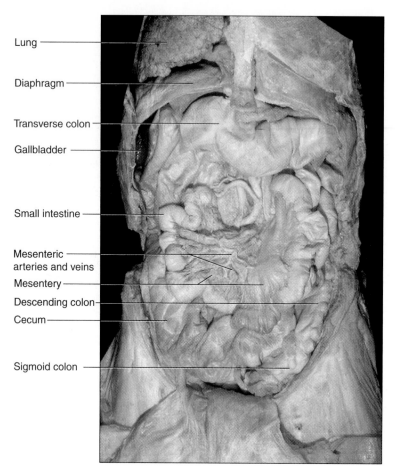

Lung

Diaphragm

Transverse colon

Gallbladder

Small intestine

Mesenteric
arteries and veins

Mesentery

Descending colon

Cecum

Sigmoid colon

Frontal View of the Abdominal Cavity
Figure A.20

© The McGraw-Hill Companies, Inc./Rebecca Gray, photographer/
Don Kincaid, dissections

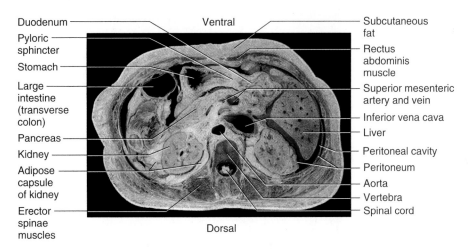

Duodenum

Pyloric sphincter

Stomach

Large intestine (transverse colon)

Pancreas

Kidney

Adipose capsule of kidney

Erector spinae muscles

Ventral

Dorsal

Subcutaneous fat

Rectus abdominis muscle

Superior mesenteric artery and vein

Inferior vena cava

Liver

Peritoneal cavity

Peritoneum

Aorta

Vertebra

Spinal cord

Transverse Section of the Abdominal Cavity
Figure A.21

© The McGraw-Hill Companies, Inc./Rebecca Gray, photographer/ Don Kincaid, dissections

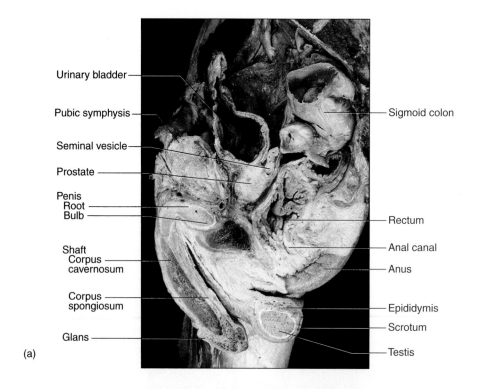

Urinary bladder

Pubic symphysis

Seminal vesicle

Prostate

Penis
Root
Bulb

Shaft
Corpus
cavernosum

Corpus
spongiosum

Glans

Sigmoid colon

Rectum

Anal canal

Anus

Epididymis

Scrotum

Testis

(a)

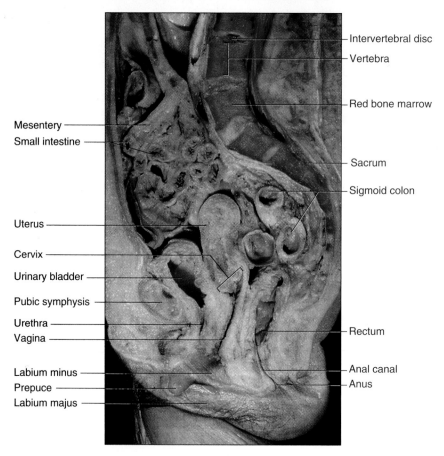

Intervertebral disc

Vertebra

Red bone marrow

Sacrum

Sigmoid colon

Mesentery

Small intestine

Uterus

Cervix

Urinary bladder

Pubic symphysis

Urethra

Vagina

Labium minus

Prepuce

Labium majus

Rectum

Anal canal

Anus

(b)

Sagittal Section of the Pelvic Cavity
Figure A.22

a: © The McGraw-Hill Companies, Inc./Dennis Strete, photographer

b: © The McGraw-Hill Companies, Inc./Rebecca Gray, photographer/
Don Kincaid, dissections

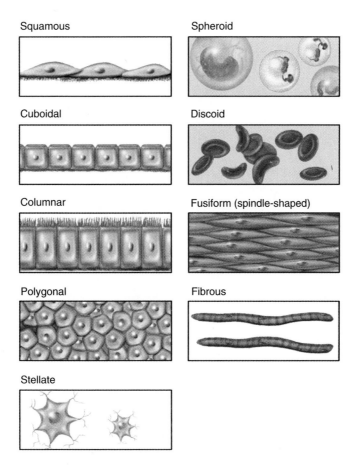

Common Cell Shapes
Figure 2.3

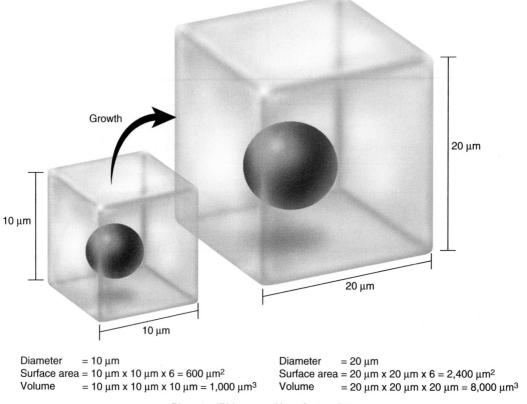

Diameter   = 10 µm
Surface area = 10 µm x 10 µm x 6 = 600 µm²
Volume   = 10 µm x 10 µm x 10 µm = 1,000 µm³

Diameter   = 20 µm
Surface area = 20 µm x 20 µm x 6 = 2,400 µm²
Volume   = 20 µm x 20 µm x 20 µm = 8,000 µm³

Diameter (D) increased by a factor of 2
Surface area increased by a factor of 4 (= D²)
Volume increased by a factor of 8 (= D³)

The Relationship Between Cell Surface Area and Volume
Figure 2.4

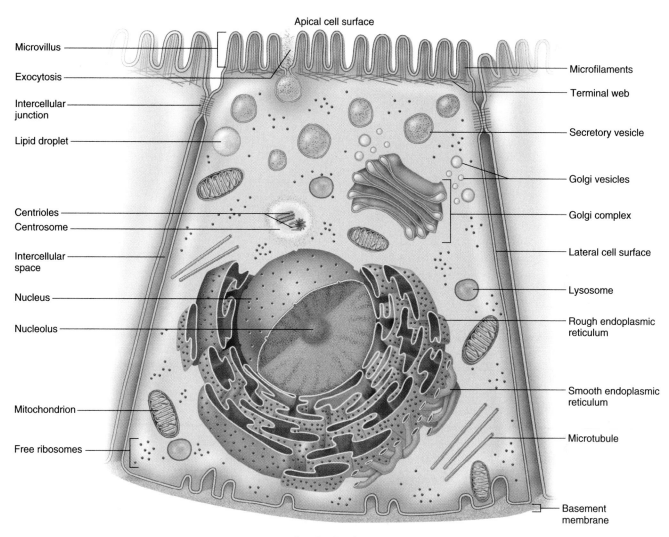

Microvillus

Exocytosis

Intercellular junction

Lipid droplet

Centrioles

Centrosome

Intercellular space

Nucleus

Nucleolus

Mitochondrion

Free ribosomes

Apical cell surface

Microfilaments

Terminal web

Secretory vesicle

Golgi vesicles

Golgi complex

Lateral cell surface

Lysosome

Rough endoplasmic reticulum

Smooth endoplasmic reticulum

Microtubule

Basement membrane

Basal cell surface

Structure of a Generalized Cell
Figure 2.5

Apical cell surface

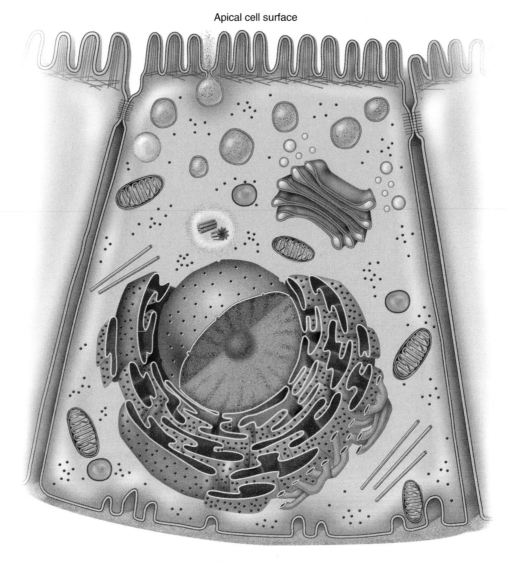

Basal cell surface

Structure of a Generalized Cell
Figure 2.5

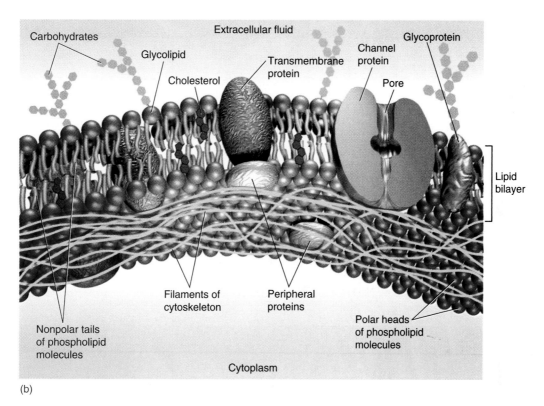

The Plasma Membrane
Figure 2.6

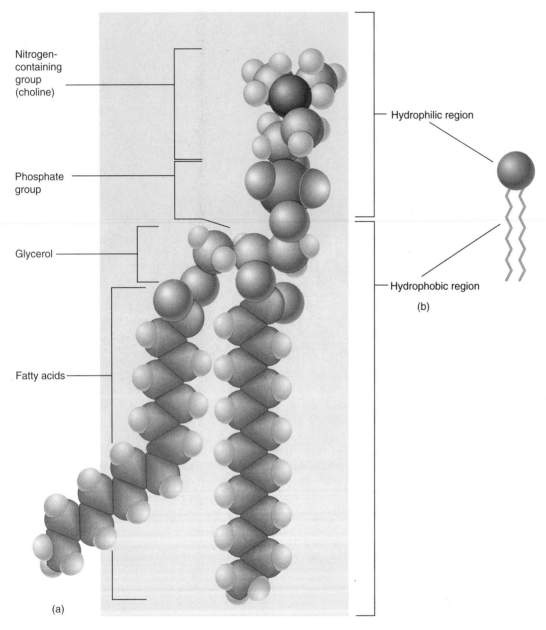

Nitrogen-
containing
group
(choline)

Phosphate
group

Glycerol

Fatty acids

Hydrophilic region

Hydrophobic region

(a)

(b)

Phospholipid Structure and Symbol
Figure 2.7

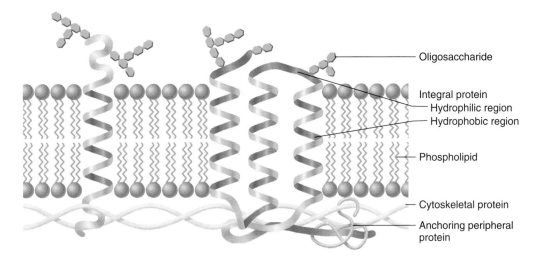

Transmembrane Proteins
Figure 2.8

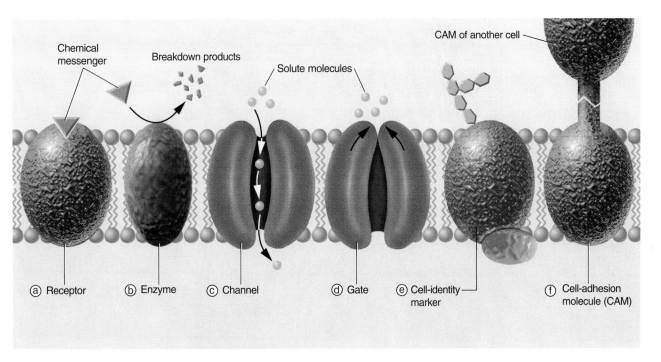

Some Functions of Plasma Membrane Proteins
Figure 2.9

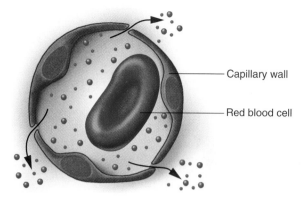

Capillary wall

Red blood cell

(a) **Filtration**

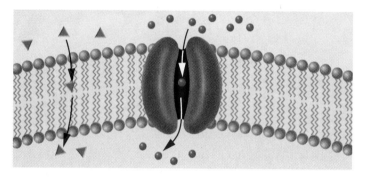

(b) **Simple diffusion**

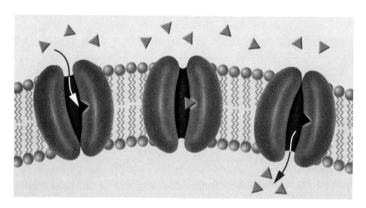

(c) **Facilitated diffusion**

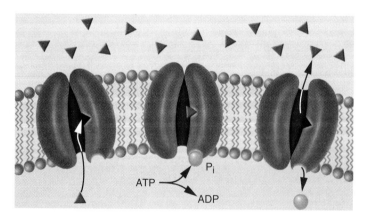

P_i

ATP

ADP

(d) **Active transport**

Some Modes of Membrane Transport
Figure 2.10

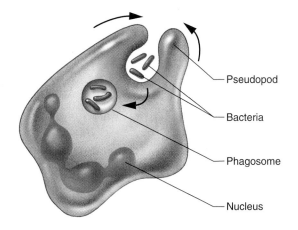

(a) **Phagocytosis**

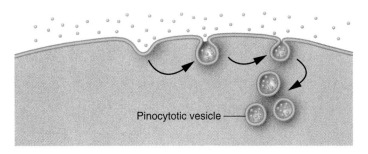

Pinocytotic vesicle

(b) **Pinocytosis**

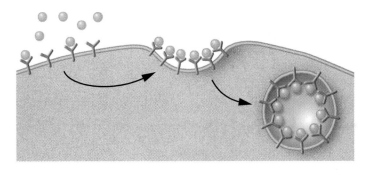

(c) **Receptor-mediated endocytosis**

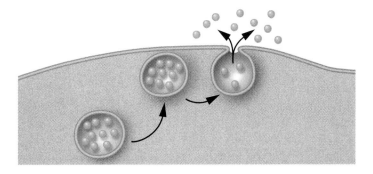

(d) **Exocytosis**

Modes of Vesicular Transport
Figure 2.11

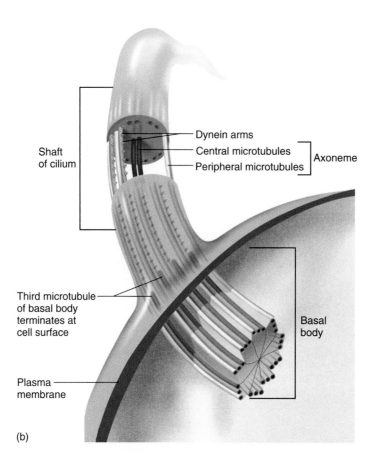

Shaft of cilium

Dynein arms
Central microtubules
Peripheral microtubules

Axoneme

Third microtubule of basal body terminates at cell surface

Basal body

Plasma membrane

(b)

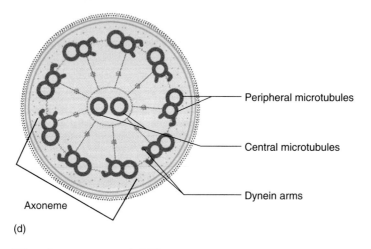

Peripheral microtubules

Central microtubules

Dynein arms

Axoneme

(d)

The Structure of Cilia
Figure 2.13

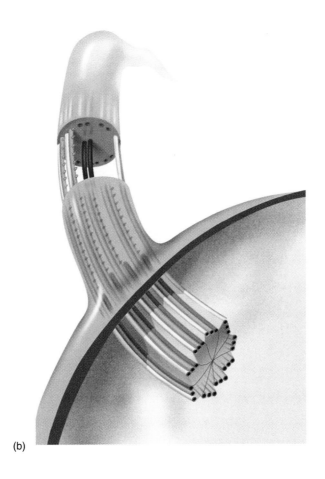

(b)

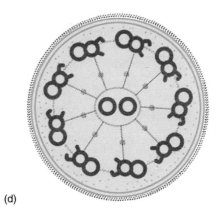

(d)

The Structure of Cilia
Figure 2.13

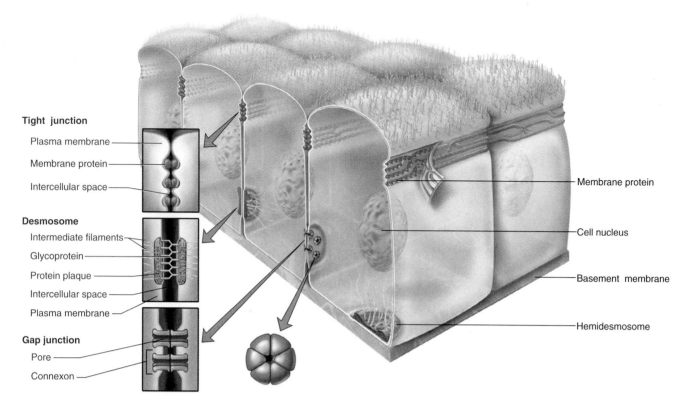

Tight junction
- Plasma membrane
- Membrane protein
- Intercellular space

Desmosome
- Intermediate filaments
- Glycoprotein
- Protein plaque
- Intercellular space
- Plasma membrane

Gap junction
- Pore
- Connexon

- Membrane protein
- Cell nucleus
- Basement membrane
- Hemidesmosome

Types of Intercellular Junctions
Figure 2.14

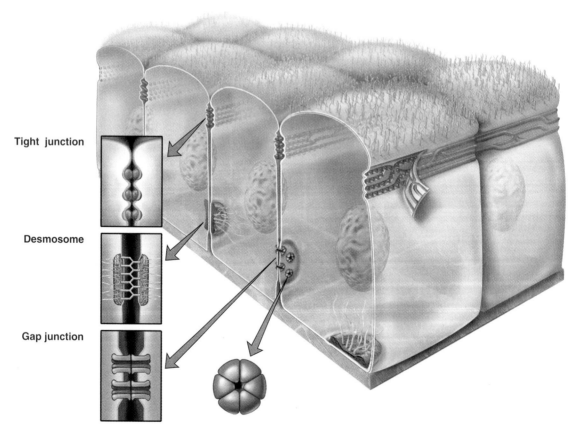

Tight junction

Desmosome

Gap junction

Types of Intercellular Junctions
Figure 2.14

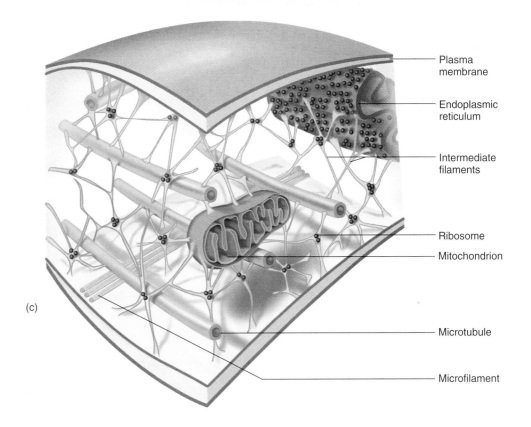

(c)

Plasma membrane

Endoplasmic reticulum

Intermediate filaments

Ribosome

Mitochondrion

Microtubule

Microfilament

The Cytoskeleton
Figure 2.15

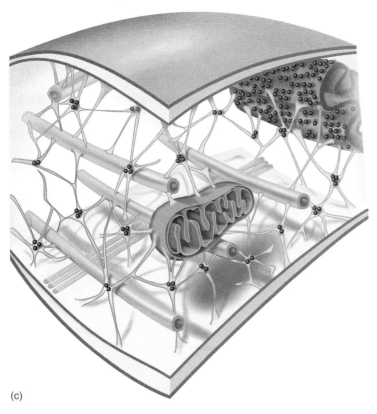

(c)

The Cytoskeleton
Figure 2.15

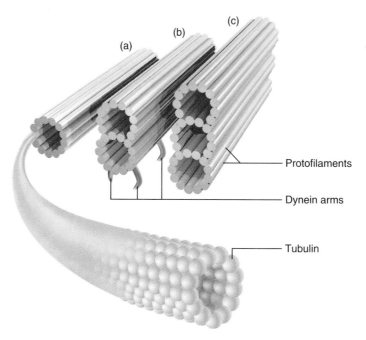

(a) (b) (c)

Protofilaments

Dynein arms

Tubulin

Microtubules
Figure 2.16

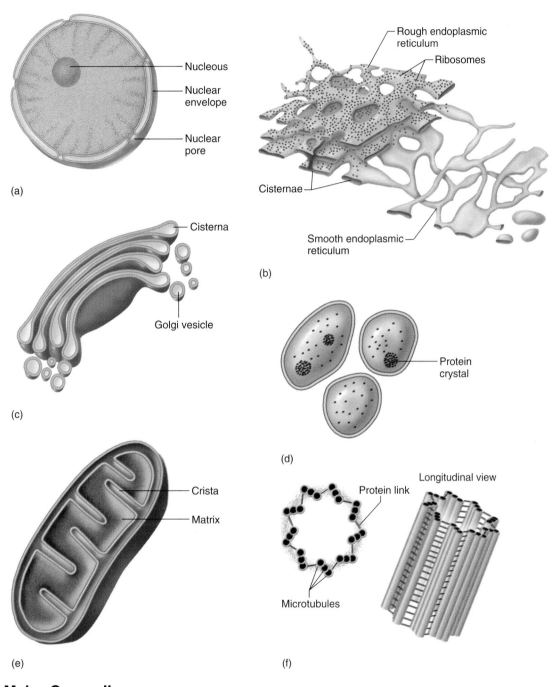

Major Organelles
Figure 2.17

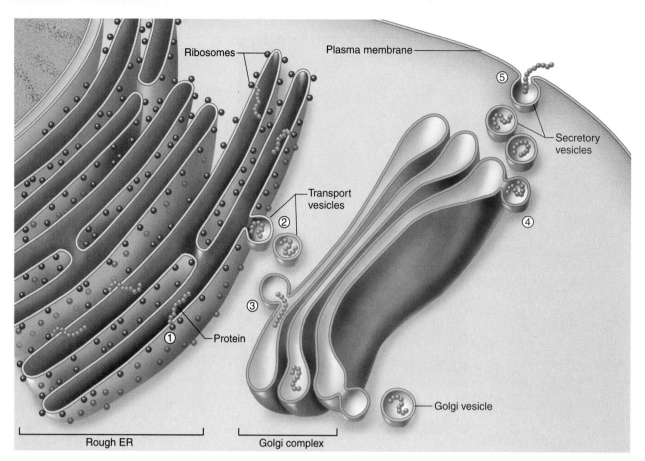

The Functional Relationship of Ribosomes
Figure 2.18

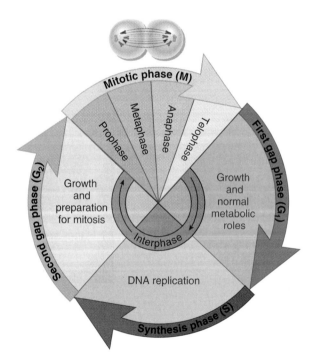

The Cell Cycle
Figure 2.19

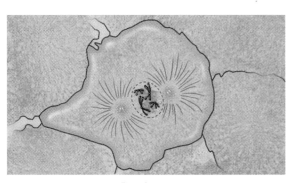

Prophase
Chromatin condenses.
Nucleoli and nuclear envelope break down.
Spindle fibers grow from centrioles.
Centrioles migrate to opposite poles of cell.

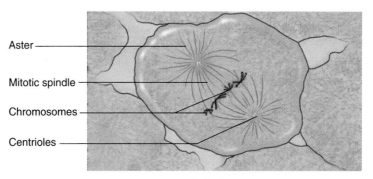

Aster
Mitotic spindle
Chromosomes
Centrioles

Metaphase
Chromosomes lie along midline of cell.
Some spindle fibers attach to kinetochores.
Fibers of aster attach to plasma membrane.

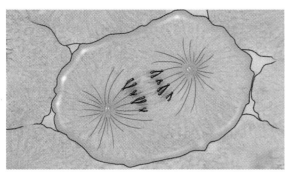

Anaphase
Centromeres divide in two.
Spindle fibers pull sister chromatids to
opposite poles of cell.
Each pole (future daughter cell) now has an
identical set of genes.

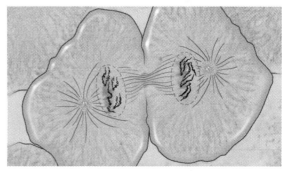

Telophase
Chromosomes gather at each pole of cell.
Chromatin decondenses.
New nuclear envelope appears at each pole.
New nucleoli appear in each nucleus.
Mitotic spindle vanishes.
(Above photo also shows cytokinesis.)

Mitosis
Figure 2.20

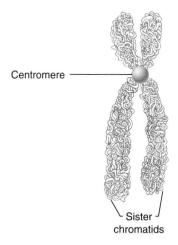

(a)

Chromosomes
Figure 2.21

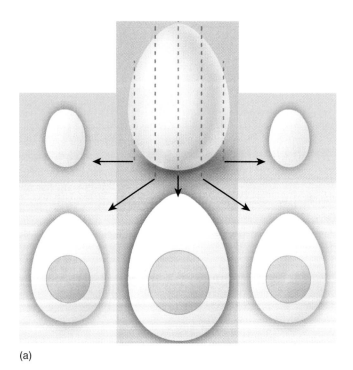

(a)

Three-Dimensional Interpretation of Two-Dimensional Images
Figure 3.1

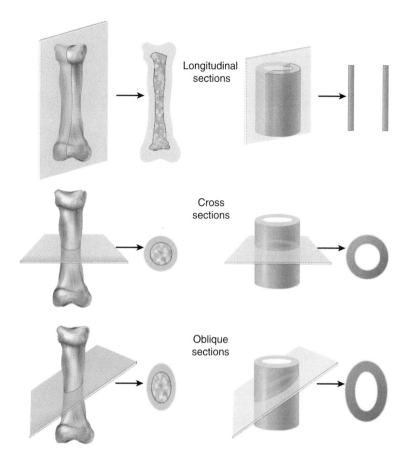

Longitudinal, Cross, and Oblique Sections
Figure 3.2

Simple epithelium

Basement membrane

Stratified epithelium

Comparison of Simple and Stratified Epithelia
Figure 3.3

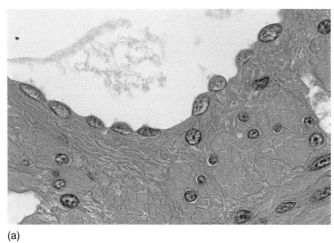

(a)

Nuclei of smooth muscle   Squamous epithelial cells

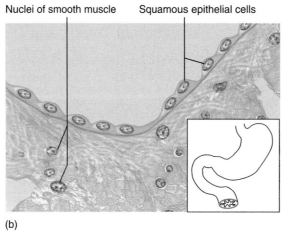

(b)

External Surface (serosa) of the Small Intestine
Figure 3.4

a: © The McGraw-Hill Companies, Inc./Dennis Strete, photographer

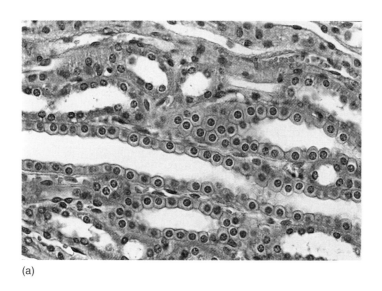

(a)

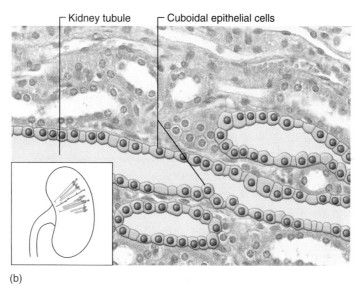

Kidney tubule   Cuboidal epithelial cells

(b)

Kidney Tubules
Figure 3.5

a: © The McGraw-Hill Companies, Inc./Dennis Strete, photographer

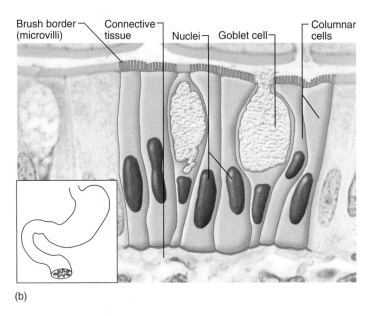

(b)

**Internal Surface (mucosa) of the Small
Intestine**
Figure 3.6

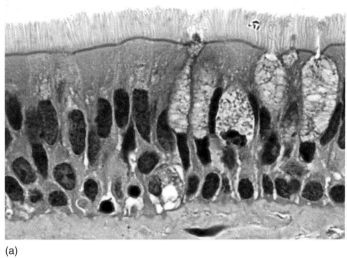

(a)

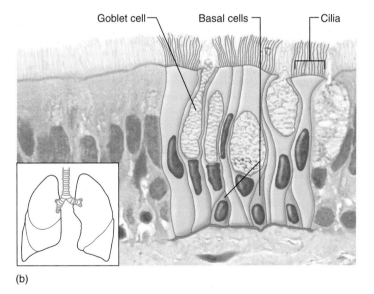

Goblet cell   Basal cells   Cilia

(b)

Mucosa of the Trachea
Figure 3.7

a: © The McGraw-Hill Companies, Inc./Dennis Strete, photographer

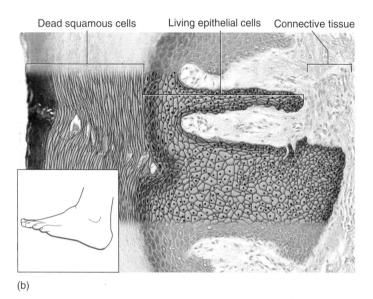

Dead squamous cells   Living epithelial cells   Connective tissue

(b)

Skin from the Sole of the Foot
Figure 3.9

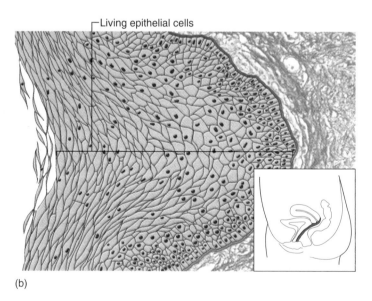

Living epithelial cells

(b)

Mucosa of the Vagina
Figure 3.10

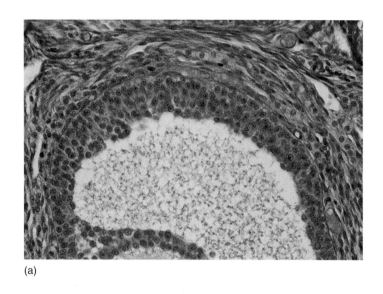

(a)

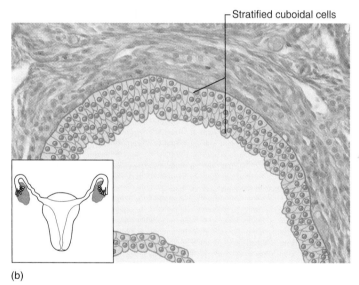

┌Stratified cuboidal cells

(b)

Wall of a Follicle in the Ovary
Figure 3.11

a: © The McGraw-Hill Companies, Inc./Dennis Strete, photographer

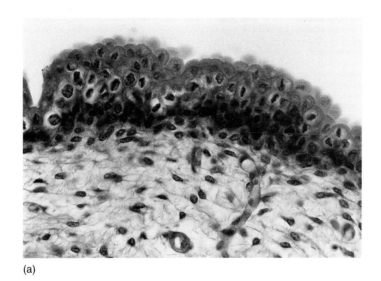

(a)

Connective tissue ─┐   ┌─Transitional epithelial cells   ┌─Blood vessels

(b)

Allantoic Duct of the Umbilical Cord
Figure 3.12

a: © The McGraw-Hill Companies, Inc./Dennis Strete, photographer

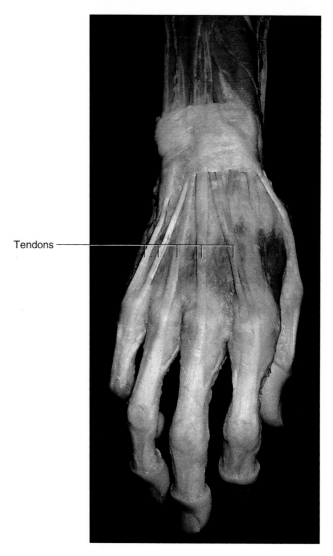

Tendons

Tendons of the Hand
Figure 3.13

© The McGraw-Hill Companies, Inc./Rebecca Gray,
photographer/Don Kincaid, dissections

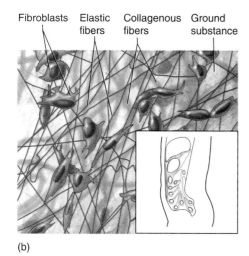

Fibroblasts Elastic   Collagenous Ground
             fibers   fibers   substance

(b)

Spread of the Mesentery
Figure 3.14

57

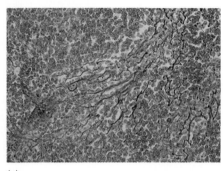

(a)

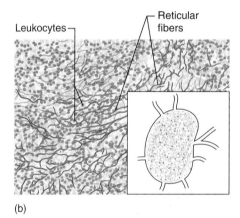

(b)

Lymph Node
Figure 3.15

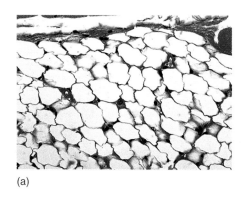

(a)

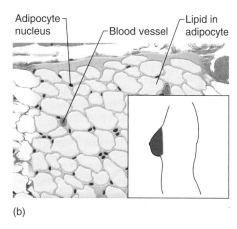

Adipocyte nucleus — ⌐Blood vessel   ⌐Lipid in adipocyte

(b)

Adipose Tissue
Figure 3.16

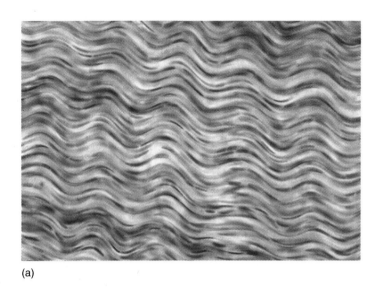

(a)

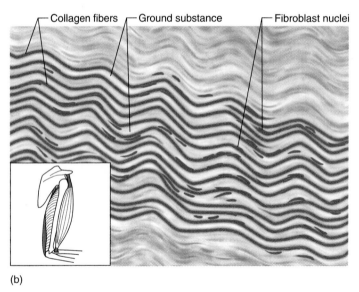

Collagen fibers   Ground substance   Fibroblast nuclei

(b)

Tendon
Figure 3.17

a: © The McGraw-Hill Companies, Inc./Dennis Strete, photographer

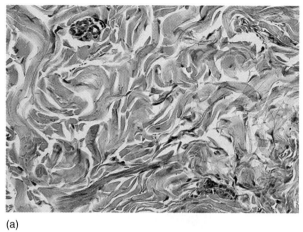

(a)

Bundles of | Gland | Fibroblast | Ground
collagen | ducts | nuclei | substance

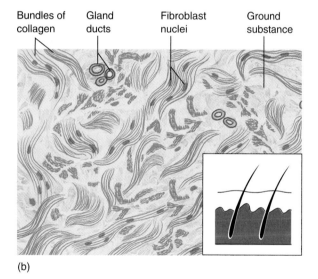

(b)

Dermis of the Skin
Figure 3.18

a: © The McGraw-Hill Companies, Inc./Dennis Strete, photographer

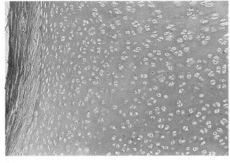

(a)

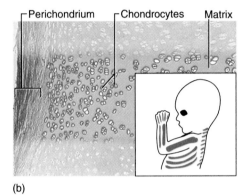

Perichondrium   Chondrocytes   Matrix

(b)

Fetal Skeleton
Figure 3.19

a: © The McGraw-Hill Companies, Inc./
Dennis Strete, photographer

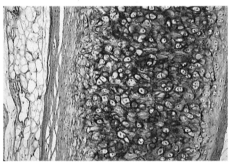

(a)

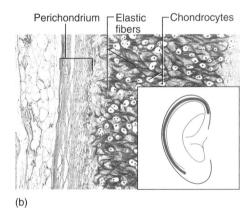

Perichondrium   Elastic   Chondrocytes
              fibers

(b)

External Ear
Figure 3.20

a: © The McGraw-Hill Companies, Inc./
Dennis Strete, photographer

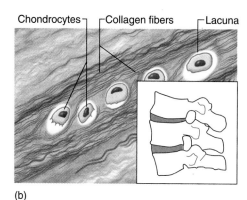
(b)

Intervertebral Disc
Figure 3.21

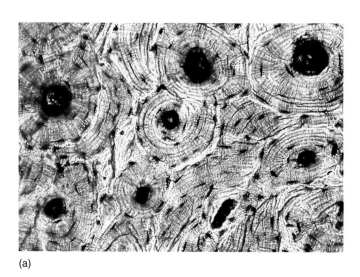
(a)

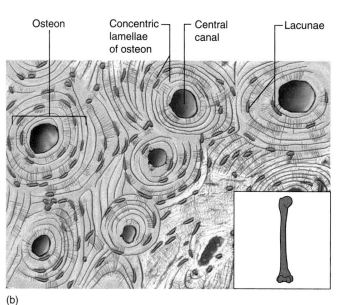
(b)

Compact Bone
Figure 3.22

a: © The McGraw-Hill Companies, Inc./Dennis Strete, photographer

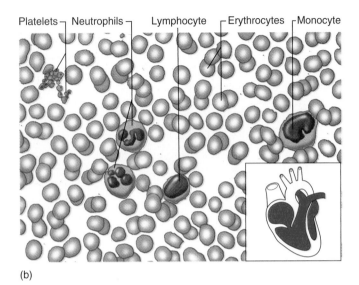

Platelets ⌐ Neutrophils ⌐ Lymphocyte ⌐ Erythrocytes ⌐ Monocyte

(b)

Blood Smear
Figure 3.23

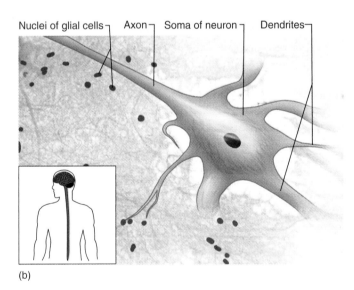

Nuclei of glial cells ⌐ Axon ⌐ Soma of neuron ⌐ Dendrites ⌐

(b)

Spinal Cord Smear
Figure 3.24

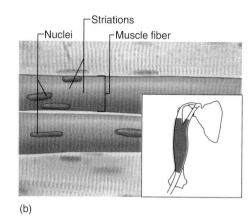

(b)

Skeletal Muscle
Figure 3.25

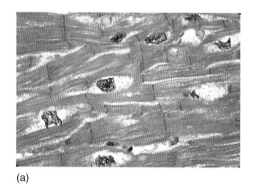

(a)

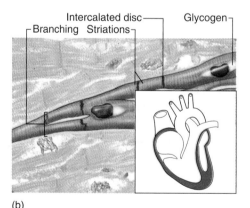

(b)

Cardiac Muscle
Figure 3.26

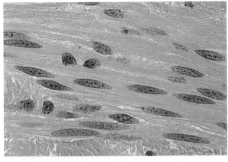

(a)

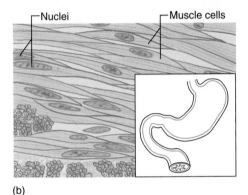

(b)

Smooth Muscle, Wall of the Small Intestine
Figure 3.27

a: © The McGraw-Hill Companies, Inc./ Dennis Strete, photographer

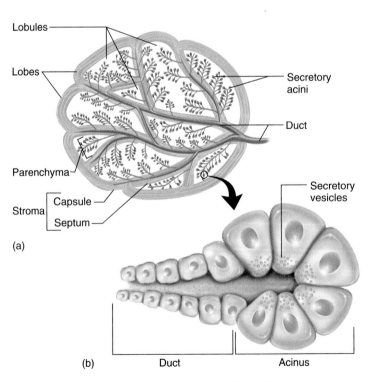

(a)

(b)

General Structure of an Exocrine Gland
Figure 3.28

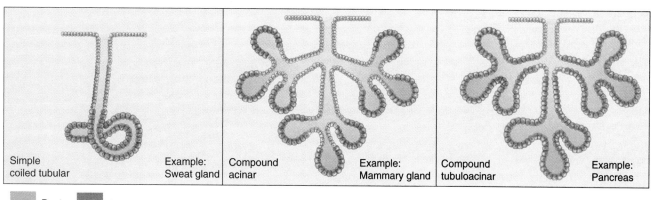

| Simple coiled tubular | Example: Sweat gland | Compound acinar | Example: Mammary gland | Compound tubuloacinar | Example: Pancreas |

Duct ▪ Secretory portion

Some Types of Exocrine Glands
Figure 3.29

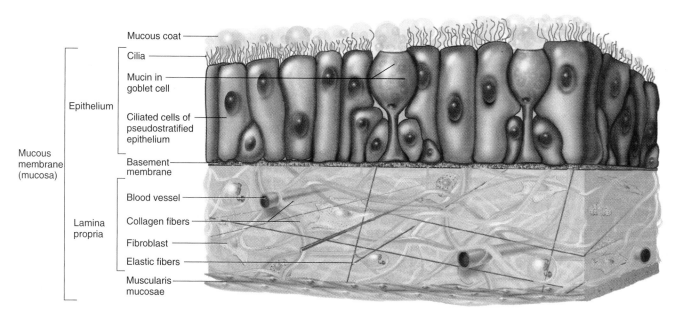

Mucous coat
Cilia
Mucin in goblet cell
Ciliated cells of pseudostratified epithelium
Basement membrane
Blood vessel
Collagen fibers
Fibroblast
Elastic fibers
Muscularis mucosae

Epithelium

Mucous membrane (mucosa)

Lamina propria

Histology of a Mucous Membrane
Figure 3.30

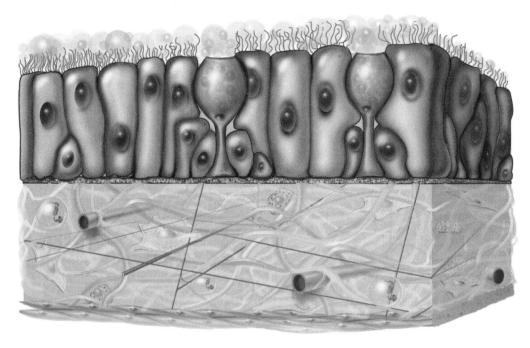

Histology of a Mucous Membrane
Figure 3.30

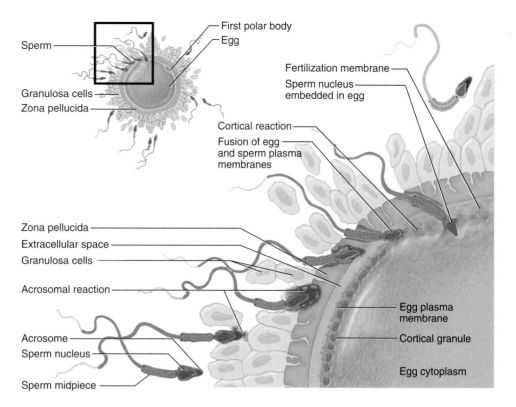

Fertilization
Figure 4.1

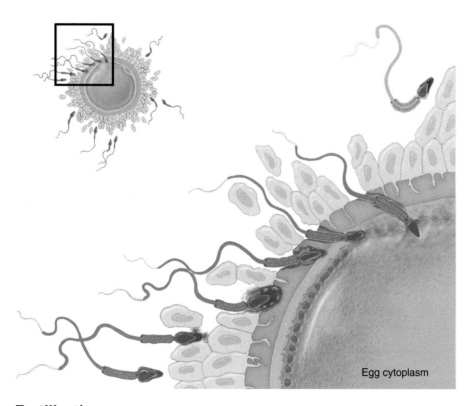

Fertilization
Figure 4.1

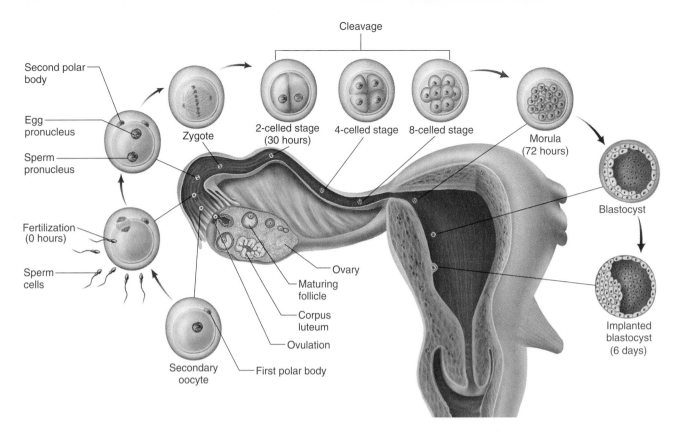

Second polar body

Egg pronucleus

Sperm pronucleus

Fertilization (0 hours)

Sperm cells

Secondary oocyte

First polar body

Ovulation

Corpus luteum

Maturing follicle

Ovary

Zygote

Cleavage

2-celled stage (30 hours)

4-celled stage

8-celled stage

Morula (72 hours)

Blastocyst

Implanted blastocyst (6 days)

Migration of the Conceptus
Figure 4.2

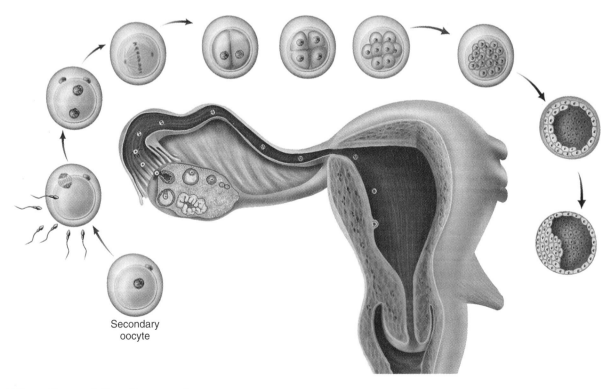

Secondary oocyte

Migration of the Conceptus
Figure 4.2

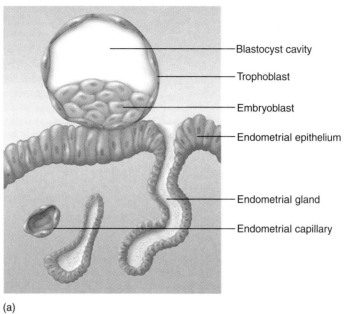

(a)

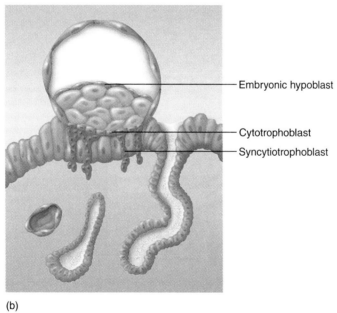

(b)

Implantation
Figure 4.3

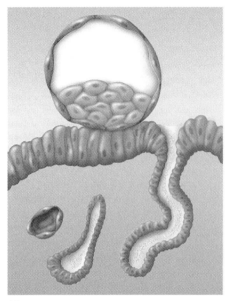

(a)

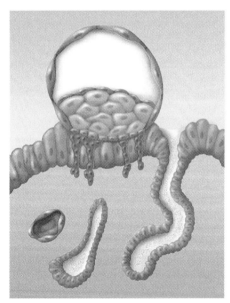

(b)

Implantation
Figure 4.3

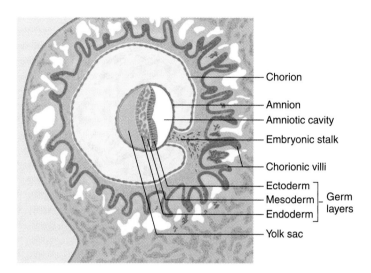

The Implanted Conceptus at 16 Days
Figure 4.4

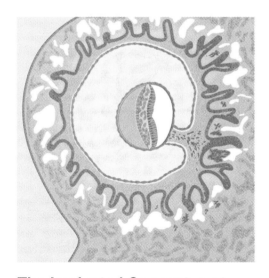

The Implanted Conceptus at 16 Days
Figure 4.4

73

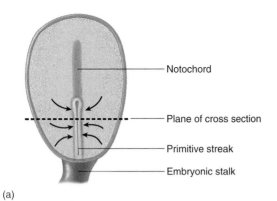

(a)

- Notochord
- Plane of cross section
- Primitive streak
- Embryonic stalk

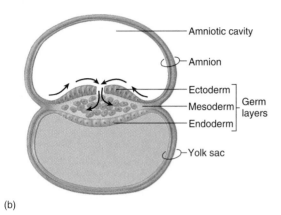

(b)

- Amniotic cavity
- Amnion
- Ectoderm
- Mesoderm — Germ layers
- Endoderm
- Yolk sac

Gastrulation
Figure 4.5

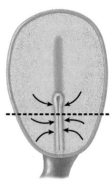

(a)

(b)

Gastrulation
Figure 4.5

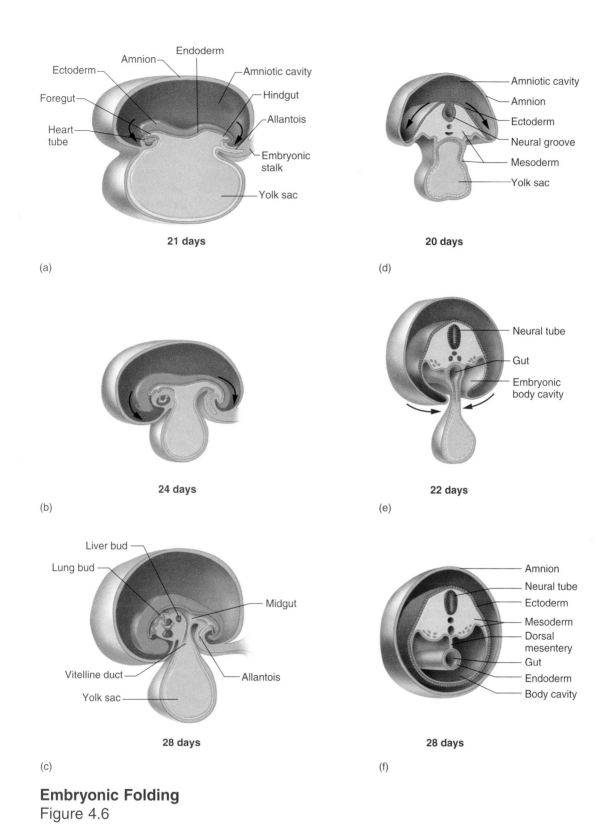

21 days

(a)

24 days

(b)

28 days

(c)

20 days

(d)

22 days

(e)

28 days

(f)

Embryonic Folding

Figure 4.6

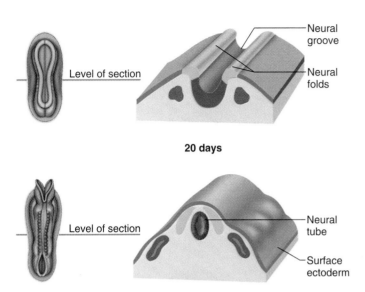

20 days

Neural groove

Neural folds

Level of section

26 days

Neural tube

Surface ectoderm

Level of section

Neurulation

Figure 4.7

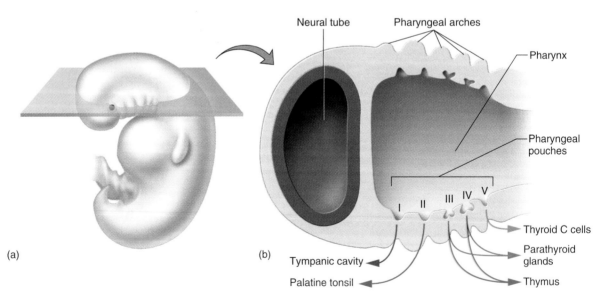

(a)

(b)

Neural tube

Pharyngeal arches

Pharynx

Pharyngeal pouches

I II III IV V

Tympanic cavity

Palatine tonsil

Thyroid C cells

Parathyroid glands

Thymus

The Pharyngeal Pouches

Figure 4.8

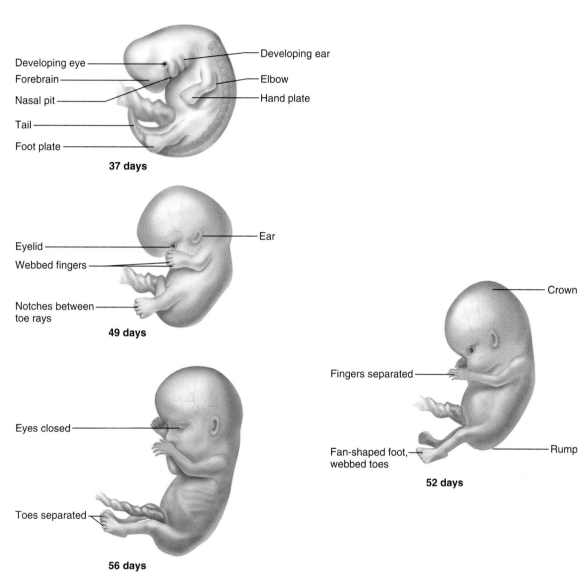

Developing eye

Forebrain

Nasal pit

Tail

Foot plate

Developing ear

Elbow

Hand plate

37 days

Eyelid

Webbed fingers

Notches between toe rays

Ear

49 days

Crown

Fingers separated

Fan-shaped foot, webbed toes

Rump

52 days

Eyes closed

Toes separated

56 days

Development of the Embryo from 37 to 56 Days
Figure 4.10

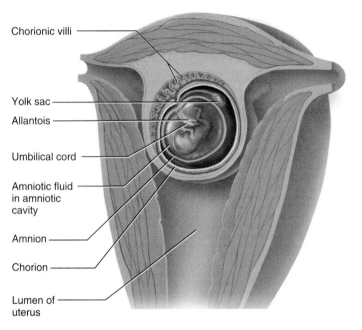

Chorionic villi

Yolk sac

Allantois

Umbilical cord

Amniotic fluid
in amniotic
cavity

Amnion

Chorion

Lumen of
uterus

The Embryonic Membranes
Figure 4.11

The Embryonic Membranes
Figure 4.11

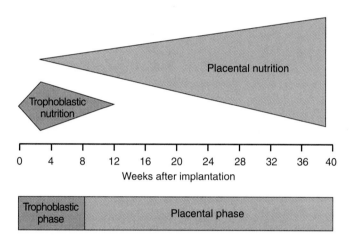

The Phases of Intrauterine Nutrition
Figure 4.12

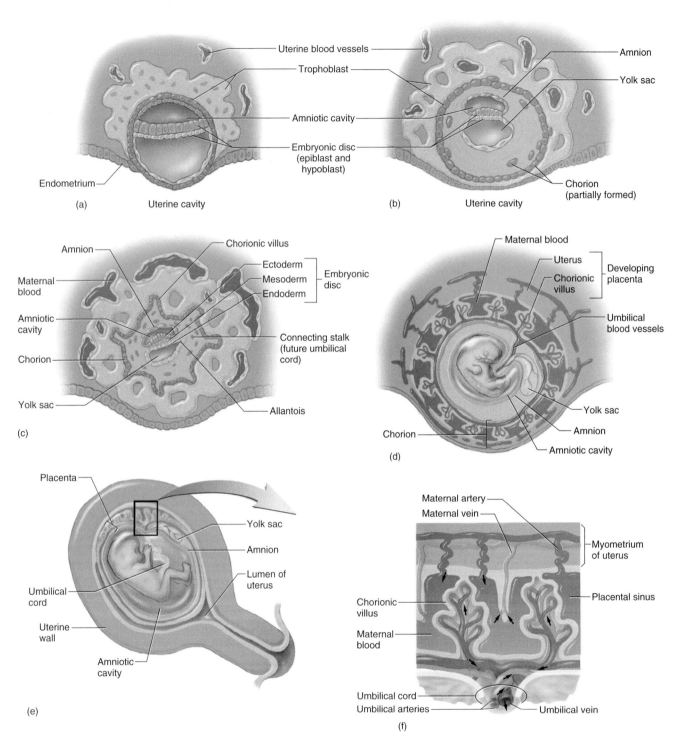

Development of the Placenta and Fetal Membranes
Figure 4.13

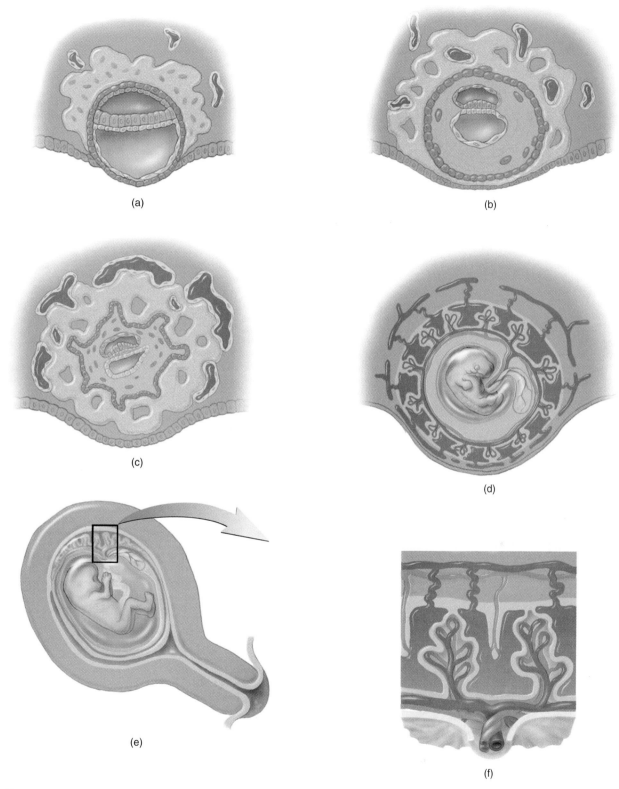

(a)

(b)

(c)

(d)

(e)

(f)

Development of the Placenta and Fetal Membranes
Figure 4.13

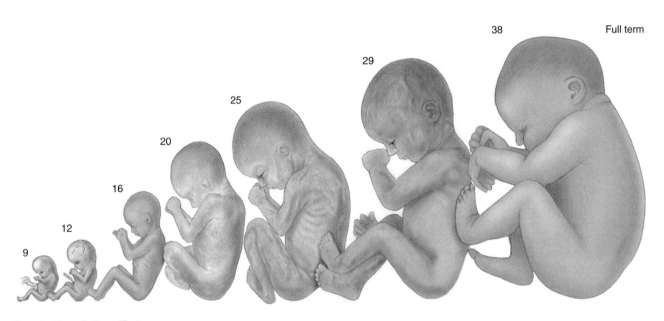

Growth of the Fetus
Figure 4.16

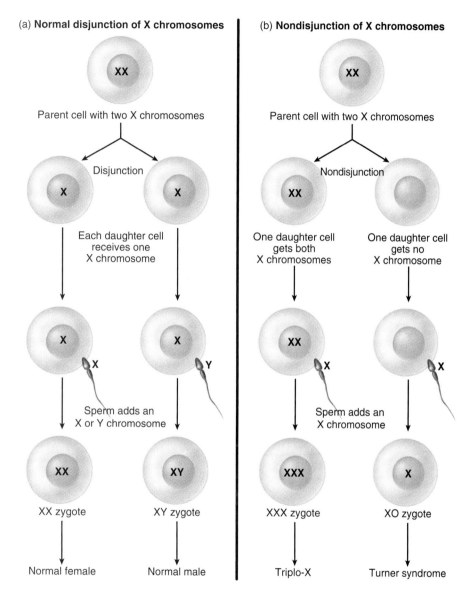

(a) **Normal disjunction of X chromosomes**

XX

Parent cell with two X chromosomes

Disjunction

X   X

Each daughter cell
receives one
X chromosome

X   X

X   Y

Sperm adds an
X or Y chromosome

XX   XY

XX zygote   XY zygote

Normal female   Normal male

(b) **Nondisjunction of X chromosomes**

XX

Parent cell with two X chromosomes

Nondisjunction

XX

One daughter cell
gets both
X chromosomes

One daughter cell
gets no
X chromosome

XX

X

X

Sperm adds an
X chromosome

XXX   X

XXX zygote   XO zygote

Triplo-X   Turner syndrome

Disjunction and Nondisjunction
Figure 4.17

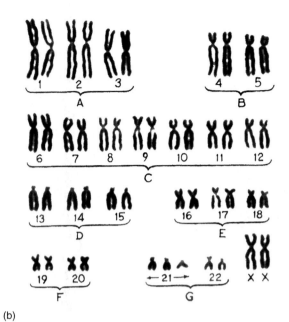

(b)

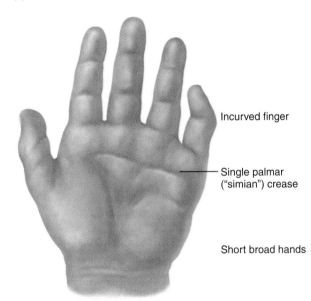

Incurved finger

Single palmar ("simian") crease

Short broad hands

(c)

(d)

Down Syndrome
Figure 4.18

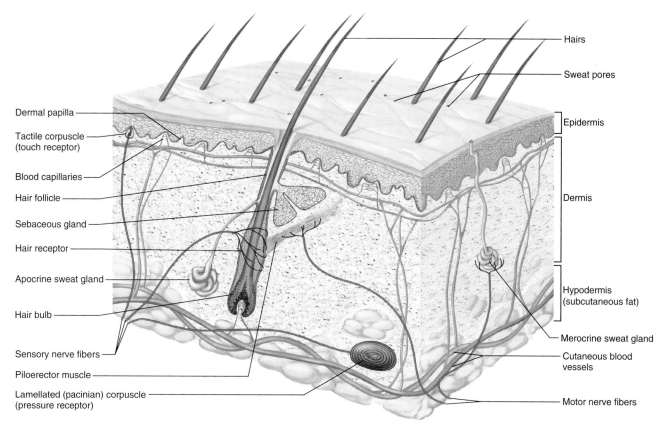

Dermal papilla

Tactile corpuscle (touch receptor)

Blood capillaries

Hair follicle

Sebaceous gland

Hair receptor

Apocrine sweat gland

Hair bulb

Sensory nerve fibers

Piloerector muscle

Lamellated (pacinian) corpuscle (pressure receptor)

Hairs

Sweat pores

Epidermis

Dermis

Hypodermis (subcutaneous fat)

Merocrine sweat gland

Cutaneous blood vessels

Motor nerve fibers

Structure of the Skin and Subcutaneous Tissue
Figure 5.1

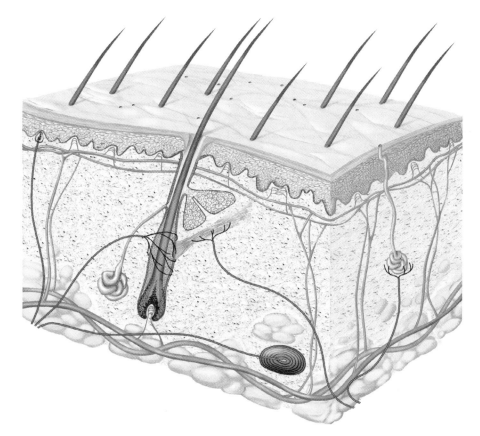

Structure of the Skin and Subcutaneous Tissue
Figure 5.1

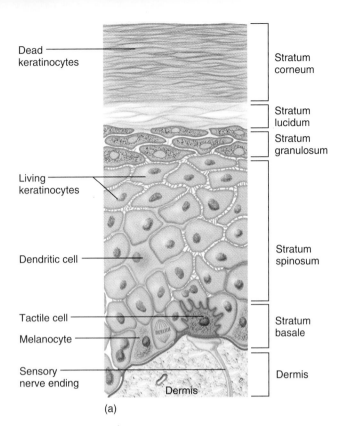

Dead keratinocytes

Stratum corneum

Stratum lucidum

Stratum granulosum

Living keratinocytes

Dendritic cell

Stratum spinosum

Tactile cell

Stratum basale

Melanocyte

Dermis

Sensory nerve ending

Dermis

(a)

The Epidermis
Figure 5.2

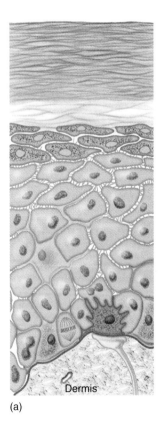

Dermis

(a)

The Epidermis
Figure 5.2

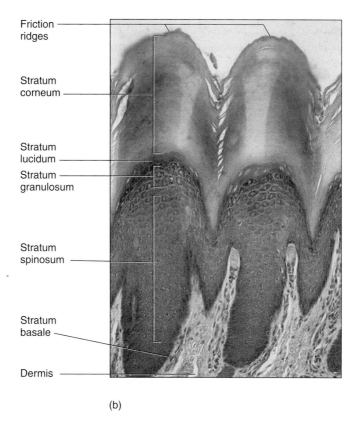

Friction ridges

Stratum corneum

Stratum lucidum

Stratum granulosum

Stratum spinosum

Stratum basale

Dermis

(b)

The Epidermis
Figure 5.2

b: © The McGraw-Hill Companies, Inc./Dennis Strete, photographer

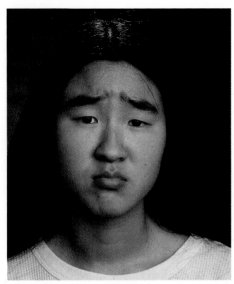

Importance of the Skin in Nonverbal Expression
Figure 5.3

© The McGraw-Hill Companies, Inc./
Joe DeGrandis, photographer

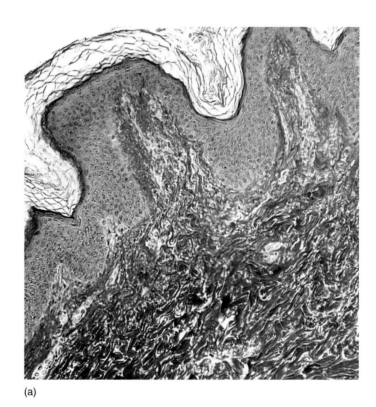

(a)

Layers of the Dermis
Figure 5.4

a: © The McGraw-Hill Companies, Inc./Dennis Strete, photographer

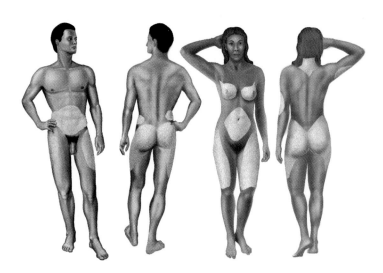

Distribution of Subcutaneous Fat in Men and Women
Figure 5.5

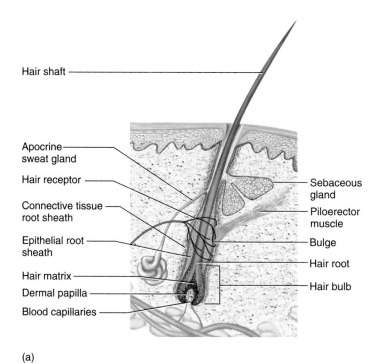

Hair shaft

Apocrine sweat gland

Hair receptor

Connective tissue root sheath

Epithelial root sheath

Hair matrix

Dermal papilla

Blood capillaries

Sebaceous gland

Piloerector muscle

Bulge

Hair root

Hair bulb

(a)

Structure of a Hair and its Follicle
Figure 5.7

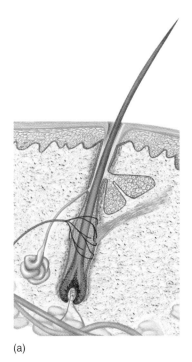

(a)

**Structure of a Hair
and Its Follicle**
Figure 5.7

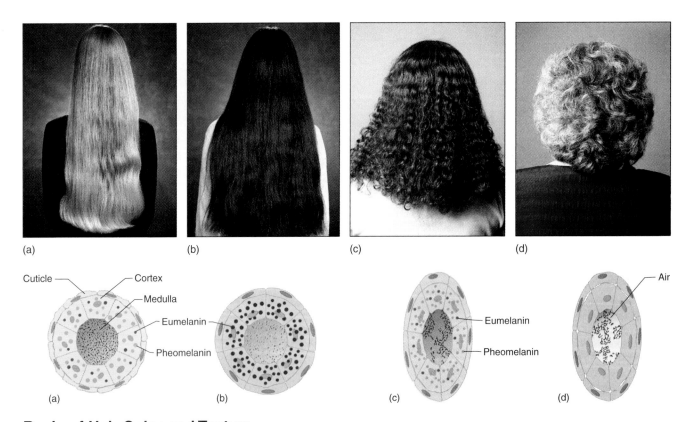

(a)   (b)   (c)   (d)

Cuticle   Cortex

Medulla

Eumelanin

Pheomelanin

(a)   (b)

Eumelanin

Pheomelanin

(c)

Air

(d)

Basis of Hair Color and Texture
Figure 5.8

a–d: © The McGraw-Hill Companies, Inc./Joe DeGrandis, photographer

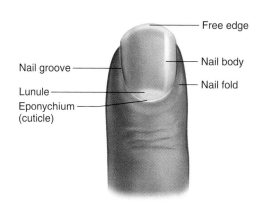

Free edge
Nail groove
Nail body
Lunule
Nail fold
Eponychium
(cuticle)

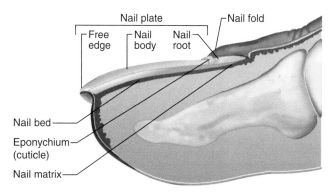

Nail plate
Free edge
Nail body
Nail root
Nail fold
Nail bed
Eponychium
(cuticle)
Nail matrix

Anatomy of a Fingernail
Figure 5.9

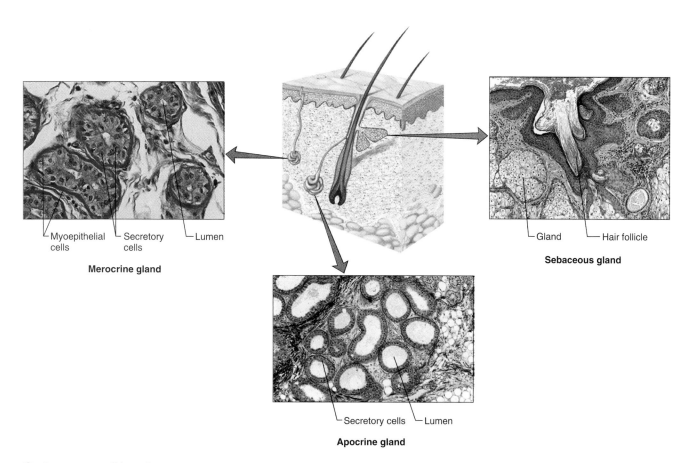

Myoepithelial cells
Secretory cells
Lumen

Merocrine gland

Gland
Hair follicle

Sebaceous gland

Secretory cells
Lumen

Apocrine gland

Cutaneous Glands
Figure 5.10

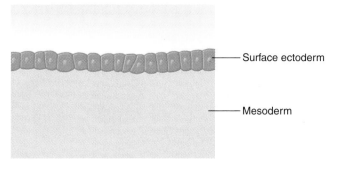

2 weeks

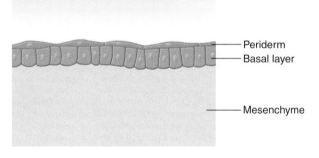

4 weeks

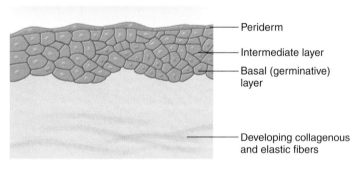

11 weeks

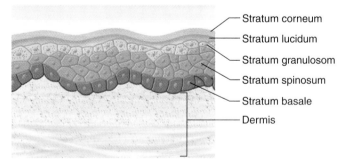

Newborn

Prenatal Development of the Epidermis and Dermis
Figure 5.11

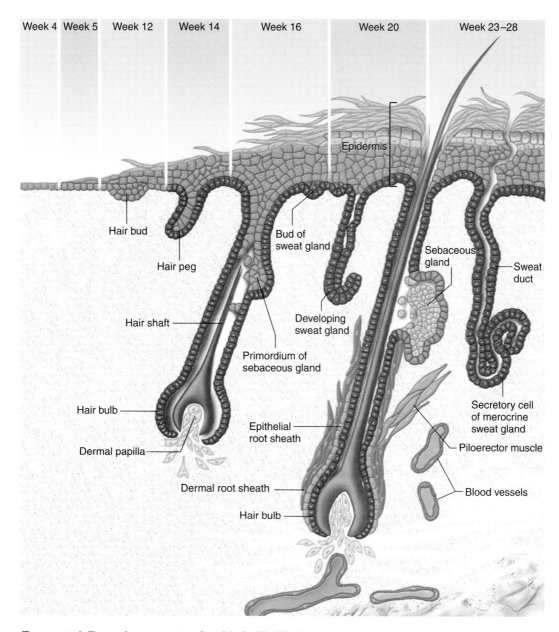

Week 4　Week 5　Week 12　Week 14　Week 16　Week 20　Week 23–28

Epidermis

Hair bud

Bud of
sweat gland

Hair peg

Sebaceous
gland

Sweat
duct

Hair shaft

Developing
sweat gland

Primordium of
sebaceous gland

Secretory cell
of merocrine
sweat gland

Hair bulb

Epithelial
root sheath

Piloerector muscle

Dermal papilla

Dermal root sheath

Blood vessels

Hair bulb

Prenatal Development of a Hair Follicle
Figure 5.12

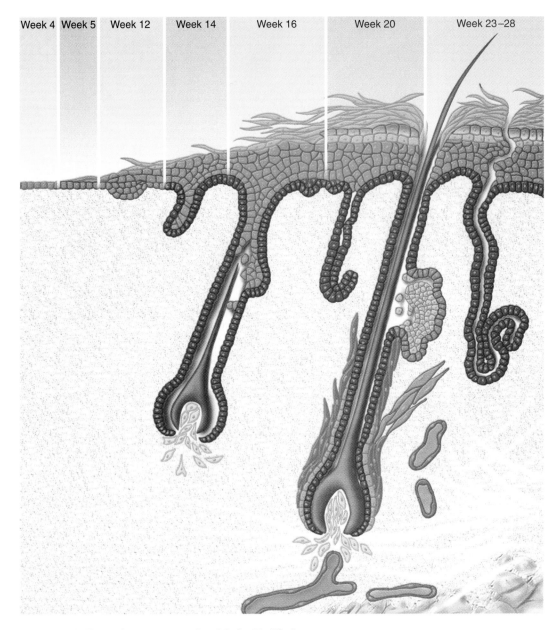

Prenatal Development of a Hair Follicle
Figure 5.12

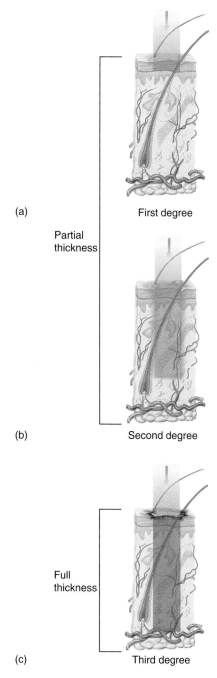

(a)

Partial
thickness

First degree

(b)

Second degree

Full
thickness

(c)

Third degree

Three Degrees of Burns
Figure 5.14

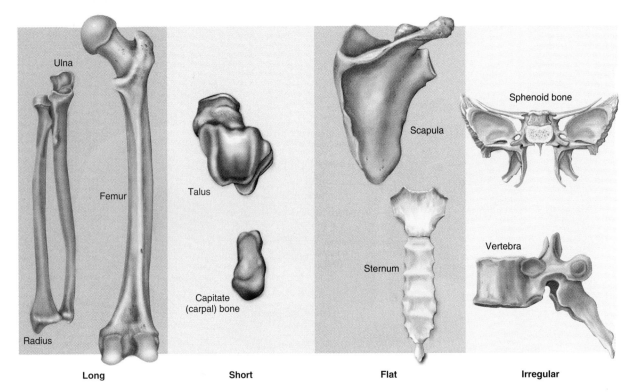

Classification of Bones by Shape
Figure 6.1

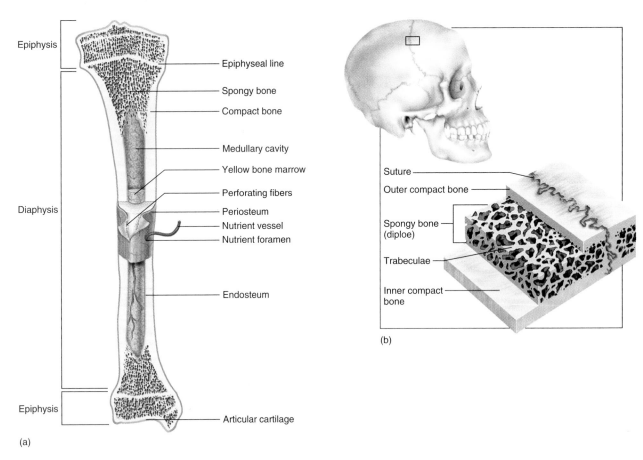

(a)

(b)

General Anatomy of Long and Flat Bones
Figure 6.2

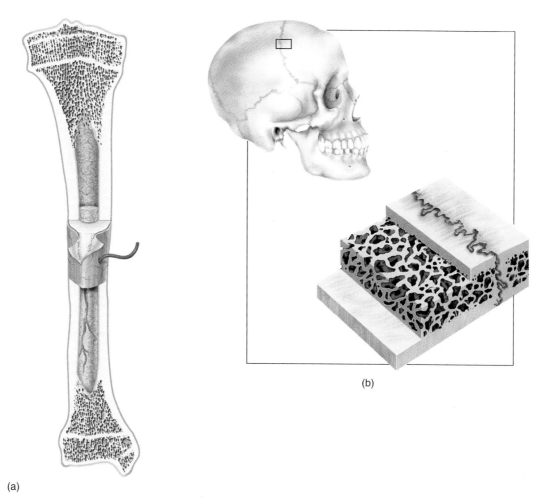

(a)

(b)

General Anatomy of Long and Flat Bones
Figure 6.2

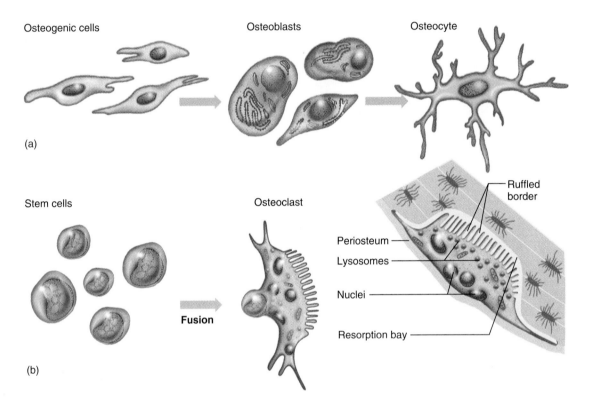

Osteogenic cells

Osteoblasts

Osteocyte

(a)

Stem cells

Osteoclast

Fusion

(b)

Ruffled border

Periosteum

Lysosomes

Nuclei

Resorption bay

Bone Cells and Their Development
Figure 6.3

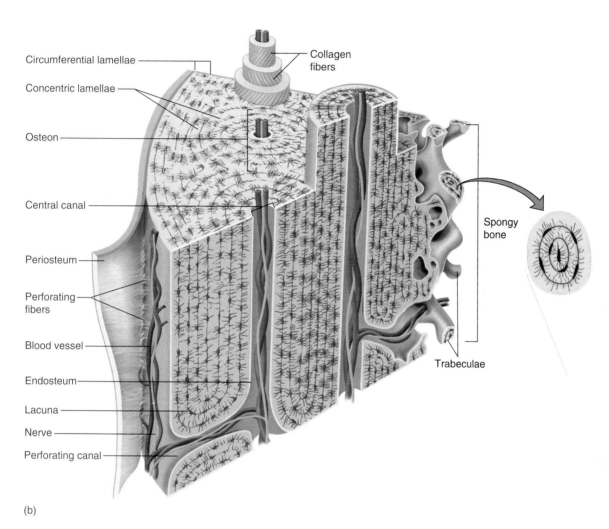

Circumferential lamellae

Concentric lamellae

Osteon

Central canal

Periosteum

Perforating fibers

Blood vessel

Endosteum

Lacuna

Nerve

Perforating canal

Collagen fibers

Spongy bone

Trabeculae

(b)

The Histology of Osseous Tissue
Figure 6.4

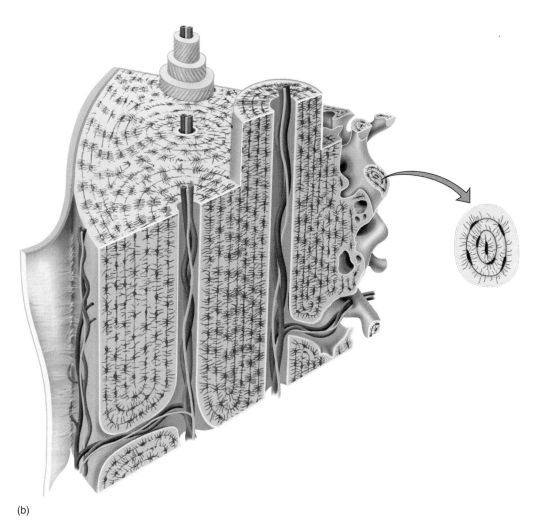

(b)

The Histology of Osseous Tissue
Figure 6.4

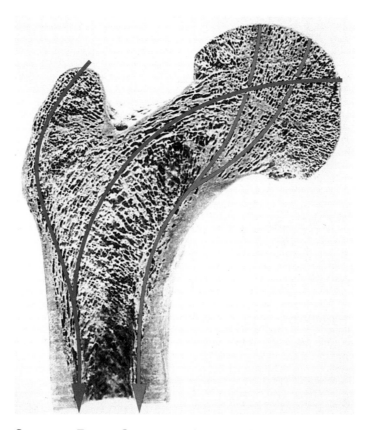

**Spongy Bone Structure in Relation to
Mechanical Stress**
Figure 6.5

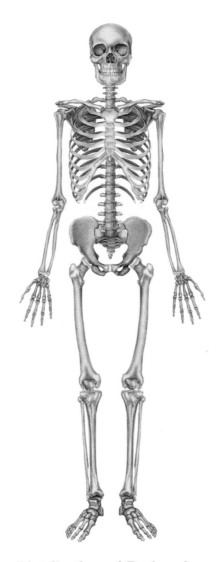

**Distribution of Red and
Yellow Bone Marrow**
Figure 6.6

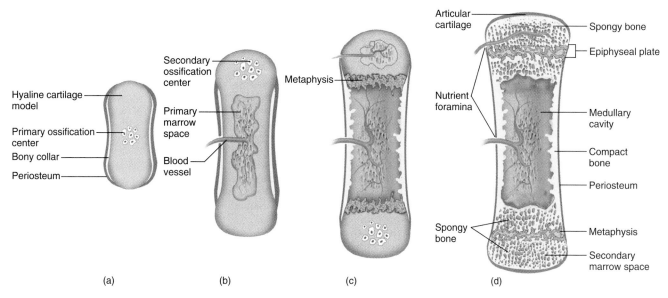

Hyaline cartilage model

Primary ossification center

Bony collar

Periosteum

Secondary ossification center

Primary marrow space

Blood vessel

Metaphysis

Articular cartilage

Nutrient foramina

Spongy bone

Spongy bone

Epiphyseal plate

Medullary cavity

Compact bone

Periosteum

Metaphysis

Secondary marrow space

(a)　　　(b)　　　(c)　　　(d)

Stages of Endochondral Ossification
Figure 6.7

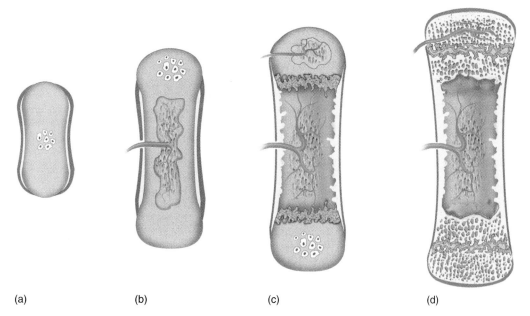

(a)　　　(b)　　　(c)　　　(d)

Stages of Endochondral Ossification
Figure 6.7

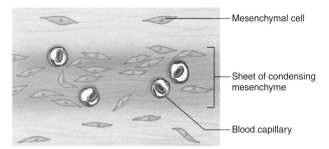

1. Embryonic mesenchyme condenses into a soft sheet permeated with blood capillaries. The mesenchymal cells in this sheet soon differentiate into osteogenic cells, which further differentiate into osteoblasts.

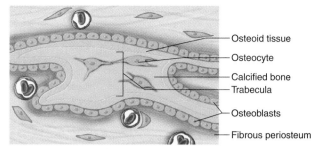

2. Osteoblasts form rows on the surface of a mesenchymal sheet, secrete a layer of osteoid tissue, and then calcify it to form bony plates, or trabeculae. Osteoblasts that become trapped in the matrix become osteocytes. A fibrous periosteum forms external to the osteoblast layer.

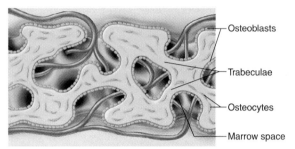

3. Continued bone deposition forms a honeycomb of bony trabeculae enclosing marrow spaces with blood vessels.

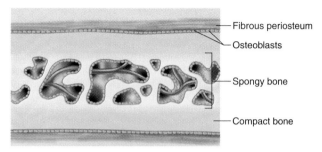

4. Further ossification at the surface of the bone fills in the spaces and produces surface plates of compact bone. Spongy bone remains in the center of the plate, forming the typical sandwichlike arrangement of a flat bone. In the skull, this middle layer of spongy bone is called the diploe.

Stages of Intramembranous Ossification
Figure 6.10

Achondroplastic Dwarfism
Figure 6.12

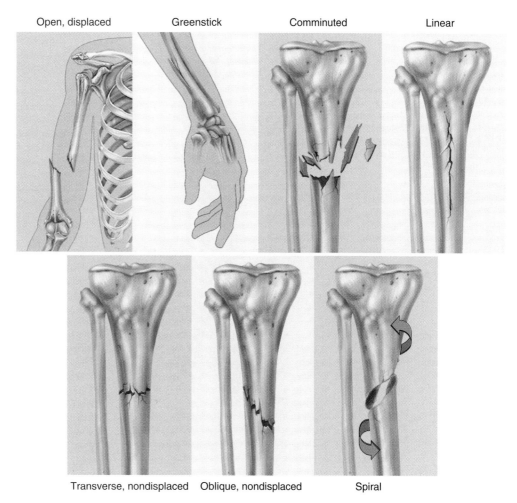

Open, displaced   Greenstick   Comminuted   Linear

Transverse, nondisplaced   Oblique, nondisplaced   Spiral

Some Types of Bone Fractures
Figure 6.13

(a)
Medullary cavity
Hematoma
Compact bone

(b)
Fibrocartilage
Soft callus
New blood vessels

(c)
Hard callus
Spongy bone

(d)

The Healing of a Bone Fracture
Figure 6.15

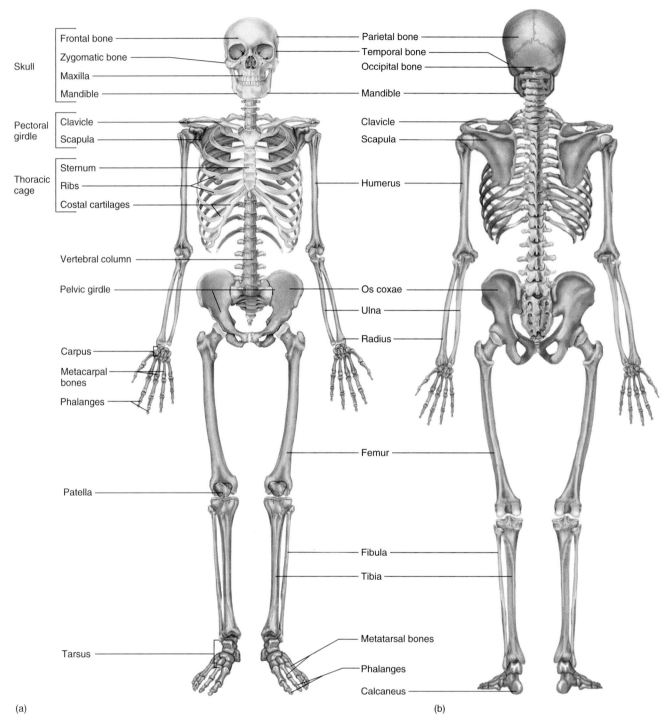

Skull
- Frontal bone
- Zygomatic bone
- Maxilla
- Mandible

- Parietal bone
- Temporal bone
- Occipital bone
- Mandible

Pectoral girdle
- Clavicle
- Scapula

- Clavicle
- Scapula

Thoracic cage
- Sternum
- Ribs
- Costal cartilages

- Humerus

- Vertebral column
- Pelvic girdle

- Os coxae
- Ulna
- Radius

- Carpus
- Metacarpal bones
- Phalanges

- Femur

- Patella

- Fibula
- Tibia

- Tarsus

- Metatarsal bones
- Phalanges
- Calcaneus

(a)   (b)

The Adult Skeleton
Figure 7.1

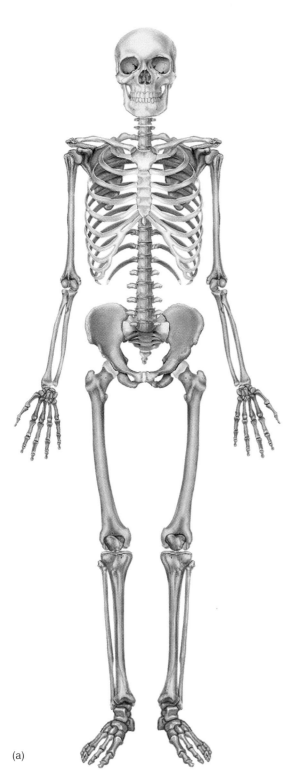

(a)

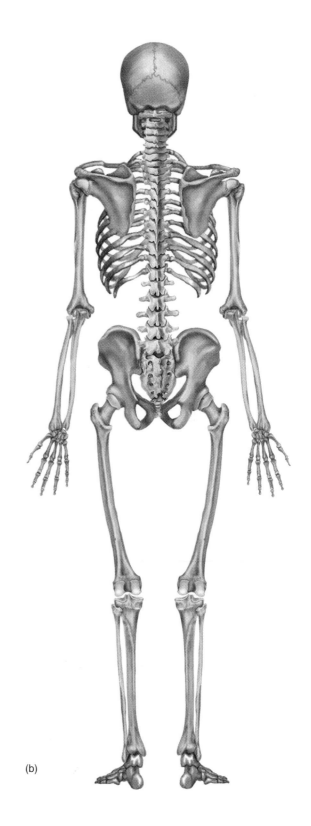

(b)

The Adult Skeleton
Figure 7.1

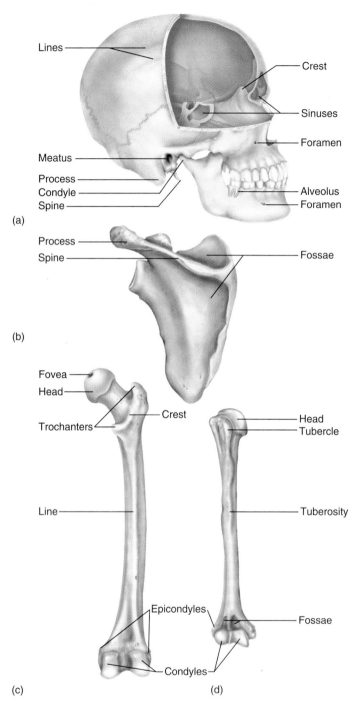

Lines

Crest

Sinuses

Foramen

Meatus
Process
Condyle
Spine

Alveolus
Foramen

(a)

Process
Spine

Fossae

(b)

Fovea
Head

Crest

Trochanters

Head
Tubercle

Line

Tuberosity

Epicondyles

Fossae

Condyles

(c)   (d)

Surface Features of Bones
Figure 7.2

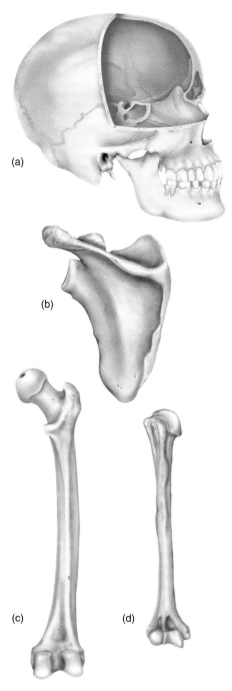

(a)

(b)

(c)   (d)

Surface Features of Bones
Figure 7.2

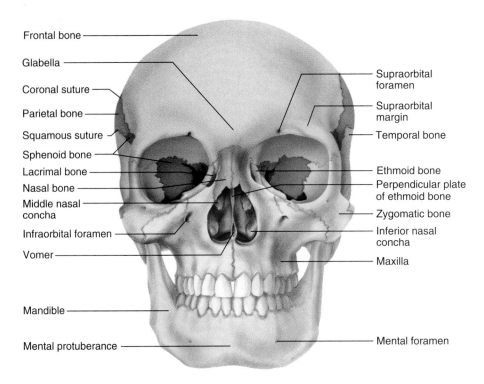

Frontal bone

Glabella

Coronal suture

Parietal bone

Squamous suture

Sphenoid bone

Lacrimal bone

Nasal bone

Middle nasal concha

Infraorbital foramen

Vomer

Mandible

Mental protuberance

Supraorbital foramen

Supraorbital margin

Temporal bone

Ethmoid bone

Perpendicular plate of ethmoid bone

Zygomatic bone

Inferior nasal concha

Maxilla

Mental foramen

The Skull, Anterior View
Figure 7.3

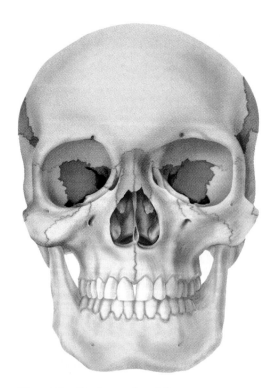

The Skull, Anterior View
Figure 7.3

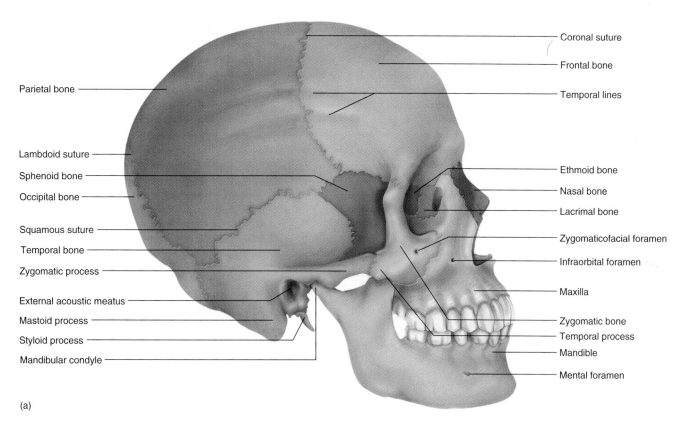

Parietal bone

Lambdoid suture
Sphenoid bone
Occipital bone

Squamous suture
Temporal bone
Zygomatic process

External acoustic meatus
Mastoid process
Styloid process
Mandibular condyle

Coronal suture
Frontal bone
Temporal lines

Ethmoid bone
Nasal bone
Lacrimal bone
Zygomaticofacial foramen
Infraorbital foramen
Maxilla
Zygomatic bone
Temporal process
Mandible
Mental foramen

(a)

The Skull
Figure 7.4

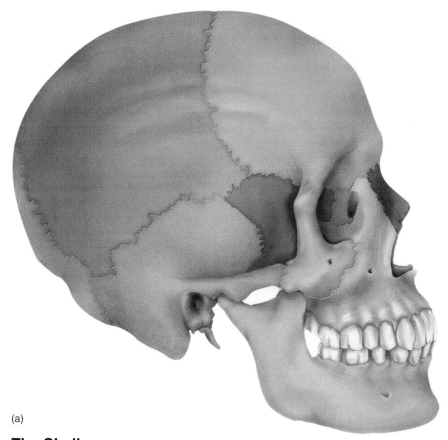

(a)

The Skull
Figure 7.4

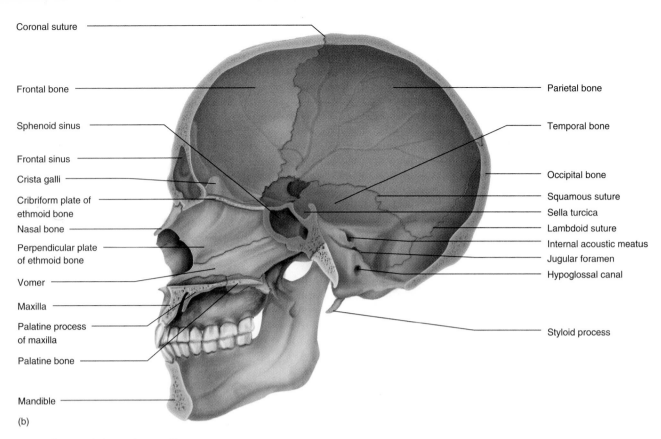

Coronal suture

Frontal bone

Sphenoid sinus

Frontal sinus

Crista galli

Cribriform plate of
ethmoid bone

Nasal bone

Perpendicular plate
of ethmoid bone

Vomer

Maxilla

Palatine process
of maxilla

Palatine bone

Mandible

Parietal bone

Temporal bone

Occipital bone

Squamous suture

Sella turcica

Lambdoid suture

Internal acoustic meatus

Jugular foramen

Hypoglossal canal

Styloid process

(b)

The Skull (*Continued*)
Figure 7.4

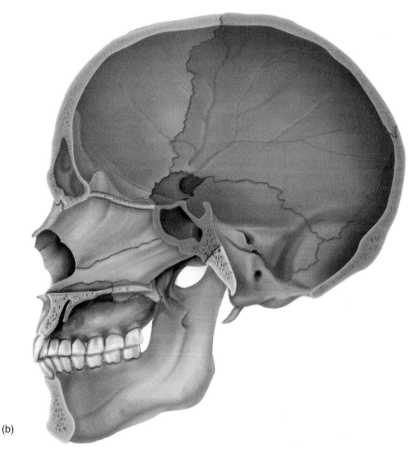

(b)

The Skull
Figure 7.4

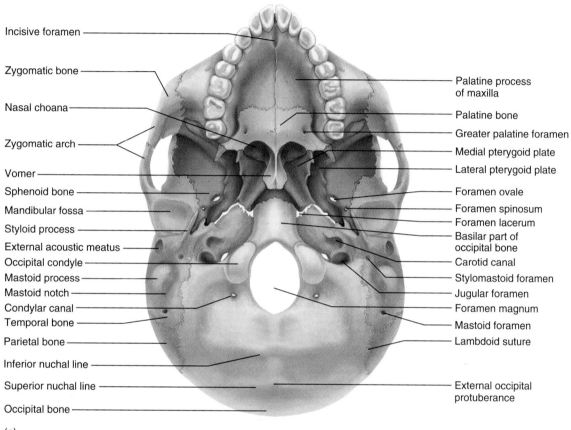

Incisive foramen

Zygomatic bone

Nasal choana

Zygomatic arch

Vomer
Sphenoid bone
Mandibular fossa
Styloid process
External acoustic meatus
Occipital condyle
Mastoid process
Mastoid notch
Condylar canal
Temporal bone
Parietal bone

Inferior nuchal line

Superior nuchal line

Occipital bone

Palatine process
of maxilla
Palatine bone
Greater palatine foramen
Medial pterygoid plate
Lateral pterygoid plate
Foramen ovale
Foramen spinosum
Foramen lacerum
Basilar part of
occipital bone
Carotid canal
Stylomastoid foramen
Jugular foramen
Foramen magnum
Mastoid foramen
Lambdoid suture

External occipital
protuberance

(a)

Base of the Skull
Figure 7.5

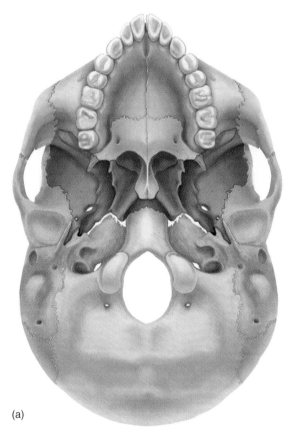

(a)

Base of the Skull
Figure 7.5

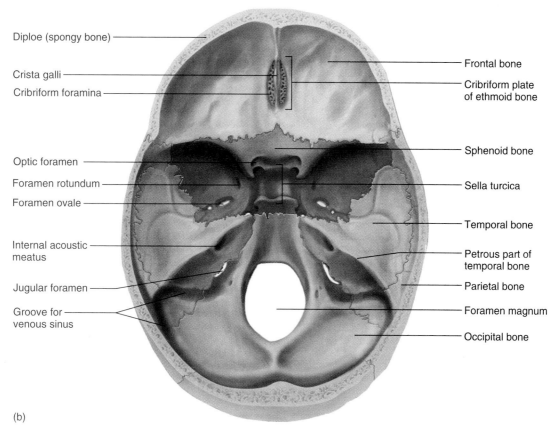

Diploe (spongy bone)

Crista galli

Cribriform foramina

Optic foramen

Foramen rotundum

Foramen ovale

Internal acoustic meatus

Jugular foramen

Groove for venous sinus

Frontal bone

Cribriform plate of ethmoid bone

Sphenoid bone

Sella turcica

Temporal bone

Petrous part of temporal bone

Parietal bone

Foramen magnum

Occipital bone

(b)

Base of the Skull (*Continued*)
Figure 7.5

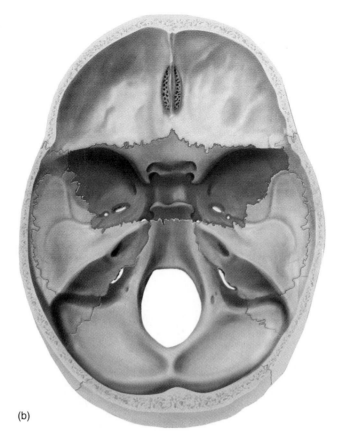

(b)

Base of the Skull
Figure 7.5

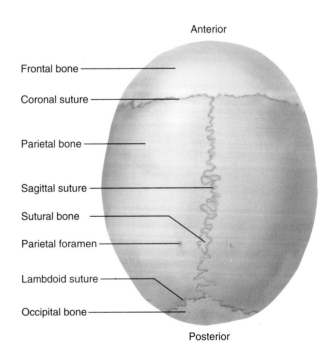

Anterior

Frontal bone

Coronal suture

Parietal bone

Sagittal suture

Sutural bone

Parietal foramen

Lambdoid suture

Occipital bone

Posterior

The Calvaria (skull)
Figure 7.6

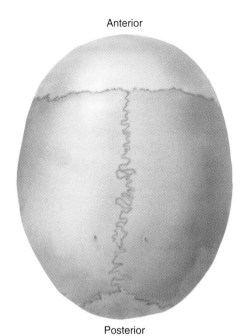

Anterior

Posterior

The Calvaria (skull)
Figure 7.6

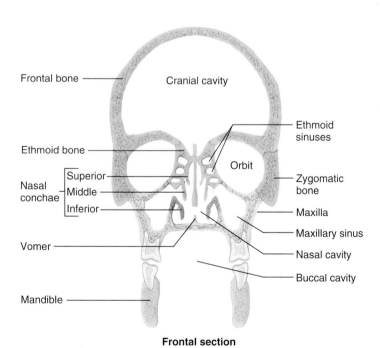

Frontal bone

Cranial cavity

Ethmoid
sinuses

Ethmoid bone

Orbit

Superior

Nasal
conchae

Middle

Inferior

Zygomatic
bone

Maxilla

Maxillary sinus

Vomer

Nasal cavity

Buccal cavity

Mandible

Frontal section

Major Cavities of the Skull
Figure 7.7

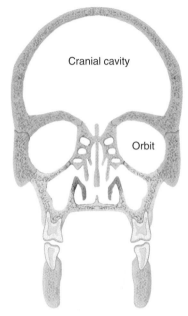

Frontal section

Major Cavities of the Skull
Figure 7.7

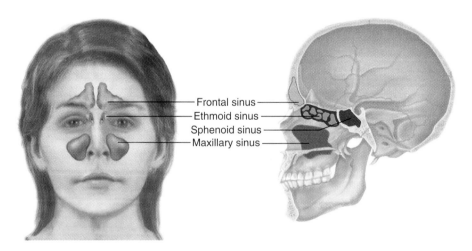

Frontal sinus
Ethmoid sinus
Sphenoid sinus
Maxillary sinus

The Paranasal Sinuses
Figure 7.8

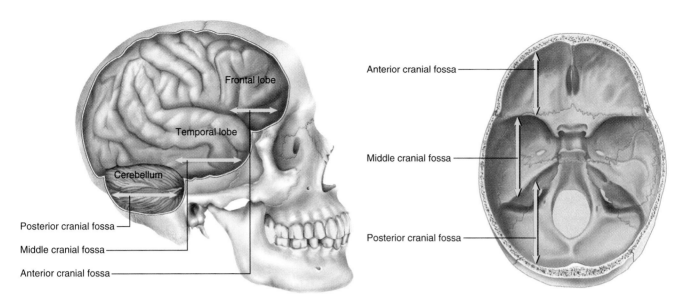

Cranial Fossae
Figure 7.9

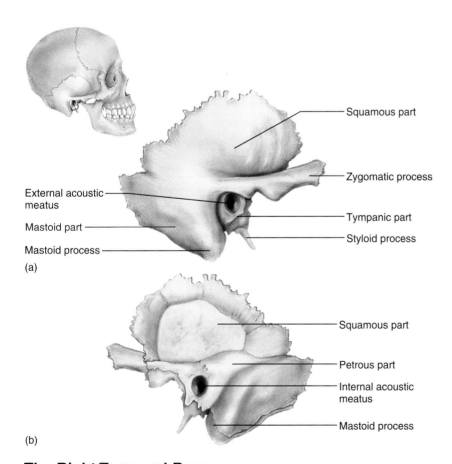

Squamous part

Zygomatic process

External acoustic meatus

Tympanic part

Mastoid part

Styloid process

Mastoid process

(a)

Squamous part

Petrous part

Internal acoustic meatus

Mastoid process

(b)

The Right Temporal Bone
Figure 7.10

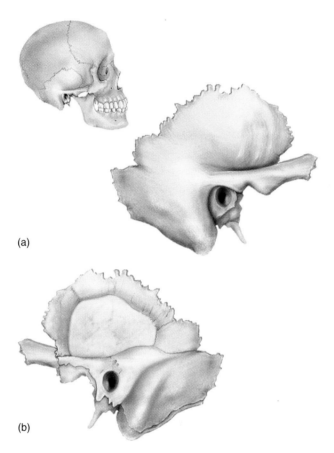

(a)

(b)

The Right Temporal Bone
Figure 7.10

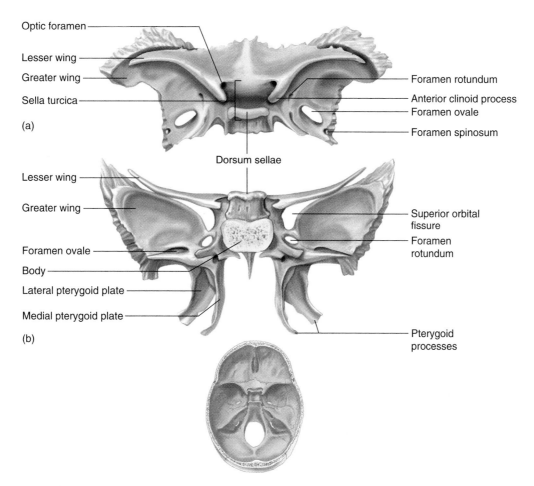

Optic foramen
Lesser wing
Greater wing
Sella turcica

(a)

Foramen rotundum
Anterior clinoid process
Foramen ovale
Foramen spinosum

Dorsum sellae

Lesser wing
Greater wing

Foramen ovale
Body
Lateral pterygoid plate
Medial pterygoid plate

(b)

Superior orbital fissure
Foramen rotundum

Pterygoid processes

The Sphenoid Bone
Figure 7.11

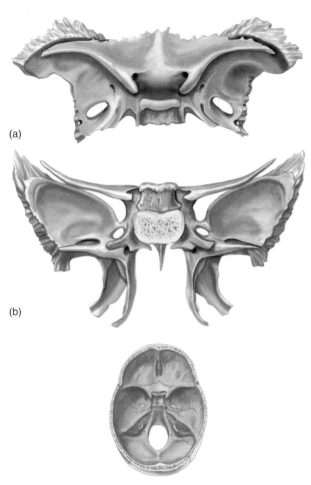

(a)

(b)

The Sphenoid Bone
Figure 7.11

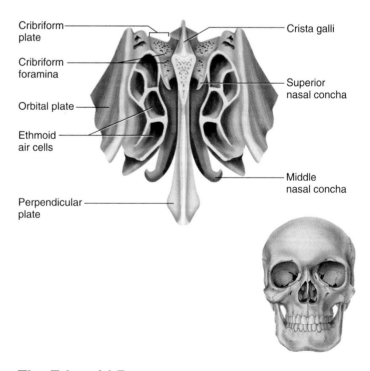

Cribriform plate

Cribriform foramina

Orbital plate

Ethmoid air cells

Perpendicular plate

Crista galli

Superior nasal concha

Middle nasal concha

The Ethmoid Bone
Figure 7.12

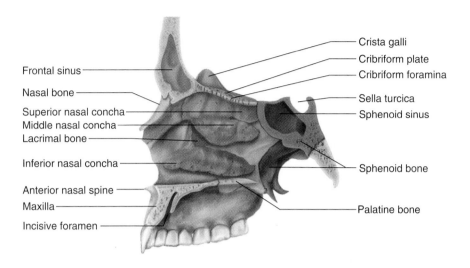

Frontal sinus

Nasal bone

Superior nasal concha

Middle nasal concha

Lacrimal bone

Inferior nasal concha

Anterior nasal spine

Maxilla

Incisive foramen

Crista galli

Cribriform plate

Cribriform foramina

Sella turcica

Sphenoid sinus

Sphenoid bone

Palatine bone

The Right Nasal Cavity
Figure 7.13

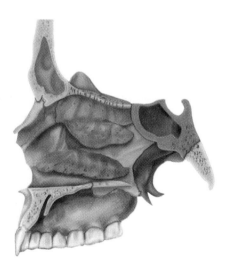

The Right Nasal Cavity
Figure 7.13

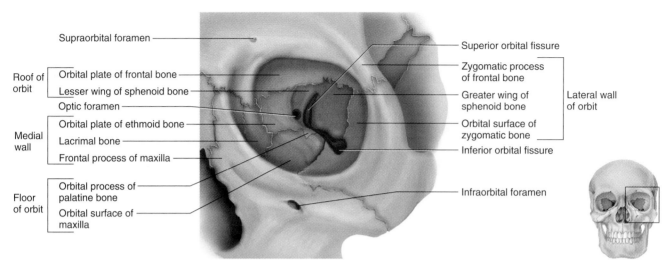

Supraorbital foramen

Roof of orbit
Orbital plate of frontal bone
Lesser wing of sphenoid bone
Optic foramen

Medial wall
Orbital plate of ethmoid bone
Lacrimal bone
Frontal process of maxilla

Floor of orbit
Orbital process of palatine bone
Orbital surface of maxilla

Superior orbital fissure

Zygomatic process of frontal bone

Greater wing of sphenoid bone

Lateral wall of orbit

Orbital surface of zygomatic bone

Inferior orbital fissure

Infraorbital foramen

The Left Orbit
Figure 7.14

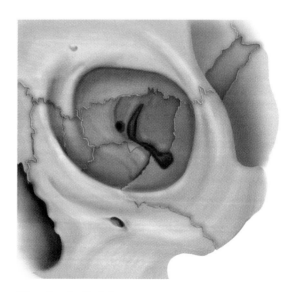

The Left Orbit
Figure 7.14

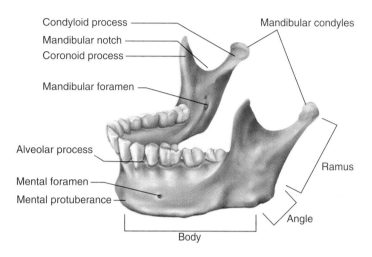

Condyloid process — Mandibular condyles
Mandibular notch —
Coronoid process —

Mandibular foramen —

Alveolar process —

Mental foramen —
Mental protuberance —

Ramus

Angle

Body

The Mandible
Figure 7.15

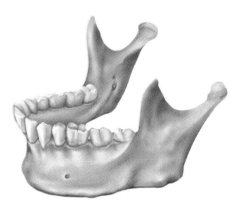

The Mandible
Figure 7.15

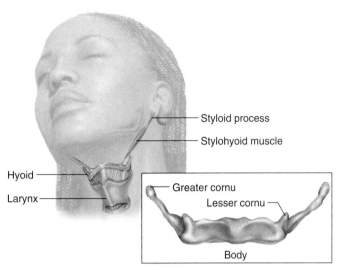

The Hyoid Bone
Figure 7.16

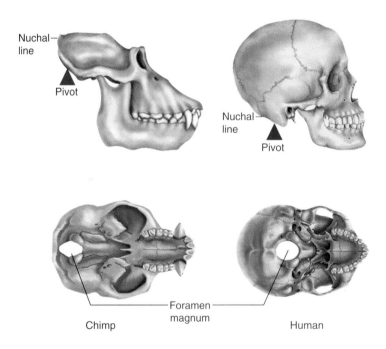

Adaptations of the Skull for Bipedalism
Figure 7.17

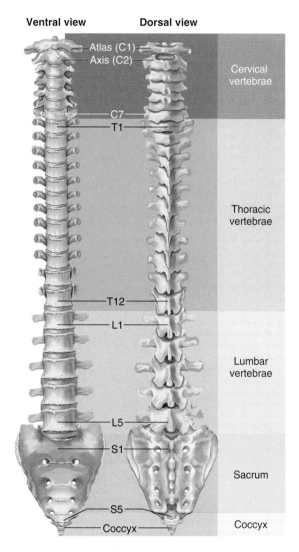

Ventral view   **Dorsal view**

Atlas (C1)
Axis (C2)

Cervical vertebrae

C7
T1

Thoracic vertebrae

T12
L1

Lumbar vertebrae

L5
S1

Sacrum

S5
Coccyx

Coccyx

The Vertebral Column
Figure 7.18

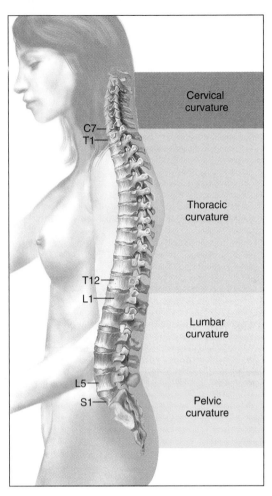

Curvatures of the Adult Vertebral Column
Figure 7.19

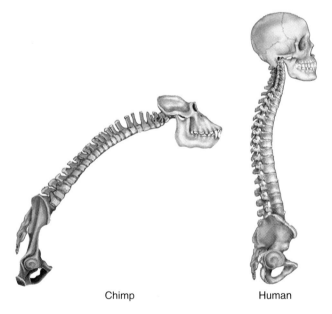

Chimp   Human

Comparison of Chimpanzee and Human Vertebral Columns
Figure 7.20

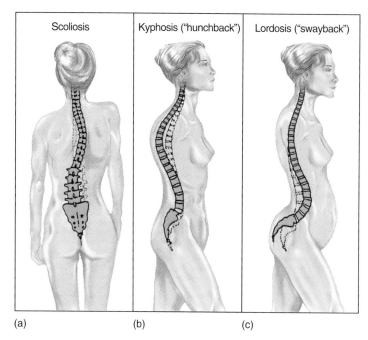

Abnormal Spinal Curvatures
Figure 7.21

2nd lumbar vertebra: superior view

Spinous process

Superior articular facet

Transverse process

Vertebral foramen

Body

Lamina

Vertebral arch

Pedicle

Intervertebral disc

Nucleus pulposus

Annulus fibrosus

A Representative Vertebra and Intervertebral Disc
Figure 7.22

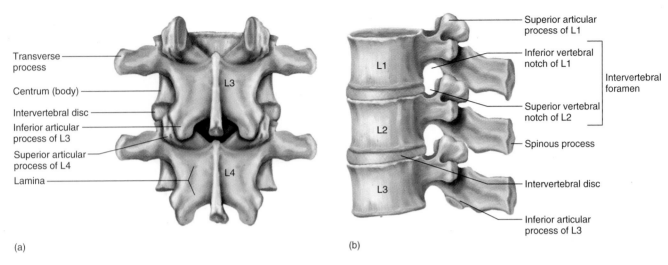

Transverse process
Centrum (body)
Intervertebral disc
Inferior articular process of L3
Superior articular process of L4
Lamina

L3
L4

(a)

Superior articular process of L1
Inferior vertebral notch of L1
Superior vertebral notch of L2
Spinous process
Intervertebral disc
Inferior articular process of L3

Intervertebral foramen

L1
L2
L3

(b)

Articulated Vertebrae
Figure 7.23

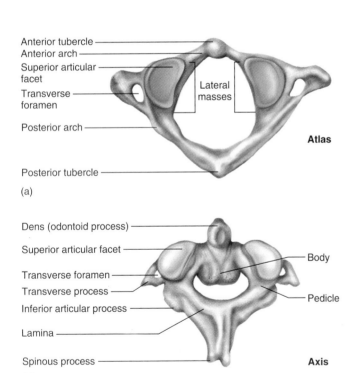

Anterior tubercle
Anterior arch
Superior articular facet
Transverse foramen
Posterior arch
Posterior tubercle

Lateral masses

Atlas

(a)

Dens (odontoid process)
Superior articular facet
Transverse foramen
Transverse process
Inferior articular process
Lamina
Spinous process

Body
Pedicle

Axis

(b)

The Atlas and Axis
Figure 7.24

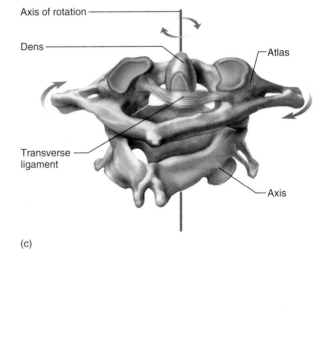

Axis of rotation
Dens
Transverse ligament

Atlas
Axis

(c)

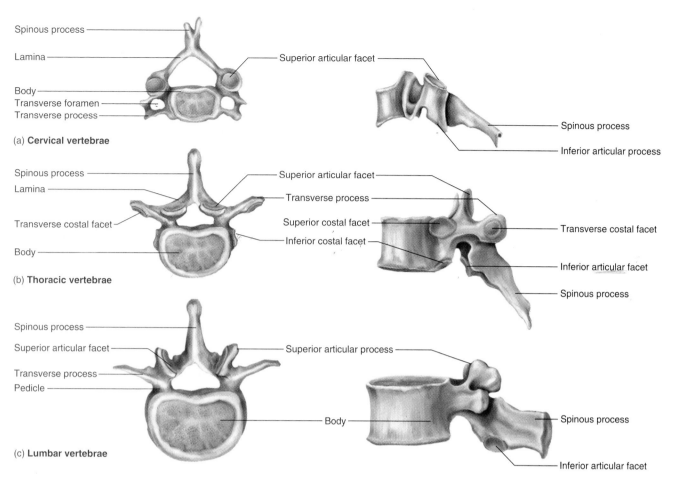

Typical Cervical, Thoracic, and Lumbar Vertebrae
Figure 7.25

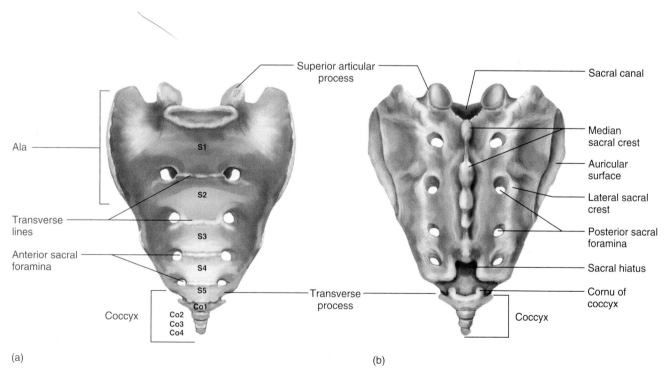

The Sacrum and Coccyx
Figure 7.26

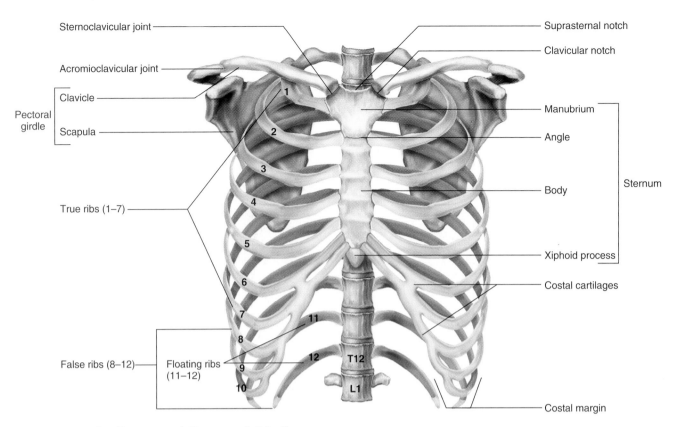

Sternoclavicular joint

Acromioclavicular joint

Pectoral girdle
Clavicle

Scapula

Suprasternal notch

Clavicular notch

Manubrium

Angle

Body
Sternum

Xiphoid process

Costal cartilages

True ribs (1–7)

False ribs (8–12)
Floating ribs (11–12)

Costal margin

The Thoracic Cage and Pectoral Girdle
Figure 7.27

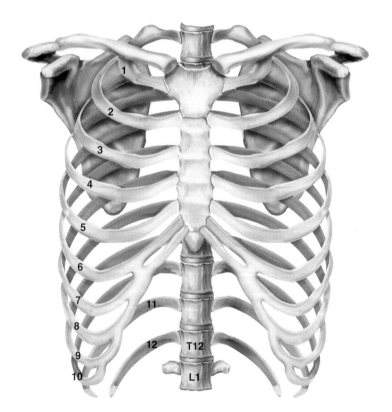

The Thoracic Cage and Pectoral Girdle
Figure 7.27

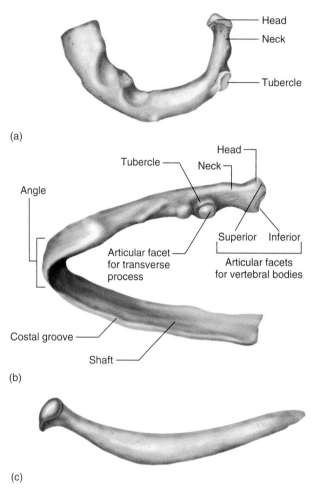

(a)

(b)

(c)

Anatomy of the Ribs
Figure 7.28

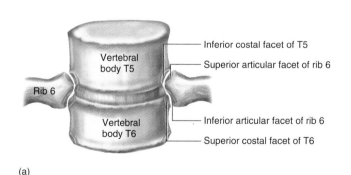

(a)

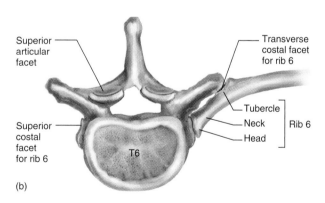

(b)

Articulation of Rib 6 with Vertebrae T5 and T6
Figure 7.29

133

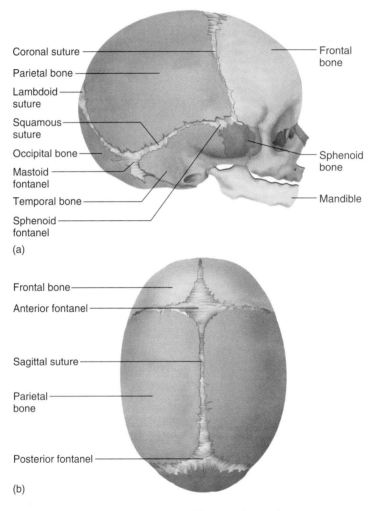

Coronal suture

Parietal bone

Lambdoid
suture

Squamous
suture

Occipital bone

Mastoid
fontanel

Temporal bone

Sphenoid
fontanel

Frontal
bone

Sphenoid
bone

Mandible

(a)

Frontal bone

Anterior fontanel

Sagittal suture

Parietal
bone

Posterior fontanel

(b)

The Fetal Skull Near the Time of Birth
Figure 7.31

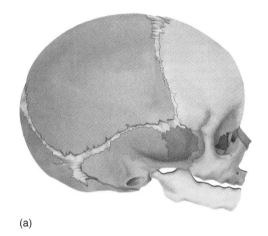

(a)

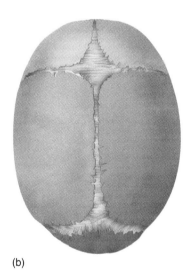

(b)

The Fetal Skull Near the Time of Birth
Figure 7.31

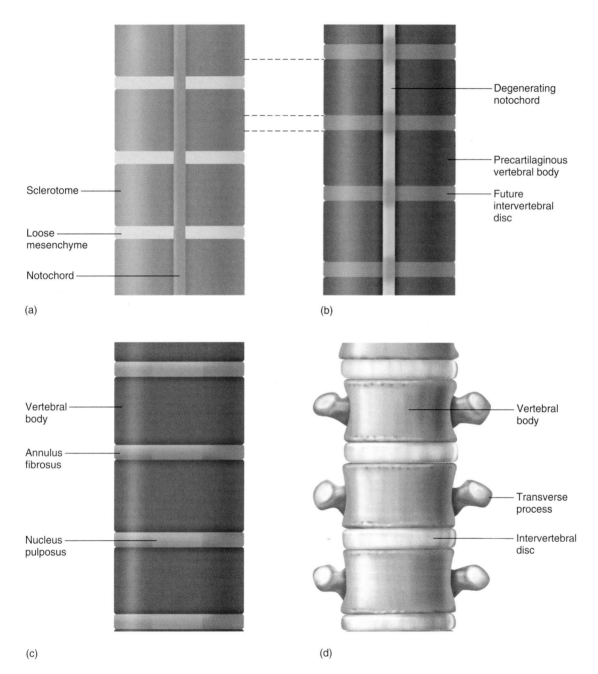

(a)

(b)

(c)

(d)

Sclerotome

Loose
mesenchyme

Notochord

Degenerating
notochord

Precartilaginous
vertebral body

Future
intervertebral
disc

Vertebral
body

Annulus
fibrosus

Nucleus
pulposus

Vertebral
body

Transverse
process

Intervertebral
disc

Development of the Vertebrae and Intervertebral Discs
Figure 7.32

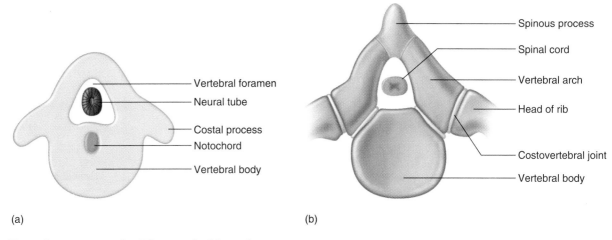

(a)   (b)

Development of a Thoracic Vertebrae
Figure 7.33

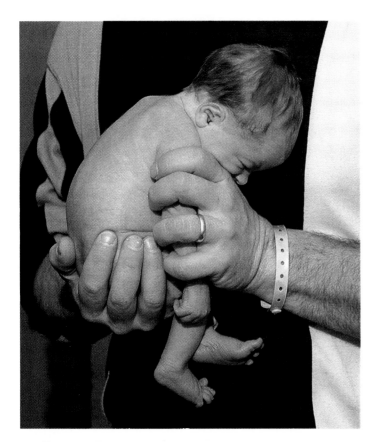

Spinal Curvature of the Newborn Infant
Figure 7.34

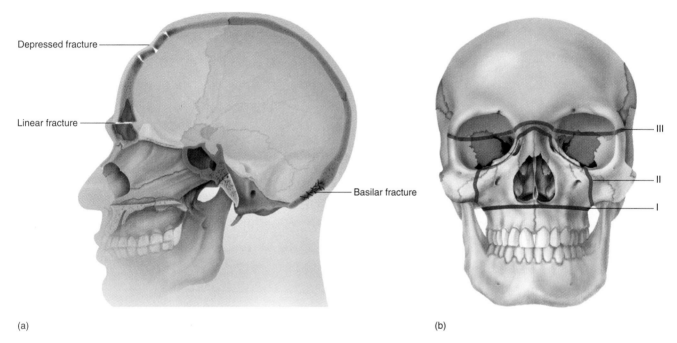

Depressed fracture

Linear fracture

Basilar fracture

(a)

III

II

I

(b)

Skull Fractures
Figure 7.35

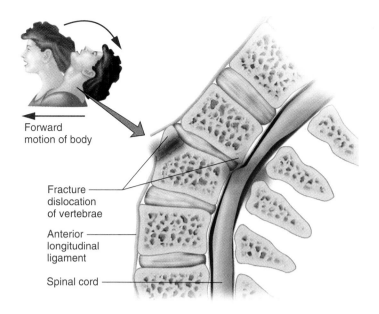

Forward
motion of body

Fracture
dislocation
of vertebrae

Anterior
longitudinal
ligament

Spinal cord

(a)

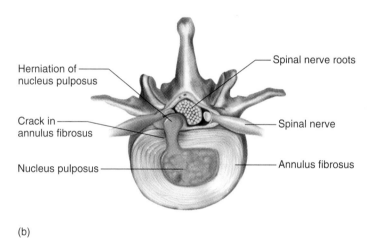

Herniation of
nucleus pulposus

Spinal nerve roots

Crack in
annulus fibrosus

Spinal nerve

Nucleus pulposus

Annulus fibrosus

(b)

Injuries to the Vertebral Column
Figure 7.36

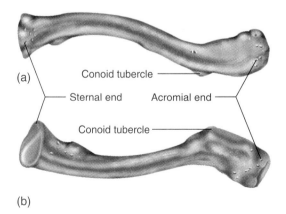

(a)

Conoid tubercle

Sternal end   Acromial end

Conoid tubercle

(b)

The Right Clavicle
Figure 8.1

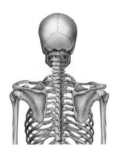

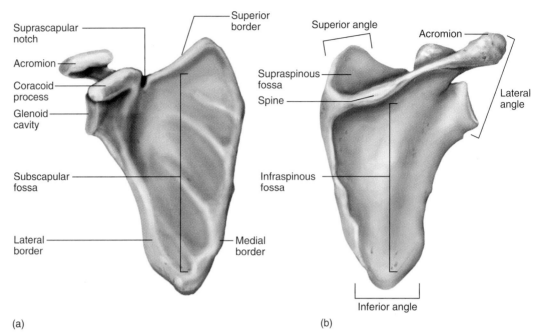

Suprascapular notch

Acromion

Coracoid process

Glenoid cavity

Subscapular fossa

Lateral border

Superior border

Medial border

Superior angle

Supraspinous fossa

Spine

Infraspinous fossa

Acromion

Lateral angle

Inferior angle

(a)

(b)

The Right Scapula
Figure 8.2

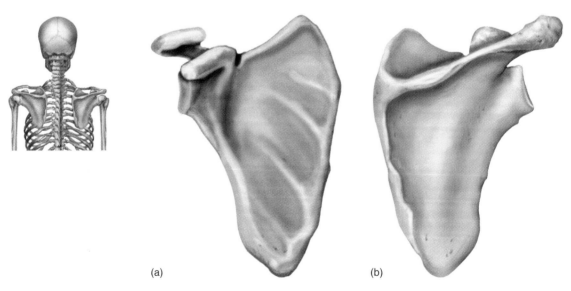

(a)

(b)

The Right Scapula
Figure 8.2

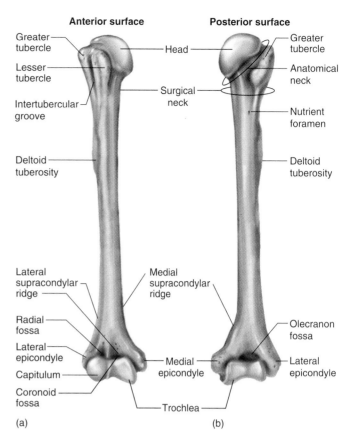

Anterior surface

Greater tubercle
Lesser tubercle
Intertubercular groove
Deltoid tuberosity

Head
Surgical neck

Posterior surface

Greater tubercle
Anatomical neck
Nutrient foramen
Deltoid tuberosity

Lateral supracondylar ridge
Radial fossa
Lateral epicondyle
Capitulum
Coronoid fossa

Medial supracondylar ridge
Medial epicondyle
Trochlea

Olecranon fossa
Lateral epicondyle

(a)

(b)

The Right Humerus
Figure 8.3

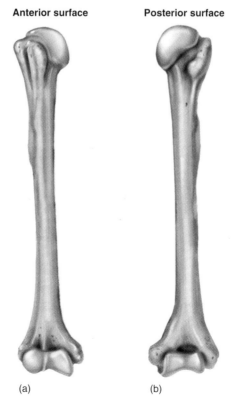

Anterior surface

Posterior surface

(a)

(b)

The Right Humerus
Figure 8.3

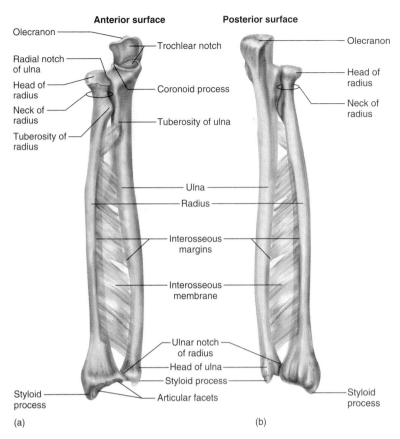

Anterior surface

- Olecranon
- Radial notch of ulna
- Head of radius
- Neck of radius
- Tuberosity of radius
- Trochlear notch
- Coronoid process
- Tuberosity of ulna
- Ulna
- Radius
- Interosseous margins
- Interosseous membrane
- Ulnar notch of radius
- Head of ulna
- Styloid process
- Articular facets
- Styloid process

Posterior surface

- Olecranon
- Head of radius
- Neck of radius
- Styloid process

(a)   (b)

The Right Radius and Ulna
Figure 8.4

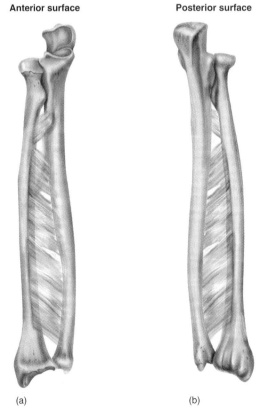

Anterior surface   **Posterior surface**

(a)   (b)

The Right Radius and Ulna
Figure 8.4

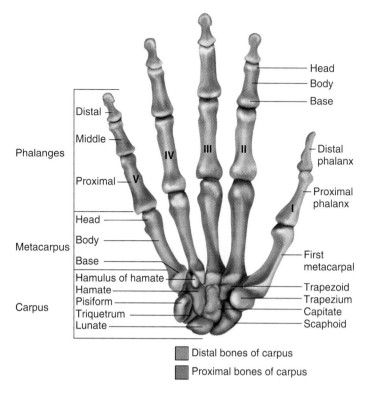

Phalanges
- Distal
- Middle
- Proximal
- Head
- Body
- Base

Metacarpus

Carpus
- Hamulus of hamate
- Hamate
- Pisiform
- Triquetrum
- Lunate

Head
Body
Base

Distal phalanx

Proximal phalanx

First metacarpal

Trapezoid
Trapezium
Capitate
Scaphoid

Distal bones of carpus
Proximal bones of carpus

The Right Wrist and Hand
Figure 8.5

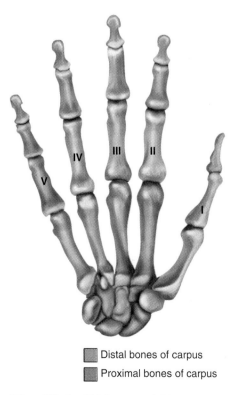

Distal bones of carpus
Proximal bones of carpus

The Right Wrist and Hand
Figure 8.5

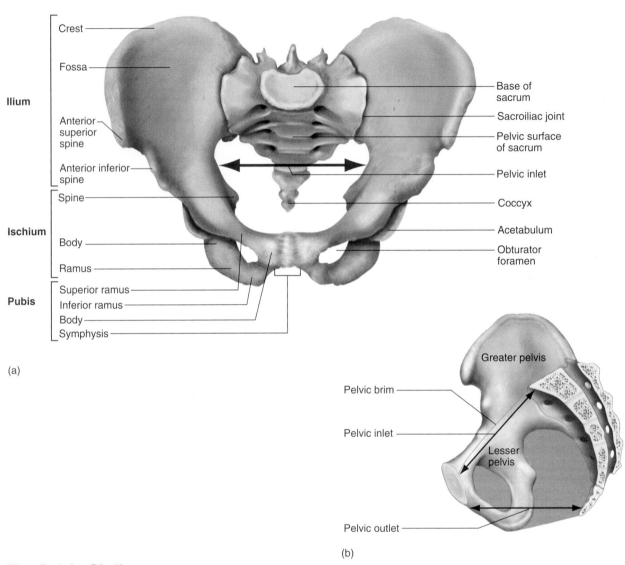

Ilium
Crest
Fossa
Anterior superior spine
Anterior inferior spine

Ischium
Spine
Body
Ramus

Pubis
Superior ramus
Inferior ramus
Body
Symphysis

Base of sacrum
Sacroiliac joint
Pelvic surface of sacrum
Pelvic inlet
Coccyx
Acetabulum
Obturator foramen

(a)

Greater pelvis
Pelvic brim
Pelvic inlet
Lesser pelvis
Pelvic outlet

(b)

The Pelvic Girdle
Figure 8.6

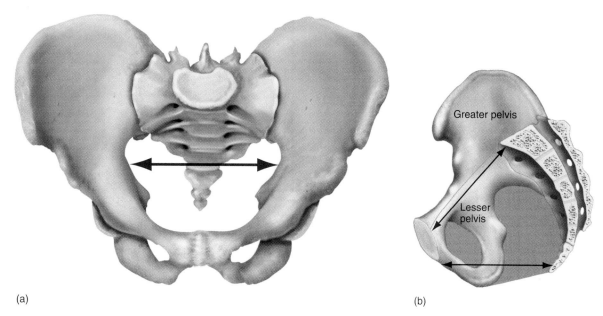

(a)

Greater pelvis

Lesser pelvis

(b)

The Pelvic Girdle
Figure 8.6

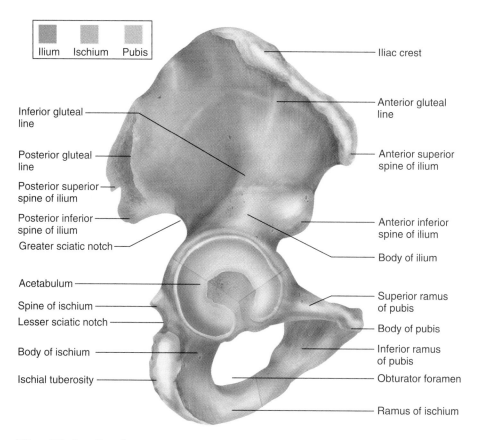

| Ilium | Ischium | Pubis |

Iliac crest

Anterior gluteal line

Inferior gluteal line

Posterior gluteal line

Posterior superior spine of ilium

Posterior inferior spine of ilium

Greater sciatic notch

Acetabulum

Spine of ischium

Lesser sciatic notch

Body of ischium

Ischial tuberosity

Anterior superior spine of ilium

Anterior inferior spine of ilium

Body of ilium

Superior ramus of pubis

Body of pubis

Inferior ramus of pubis

Obturator foramen

Ramus of ischium

The Right Os Coxae
Figure 8.7

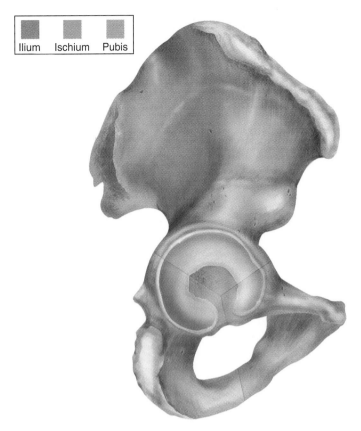

Ilium Ischium Pubis

The Right Os Coxae
Figure 8.7

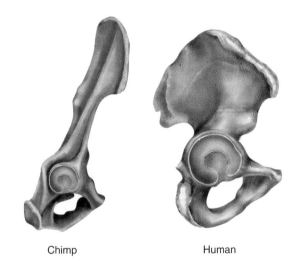

Chimp   Human

Chimpanzee and Human Os Coxae
Figure 8.8

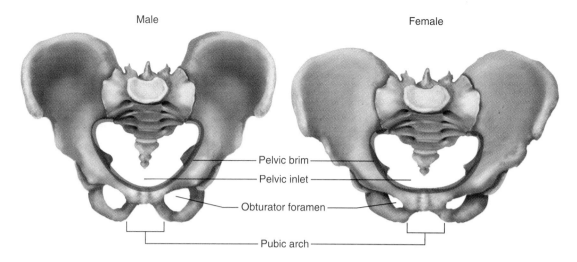

Comparison of the Male and Female Pelvic Girdles
Figure 8.9

- Pelvic brim
- Pelvic inlet
- Obturator foramen
- Pubic arch

Male

Female

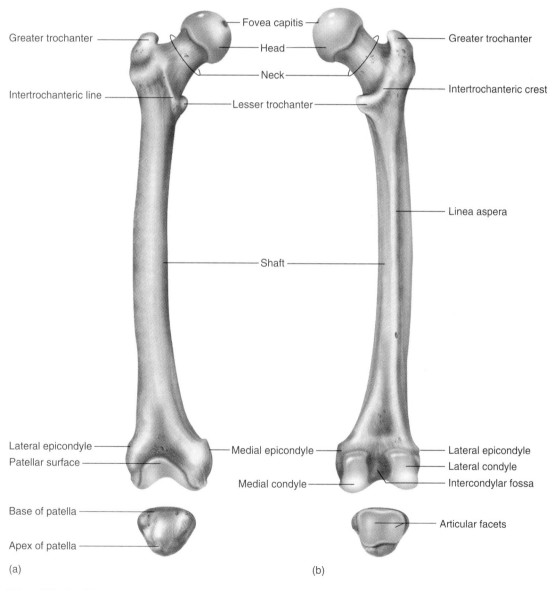

- Fovea capitis
- Greater trochanter
- Head
- Neck
- Greater trochanter
- Intertrochanteric line
- Intertrochanteric crest
- Lesser trochanter
- Linea aspera
- Shaft
- Lateral epicondyle
- Medial epicondyle
- Lateral epicondyle
- Patellar surface
- Lateral condyle
- Medial condyle
- Intercondylar fossa
- Base of patella
- Apex of patella
- Articular facets

(a)

(b)

The Right Femur and Patella
Figure 8.10

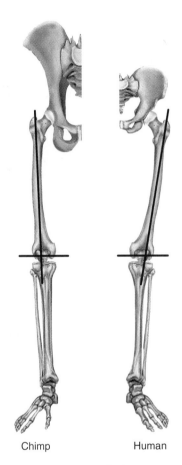

Chimp   Human

**Adaptation of the
Lower Limb for
Bipedalism**
Figure 8.11

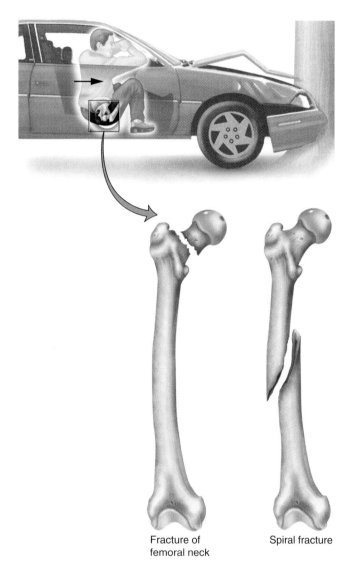

Fracture of
femoral neck

Spiral fracture

Fractures of the Femur
Figure 8.12

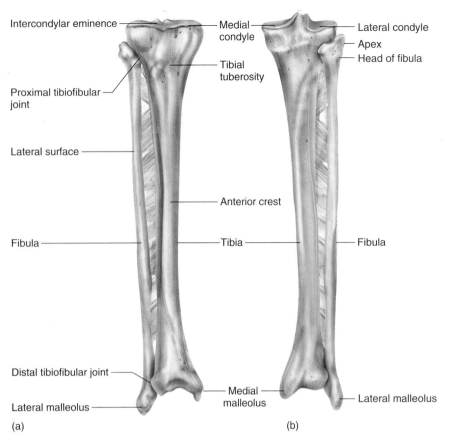

Intercondylar eminence — Medial condyle — Lateral condyle — Apex — Head of fibula

Proximal tibiofibular joint — Tibial tuberosity

Lateral surface — Anterior crest

Fibula — Tibia — Fibula

Distal tibiofibular joint — Medial malleolus — Lateral malleolus

Lateral malleolus

(a) (b)

The Right Tibia and Fibula
Figure 8.13

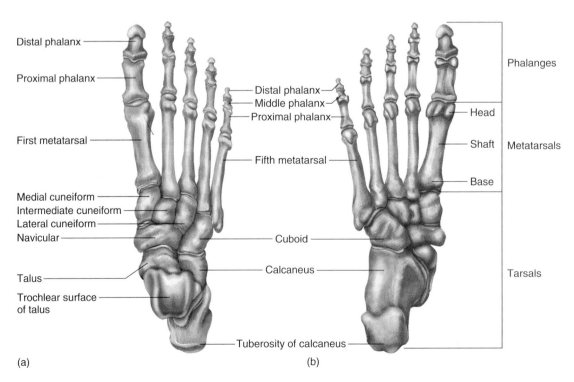

Distal phalanx — Distal phalanx — Phalanges

Proximal phalanx — Middle phalanx — Head

Proximal phalanx — Shaft — Metatarsals

First metatarsal — Fifth metatarsal — Base

Medial cuneiform — Cuboid — Tarsals

Intermediate cuneiform

Lateral cuneiform

Navicular — Calcaneus

Talus

Trochlear surface of talus

Tuberosity of calcaneus

(a) (b)

The Right Foot
Figure 8.14

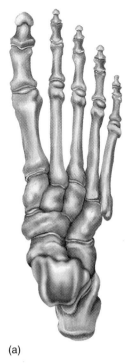
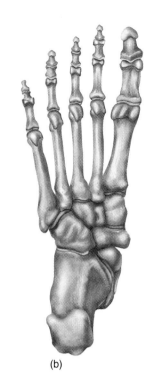

(a)   (b)

The Right Foot
Figure 8.14

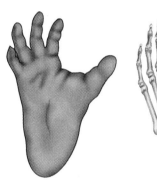
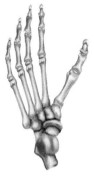

Chimp

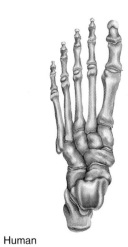

Human

**Some Adaptations of the
Foot for Bipedalism**
Figure 8.15

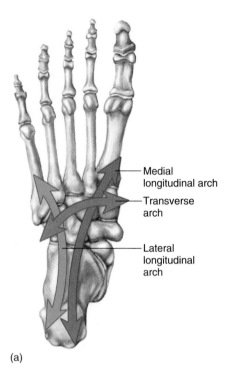

(a)

Arches of the Foot
Figure 8.16

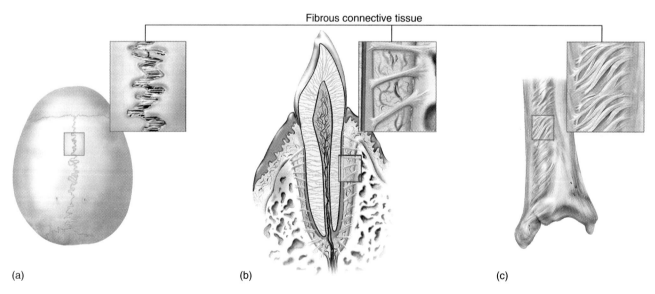

Fibrous connective tissue

(a)   (b)   (c)

Types of Fibrous Joints
Figure 9.1

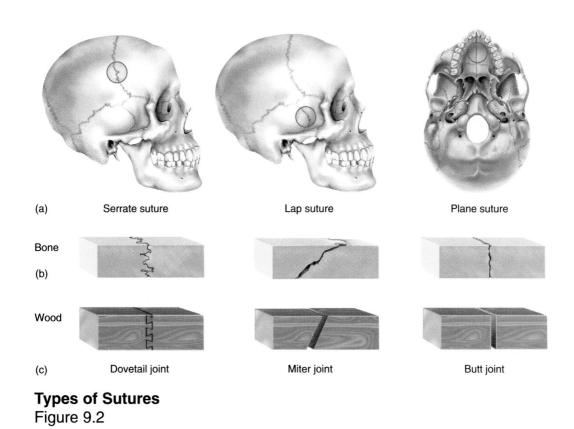

(a)　　Serrate suture　　　　　　　　Lap suture　　　　　　　　Plane suture

Bone

(b)

Wood

(c)　　Dovetail joint　　　　　　　　Miter joint　　　　　　　　Butt joint

Types of Sutures
Figure 9.2

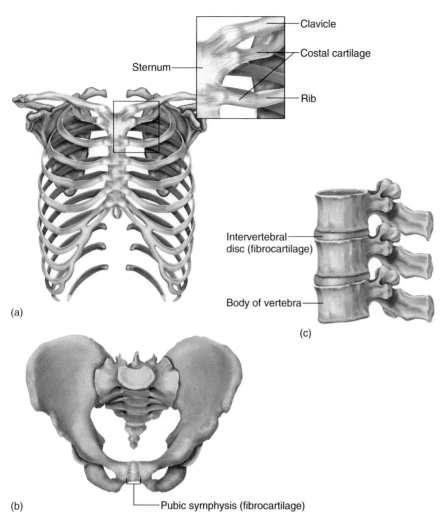

Clavicle
Costal cartilage
Sternum
Rib

(a)

Intervertebral disc (fibrocartilage)

Body of vertebra

(c)

Pubic symphysis (fibrocartilage)

(b)

Cartilaginous Joints
Figure 9.3

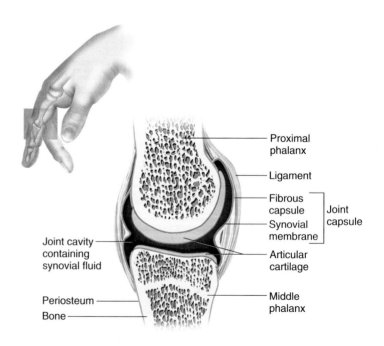

Proximal phalanx

Ligament

Fibrous capsule — Joint capsule
Synovial membrane

Articular cartilage

Middle phalanx

Joint cavity containing synovial fluid

Periosteum

Bone

Structure of a Simple Synovial Joint
Figure 9.4

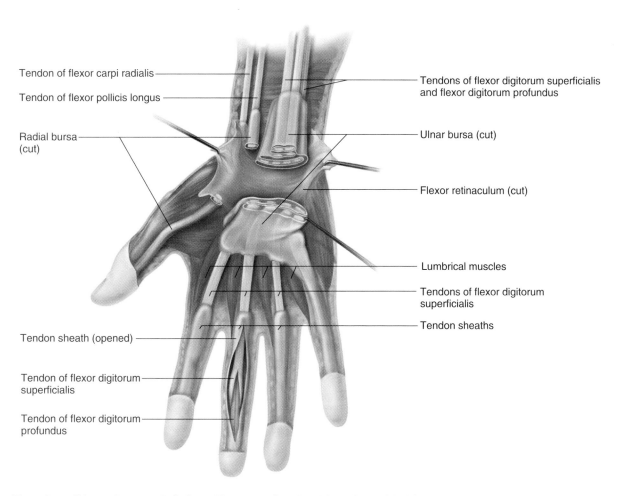

Tendon of flexor carpi radialis

Tendon of flexor pollicis longus

Radial bursa (cut)

Tendons of flexor digitorum superficialis and flexor digitorum profundus

Ulnar bursa (cut)

Flexor retinaculum (cut)

Lumbrical muscles

Tendons of flexor digitorum superficialis

Tendon sheaths

Tendon sheath (opened)

Tendon of flexor digitorum superficialis

Tendon of flexor digitorum profundus

Tendon Sheaths and Other Bursae in the Hand and Wrist
Figure 9.5

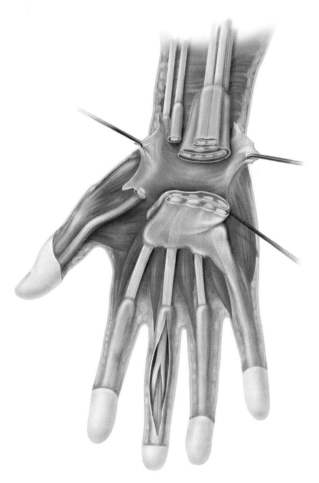

Tendon Sheaths and Other Bursae in the Hand and Wrist
Figure 9.5

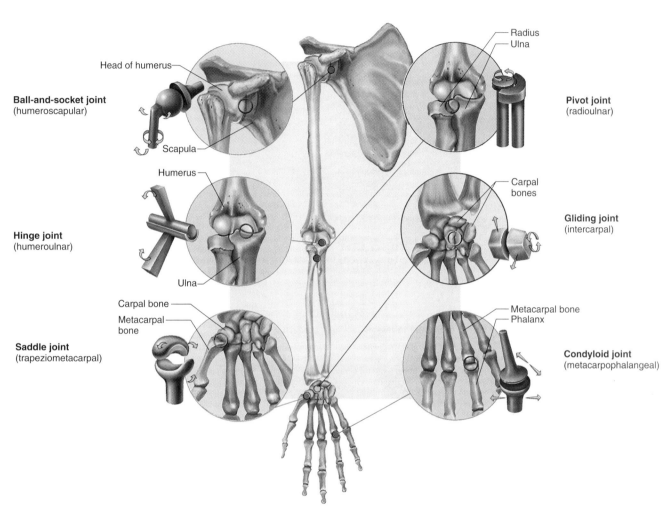

Ball-and-socket joint
(humeroscapular)

Head of humerus

Scapula

Hinge joint
(humeroulnar)

Humerus

Ulna

Saddle joint
(trapeziometacarpal)

Carpal bone

Metacarpal
bone

Radius
Ulna

Pivot joint
(radioulnar)

Carpal
bones

Gliding joint
(intercarpal)

Metacarpal bone
Phalanx

Condyloid joint
(metacarpophalangeal)

The Six Types of Synovial Joints
Figure 9.6

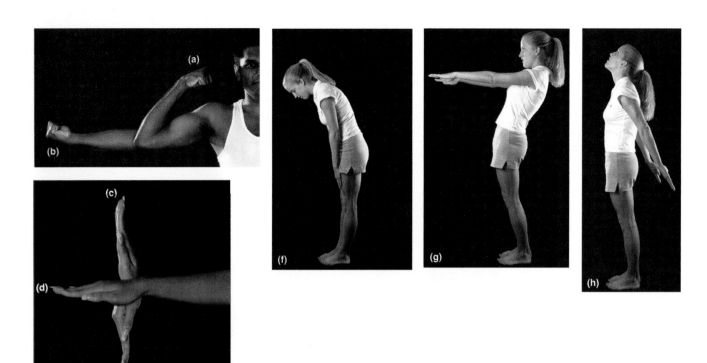

Joint Flexion and Extension
Figure 9.7

a–h: © The McGraw-Hill Companies, Inc./Rebecca Gray, photographer/Don Kincaid, dissections

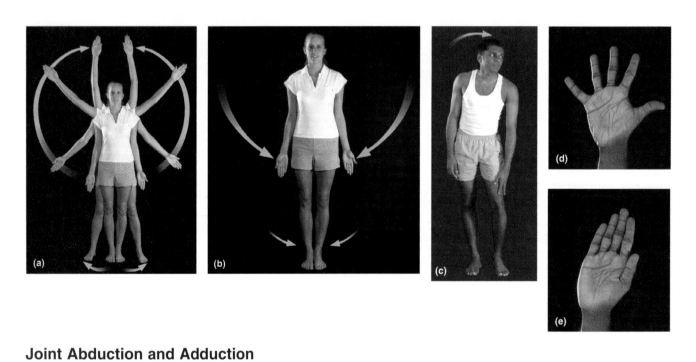

Joint Abduction and Adduction
Figure 9.8

a–e: © The McGraw-Hill Companies, Inc./Rebecca Gray, photographer/Don Kincaid, dissections

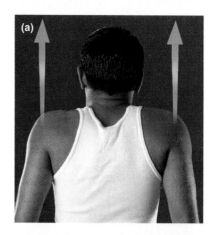

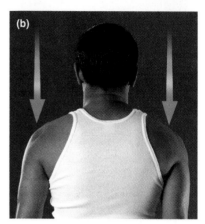

Elevation and Depression
Figure 9.9

a,b: © The McGraw-Hill Companies, Inc./
Rebecca Gray, photographer/
Don Kincaid, dissections

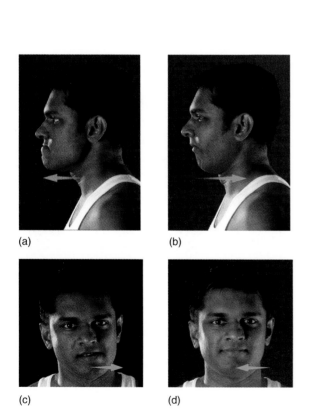

(a)

(b)

(c)

(d)

Some Horizontal Joint Movements
Figure 9.10

a–d: © The McGraw-Hill Companies, Inc./Rebecca
Gray, photographer/Don Kincaid, dissections

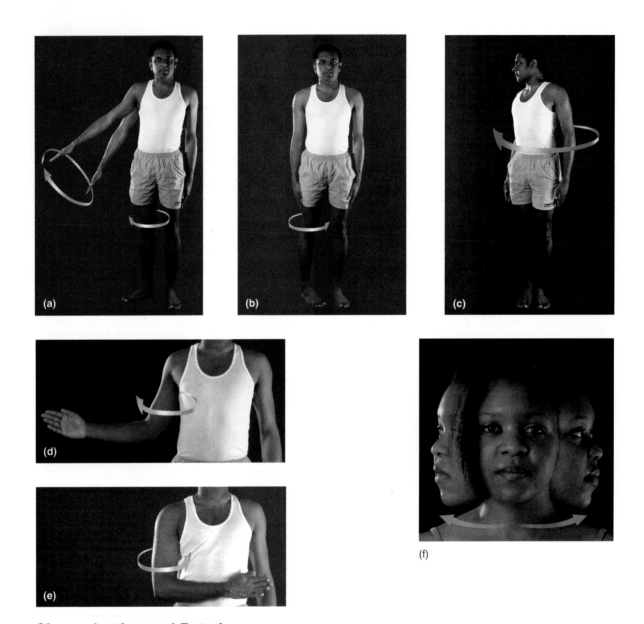

Circumduction and Rotation
Figure 9.11

a–f: © The McGraw-Hill Companies, Inc./Rebecca Gray, photographer/Don Kincaid, dissections

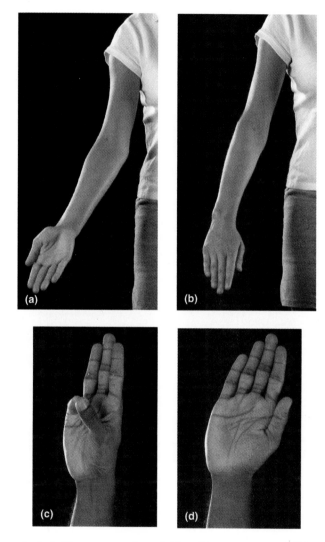

Joint Movements of the Forearm and Thumb
Figure 9.12

a–d: © The McGraw-Hill Companies, Inc./Rebecca Gray,
photographer/Don Kincaid, dissections

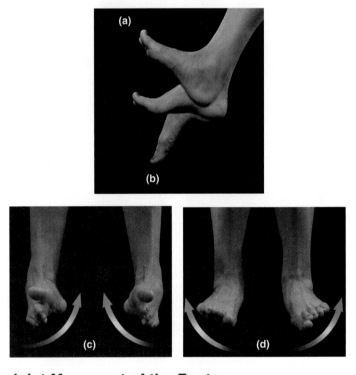

Joint Movement of the Foot
Figure 9.13

b–d: © The McGraw-Hill Companies, Inc./Rebecca Gray, photographer/Don Kincaid, dissections

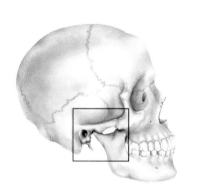

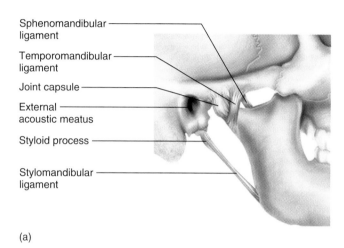

Sphenomandibular ligament

Temporomandibular ligament

Joint capsule

External acoustic meatus

Styloid process

Stylomandibular ligament

(a)

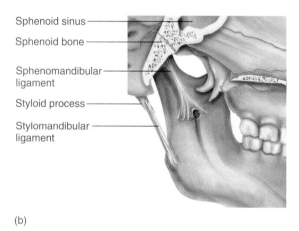

Sphenoid sinus

Sphenoid bone

Sphenomandibular ligament

Styloid process

Stylomandibular ligament

(b)

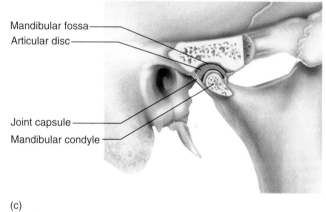

Mandibular fossa

Articular disc

Joint capsule

Mandibular condyle

(c)

The Temporomandibular Joint (TMJ)
Figure 9.14

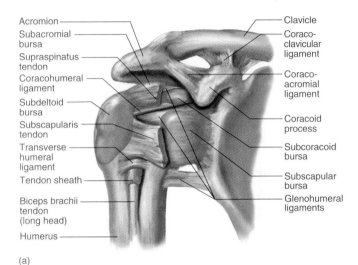

Acromion
Subacromial bursa
Supraspinatus tendon
Coracohumeral ligament
Subdeltoid bursa
Subscapularis tendon
Transverse humeral ligament
Tendon sheath
Biceps brachii tendon (long head)
Humerus

Clavicle
Coraco-clavicular ligament
Coraco-acromial ligament
Coracoid process
Subcoracoid bursa
Subscapular bursa
Glenohumeral ligaments

(a)

The Shoulder (humeroscapular) Joint
Figure 9.15

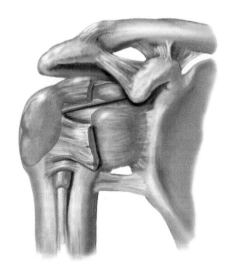

(a)

The Shoulder (humeroscapular) Joint
Figure 9.15

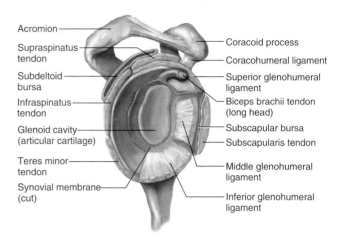

Acromion

Supraspinatus
tendon

Subdeltoid
bursa

Infraspinatus
tendon

Glenoid cavity
(articular cartilage)

Teres minor
tendon

Synovial membrane
(cut)

Coracoid process

Coracohumeral ligament

Superior glenohumeral
ligament

Biceps brachii tendon
(long head)

Subscapular bursa

Subscapularis tendon

Middle glenohumeral
ligament

Inferior glenohumeral
ligament

(b)

The Shoulder (humeroscapular) Joint
Figure 9.15

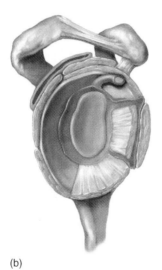

(b)

**The Shoulder
(humeroscapular) Joint**
Figure 9.15

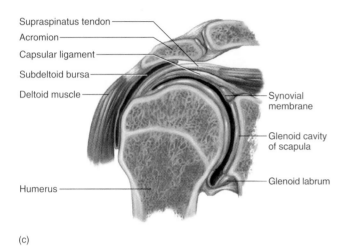

Supraspinatus tendon
Acromion
Capsular ligament
Subdeltoid bursa
Deltoid muscle

Synovial membrane

Glenoid cavity of scapula

Glenoid labrum

Humerus

(c)

The Shoulder (humeroscapular) Joint
Figure 9.15

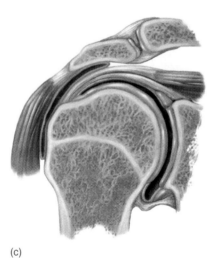

(c)

The Shoulder (humeroscapular) Joint
Figure 9.15

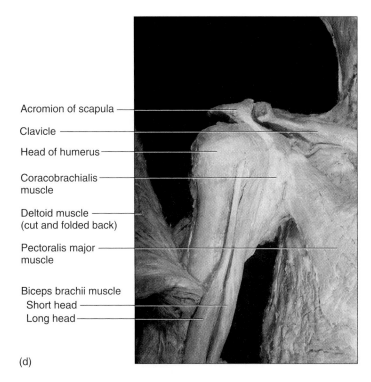

Acromion of scapula

Clavicle

Head of humerus

Coracobrachialis
muscle

Deltoid muscle
(cut and folded back)

Pectoralis major
muscle

Biceps brachii muscle
Short head
Long head

(d)

The Shoulder (humeroscapular) Joint (*Continued*)
Figure 9.15

d: © The McGraw-Hill Companies, Inc./Rebecca Gray, photographer/Don Kincaid, dissections

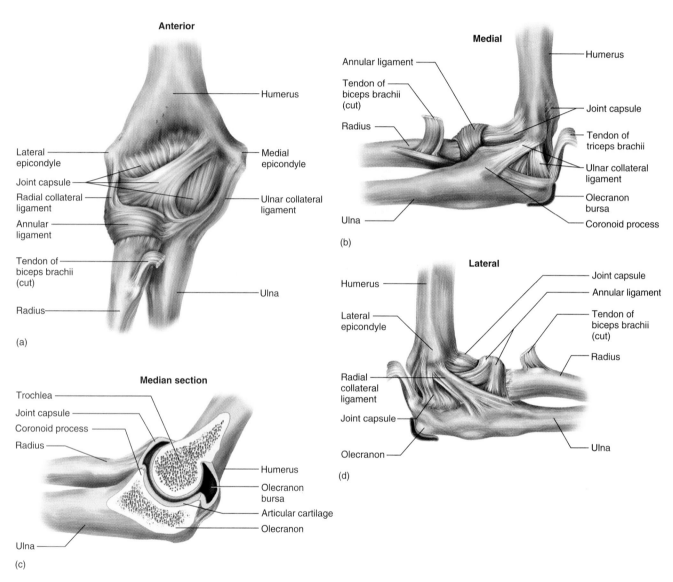

Anterior

Humerus

Lateral epicondyle

Joint capsule

Radial collateral ligament

Annular ligament

Tendon of biceps brachii (cut)

Radius

Medial epicondyle

Ulnar collateral ligament

Ulna

(a)

Median section

Trochlea

Joint capsule

Coronoid process

Radius

Humerus

Olecranon bursa

Articular cartilage

Olecranon

Ulna

(c)

Medial

Annular ligament

Tendon of biceps brachii (cut)

Radius

Ulna

Humerus

Joint capsule

Tendon of triceps brachii

Ulnar collateral ligament

Olecranon bursa

Coronoid process

(b)

Lateral

Humerus

Lateral epicondyle

Radial collateral ligament

Joint capsule

Olecranon

Joint capsule

Annular ligament

Tendon of biceps brachii (cut)

Radius

Ulna

(d)

The Elbow Joint
Figure 9.16

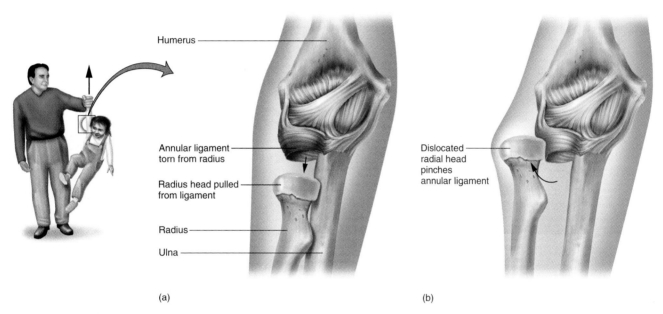

Humerus

Annular ligament
torn from radius

Radius head pulled
from ligament

Radius

Ulna

(a)

Dislocated
radial head
pinches
annular ligament

(b)

Pulled Elbow
Figure 9.17

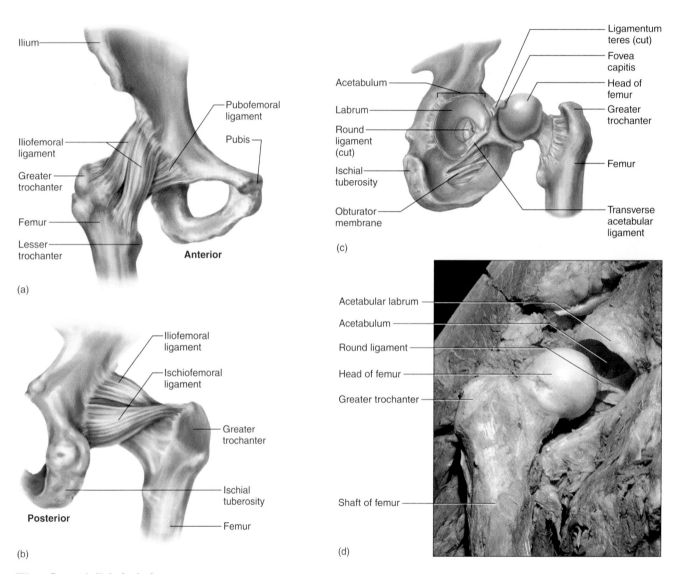

(a)

Ilium

Iliofemoral ligament

Greater trochanter

Femur

Lesser trochanter

Pubofemoral ligament

Pubis

Anterior

(b)

Iliofemoral ligament

Ischiofemoral ligament

Greater trochanter

Ischial tuberosity

Femur

Posterior

(c)

Acetabulum

Labrum

Round ligament (cut)

Ischial tuberosity

Obturator membrane

Ligamentum teres (cut)

Fovea capitis

Head of femur

Greater trochanter

Femur

Transverse acetabular ligament

(d)

Acetabular labrum

Acetabulum

Round ligament

Head of femur

Greater trochanter

Shaft of femur

The Coxal (hip) Joint
Figure 9.18

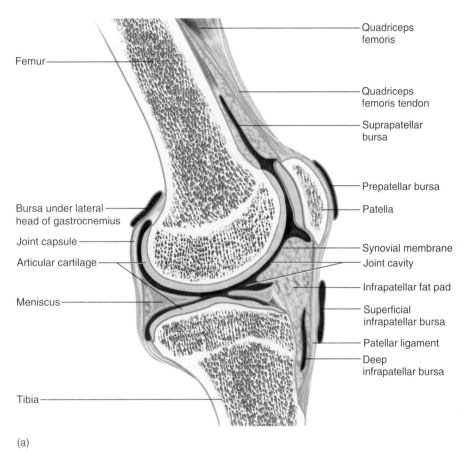

Femur

Bursa under lateral head of gastrocnemius

Joint capsule

Articular cartilage

Meniscus

Tibia

Quadriceps femoris

Quadriceps femoris tendon

Suprapatellar bursa

Prepatellar bursa

Patella

Synovial membrane

Joint cavity

Infrapatellar fat pad

Superficial infrapatellar bursa

Patellar ligament

Deep infrapatellar bursa

(a)

The Knee Joint
Figure 9.19

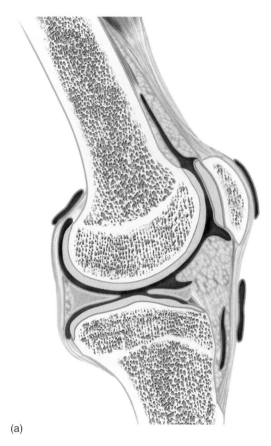

(a)

The Knee Joint
Figure 9.19

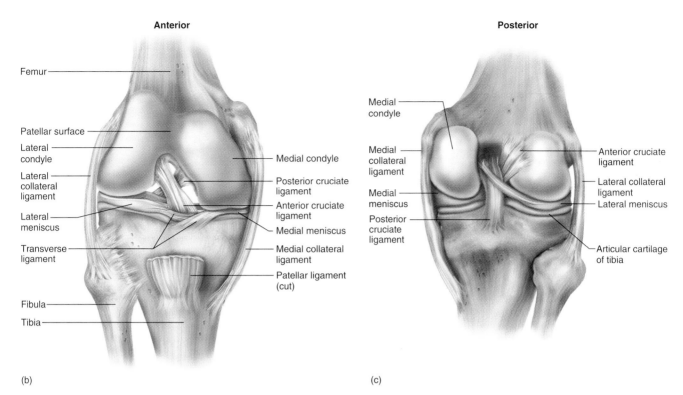

The Knee Joint
Figure 9.19

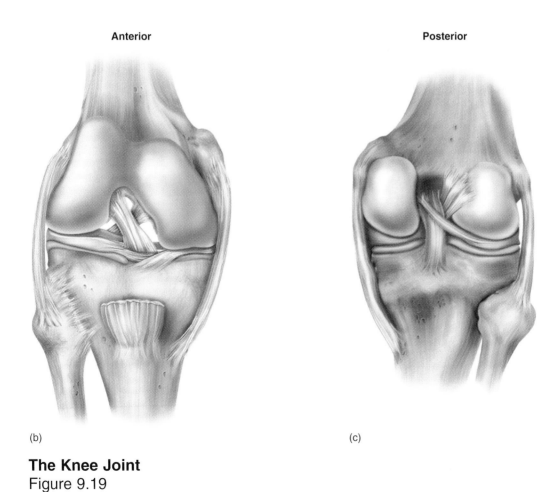

The Knee Joint
Figure 9.19

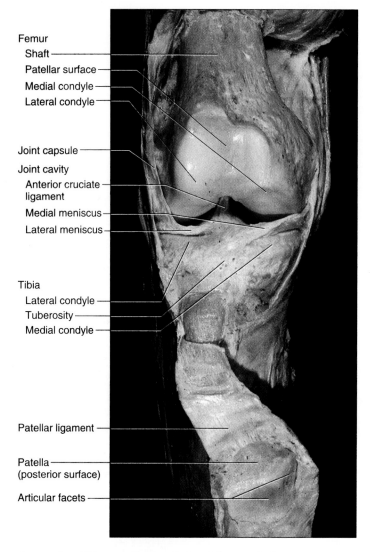

Femur
Shaft
Patellar surface
Medial condyle
Lateral condyle

Joint capsule

Joint cavity
Anterior cruciate ligament
Medial meniscus
Lateral meniscus

Tibia
Lateral condyle
Tuberosity
Medial condyle

Patellar ligament

Patella (posterior surface)

Articular facets

Anterior Dissection of the Knee Joint
Figure 9.20

© The McGraw-Hill Companies, Inc./Rebecca Gray, photographer/Don Kincaid, dissections

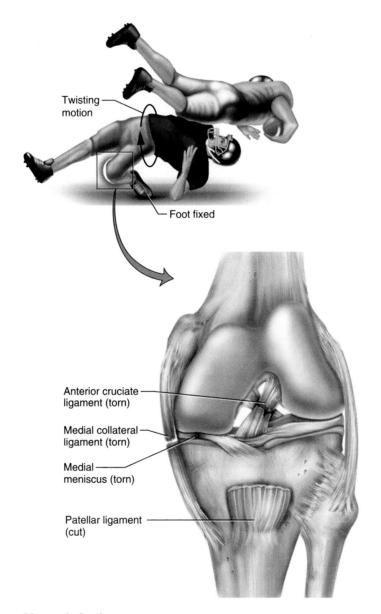

Twisting
motion

Foot fixed

Anterior cruciate
ligament (torn)

Medial collateral
ligament (torn)

Medial
meniscus (torn)

Patellar ligament
(cut)

Knee Injuries
Figure 9.21

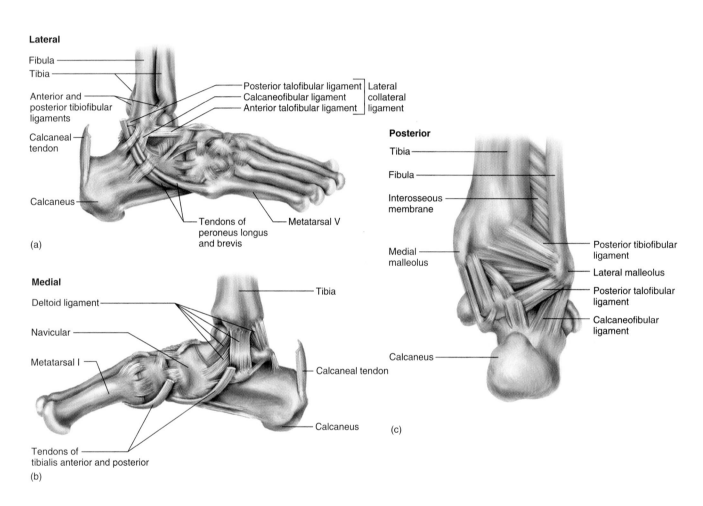

Lateral

Fibula

Tibia

Anterior and posterior tibiofibular ligaments

Calcaneal tendon

Calcaneus

Posterior talofibular ligament ⎤ Lateral
Calcaneofibular ligament ⎬ collateral
Anterior talofibular ligament ⎦ ligament

Tendons of peroneus longus and brevis

Metatarsal V

(a)

Medial

Deltoid ligament

Navicular

Metatarsal I

Tibia

Calcaneal tendon

Calcaneus

Tendons of tibialis anterior and posterior

(b)

Posterior

Tibia

Fibula

Interosseous membrane

Medial malleolus

Calcaneus

Posterior tibiofibular ligament

Lateral malleolus

Posterior talofibular ligament

Calcaneofibular ligament

(c)

The Talocrural (ankle) Joint and Ligaments of the Right Foot
Figure 9.22

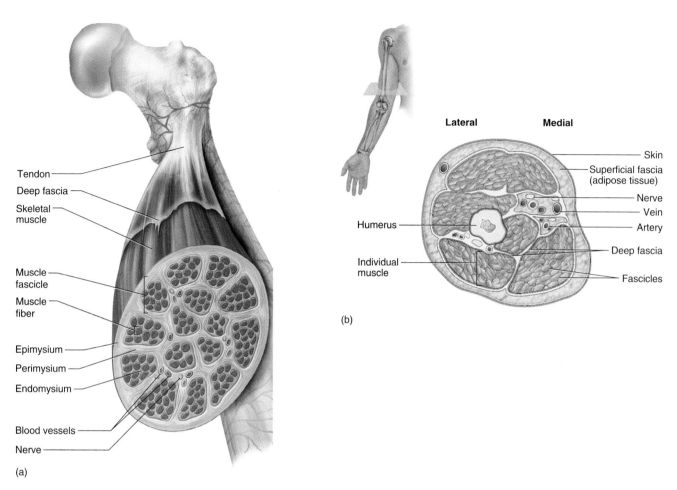

Labels on figure (a):
Tendon
Deep fascia
Skeletal muscle
Muscle fascicle
Muscle fiber
Epimysium
Perimysium
Endomysium
Blood vessels
Nerve

(a)

Labels on figure (b):
Lateral
Medial
Skin
Superficial fascia (adipose tissue)
Nerve
Vein
Artery
Deep fascia
Fascicles
Humerus
Individual muscle

(b)

The Connective Tissues of a Skeletal Muscle
Figure 10.2

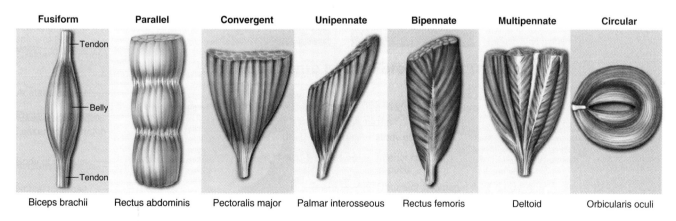

Fusiform	Parallel	Convergent	Unipennate	Bipennate	Multipennate	Circular
Tendon / Belly / Tendon						
Biceps brachii	Rectus abdominis	Pectoralis major	Palmar interosseous	Rectus femoris	Deltoid	Orbicularis oculi

Classification of Muscles According to Fascicle Orientation
Figure 10.3

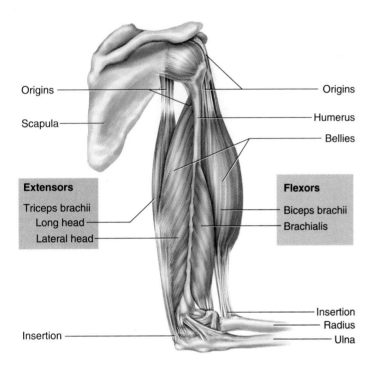

A Muscle Group Acting on the Elbow
Figure 10.4

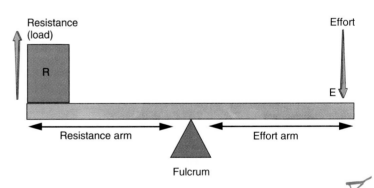

Basic Components of a Lever
Figure 10.5

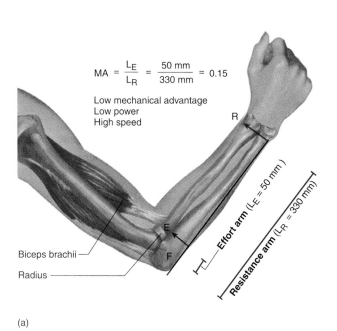

$$MA = \frac{L_E}{L_R} = \frac{50 \text{ mm}}{330 \text{ mm}} = 0.15$$

Low mechanical advantage
Low power
High speed

R

E

F

Effort arm (L$_E$ = 50 mm)

Resistance arm (L$_R$ = 330 mm)

Biceps brachii

Radius

(a)

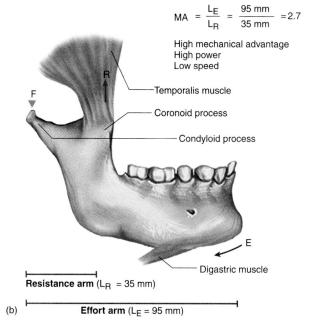

$$MA = \frac{L_E}{L_R} = \frac{95 \text{ mm}}{35 \text{ mm}} = 2.7$$

High mechanical advantage
High power
Low speed

R

F

Temporalis muscle

Coronoid process

Condyloid process

E

Digastric muscle

Resistance arm (L$_R$ = 35 mm)

Effort arm (L$_E$ = 95 mm)

(b)

Mechanical Advantage (MA)
Figure 10.6

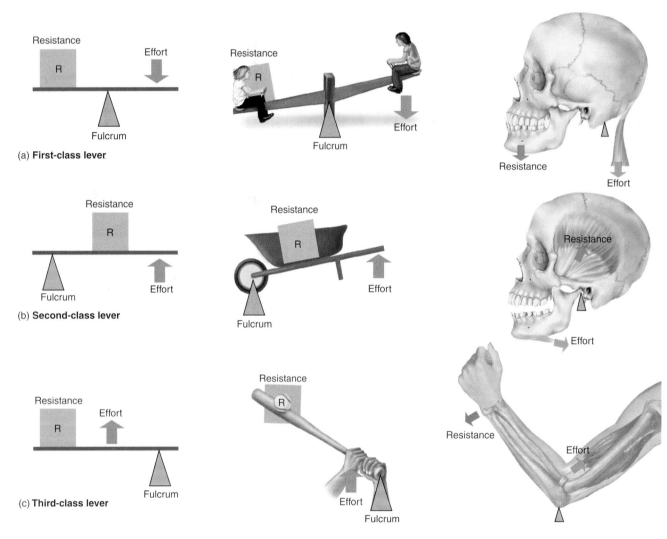

(a) **First-class lever**

(b) **Second-class lever**

(c) **Third-class lever**

The Three Classes of Levers
Figure 10.7

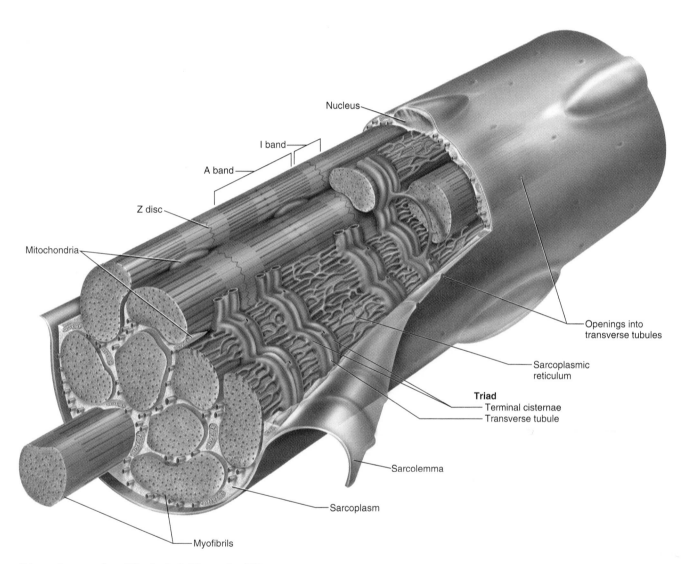

Structure of a Skeletal Muscle Fiber
Figure 10.8

Nucleus

I band

A band

Z disc

Mitochondria

Openings into
transverse tubules

Sarcoplasmic
reticulum

Triad
Terminal cisternae
Transverse tubule

Sarcolemma

Sarcoplasm

Myofibrils

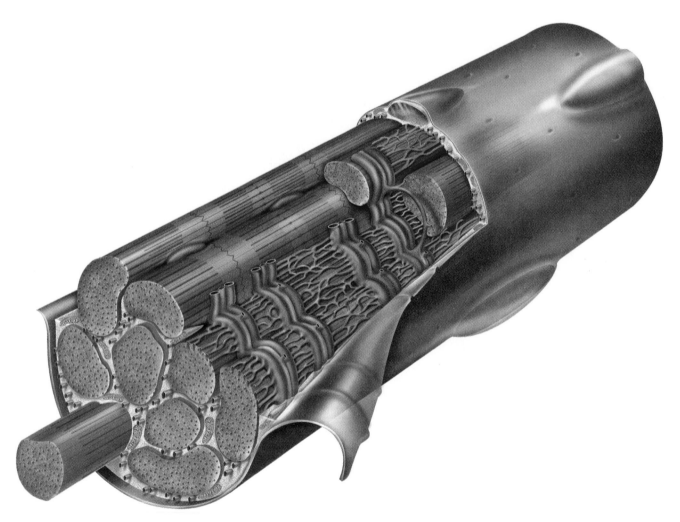

Structure of a Skeletal Muscle Fiber
Figure 10.8

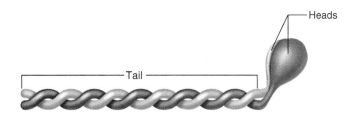

Myosin molecule

(a)

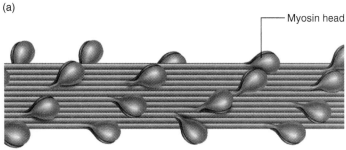

Thick filament

(b)

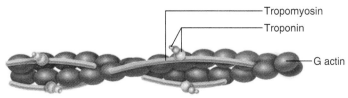

Thin filament

(c)

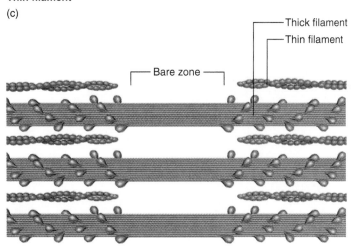

Portion of a sarcomere showing the overlap of thick and thin filaments

(d)

Molecular Structure of Thick and Thin Filaments

Figure 10.9

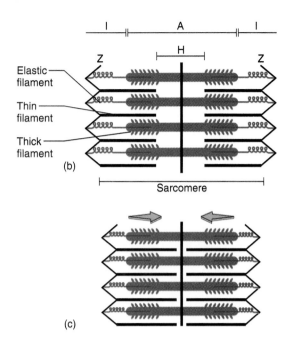

(b)

Elastic filament
Thin filament
Thick filament

Sarcomere

(c)

Muscle Striations and Their Molecular Basis
Figure 10.10

Motor nerve fiber (axon)

Schwann cells

Synaptic knob

Synaptic vesicles (containing ACh)

Basal lamina (containing AChE)

Sarcolemma

Motor end plate

Nucleus of muscle fiber

Synaptic cleft

Junctional folds

A Neuromuscular Junction
Figure 10.12

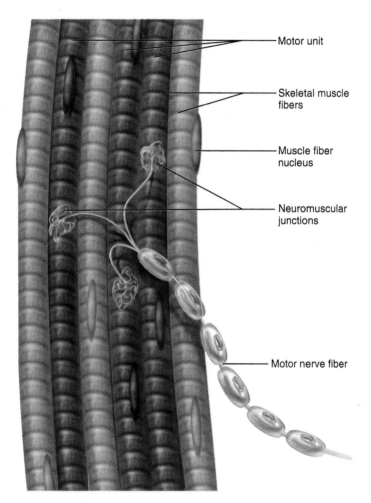

Motor unit

Skeletal muscle fibers

Muscle fiber nucleus

Neuromuscular junctions

Motor nerve fiber

A Motor Unit
Figure 10.13

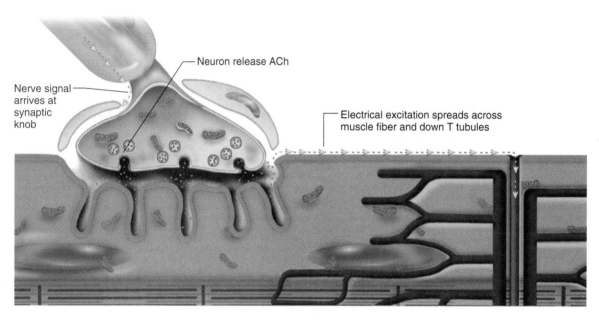

Nerve signal arrives at synaptic knob

Neuron release ACh

Electrical excitation spreads across muscle fiber and down T tubules

1. Excitation

Excitation of T tubule stimulates SR to release calcium ions

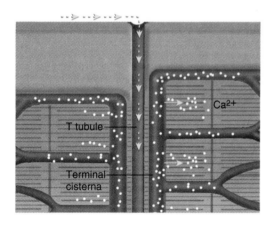

Ca²⁺

T tubule

Terminal cisterna

2. Excitation-Contraction Coupling

Calcium binds to troponin; tropomyosin shifts and exposes active sites of actin

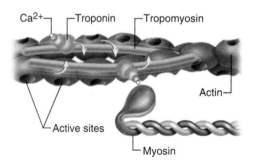

Ca²⁺   Troponin   Tropomyosin

Active sites   Actin

Myosin

The Principal Events in Muscle Contraction and Relaxation
Figure 10.15

Myosin binds ATP, breaks it down to ADP and phosphate (P_i), and binds to actin

Myosin releases ADP and P_i, flexes, and pulls on thin filament

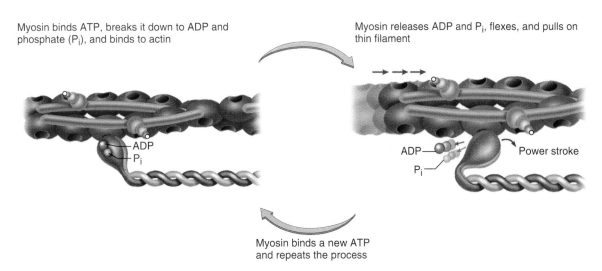

ADP
P_i

ADP
P_i

Power stroke

Myosin binds a new ATP and repeats the process

3. Contraction

Nerve signal ceases, ACh in synapse breaks down, and muscle excitation ceases

Ca^{2+} returns to SR, tropomyosin shifts and blocks active sites, myosin-actin links cannot form

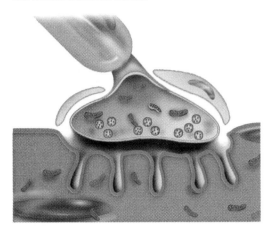

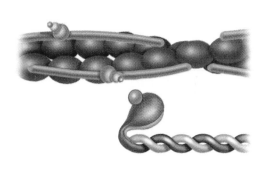

4. Relaxation

The Principal Events in Muscle Contraction and Relaxation (*Continued*)
Figure 10.15

Cardiac muscle

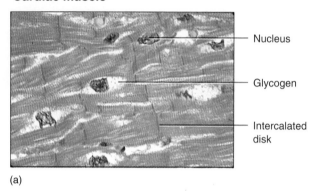

Nucleus

Glycogen

Intercalated disk

(a)

Smooth muscle

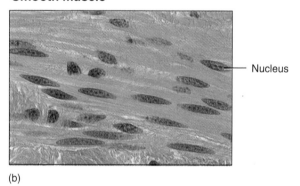

Nucleus

(b)

Cardiac and Smooth Muscle
Figure 10.17

© The McGraw-Hill Companies, Inc./Dennis Strete, photographer

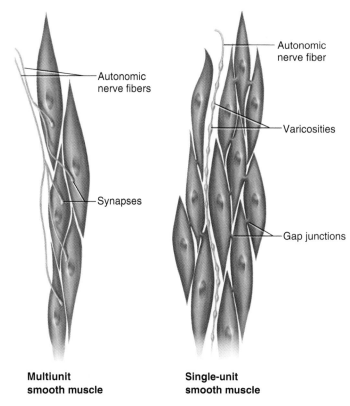

- Autonomic nerve fibers
- Autonomic nerve fiber
- Varicosities
- Synapses
- Gap junctions

Multiunit smooth muscle

Single-unit smooth muscle

Smooth Muscle Types
Figure 10.18

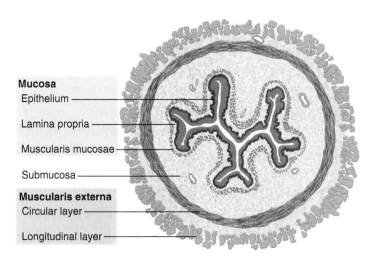

Mucosa
- Epithelium
- Lamina propria
- Muscularis mucosae

- Submucosa

Muscularis externa
- Circular layer
- Longitudinal layer

Layers of Visceral Muscle in the Wall of the Esophagus
Figure 10.19

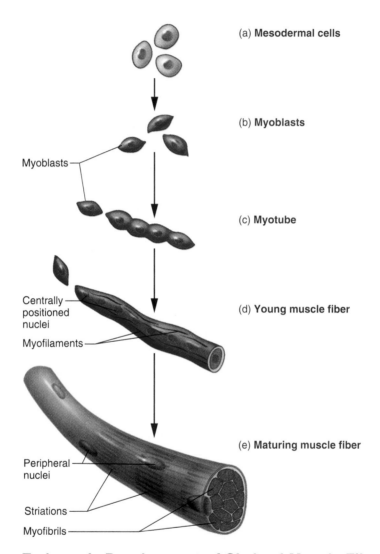

(a) **Mesodermal cells**

(b) **Myoblasts**

Myoblasts

(c) **Myotube**

Centrally
positioned
nuclei

Myofilaments

(d) **Young muscle fiber**

(e) **Maturing muscle fiber**

Peripheral
nuclei

Striations

Myofibrils

Embryonic Development of Skeletal Muscle Fibers
Figure 10.20

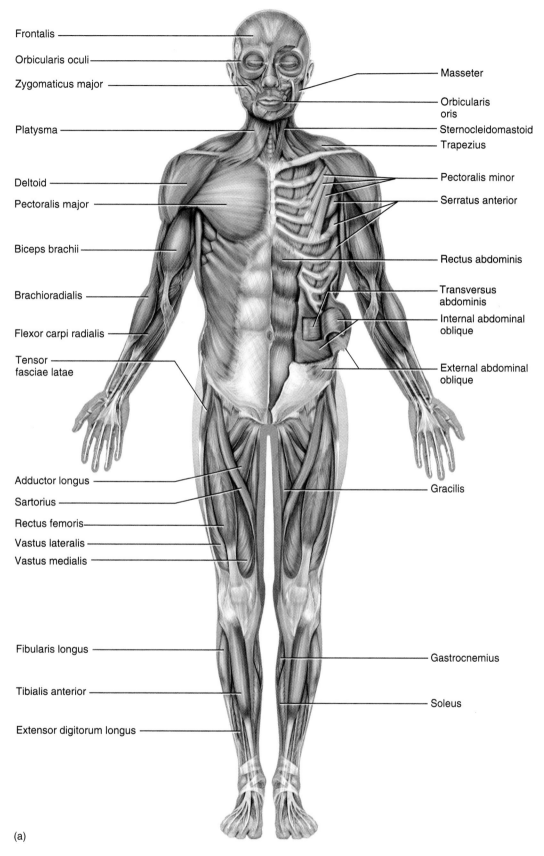

Frontalis

Orbicularis oculi

Zygomaticus major

Platysma

Deltoid

Pectoralis major

Biceps brachii

Brachioradialis

Flexor carpi radialis

Tensor
fasciae latae

Adductor longus

Sartorius

Rectus femoris

Vastus lateralis

Vastus medialis

Fibularis longus

Tibialis anterior

Extensor digitorum longus

Masseter

Orbicularis
oris

Sternocleidomastoid

Trapezius

Pectoralis minor

Serratus anterior

Rectus abdominis

Transversus
abdominis

Internal abdominal
oblique

External abdominal
oblique

Gracilis

Gastrocnemius

Soleus

(a)

The Muscular System
Figure 11.1

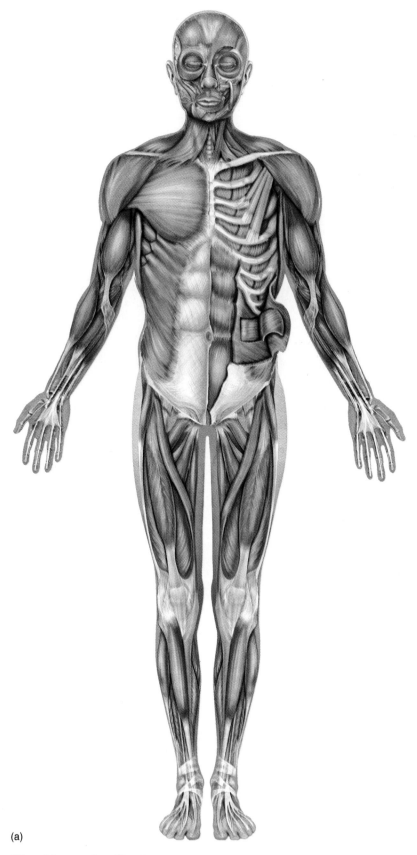

(a)

The Muscular System
Figure 11.1

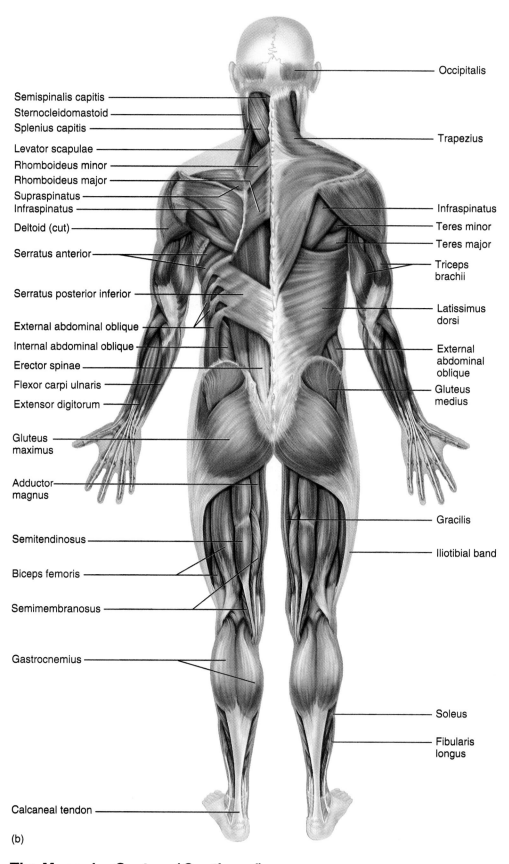

Occipitalis

Semispinalis capitis
Sternocleidomastoid
Splenius capitis

Trapezius

Levator scapulae
Rhomboideus minor
Rhomboideus major
Supraspinatus
Infraspinatus

Infraspinatus
Teres minor
Teres major

Deltoid (cut)

Triceps
brachii

Serratus anterior

Serratus posterior inferior

Latissimus
dorsi

External abdominal oblique

External
abdominal
oblique

Internal abdominal oblique

Erector spinae

Flexor carpi ulnaris

Gluteus
medius

Extensor digitorum

Gluteus
maximus

Adductor
magnus

Gracilis

Semitendinosus

Iliotibial band

Biceps femoris

Semimembranosus

Gastrocnemius

Soleus

Fibularis
longus

Calcaneal tendon

(b)

The Muscular System (*Continued*)
Figure 11.1

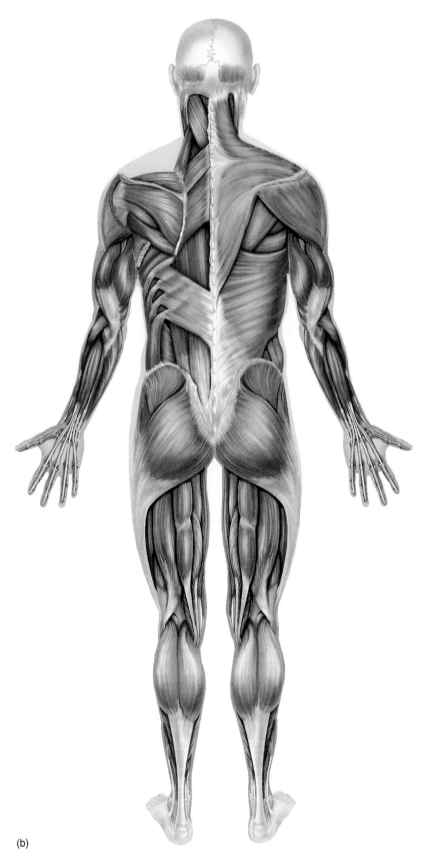

(b)

The Muscular System
Figure 11.1

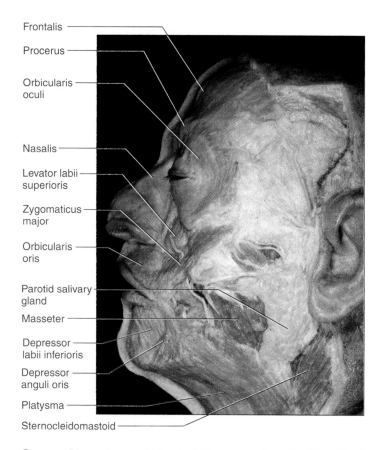

Frontalis

Procerus

Orbicularis
oculi

Nasalis

Levator labii
superioris

Zygomaticus
major

Orbicularis
oris

Parotid salivary
gland

Masseter

Depressor
labii inferioris

Depressor
anguli oris

Platysma

Sternocleidomastoid

Some Muscles of Facial Expression in the Cadaver
Figure 11.2

© The McGraw-Hill Companies, Inc./Rebecca Gray, photographer/
Don Kincaid, dissections

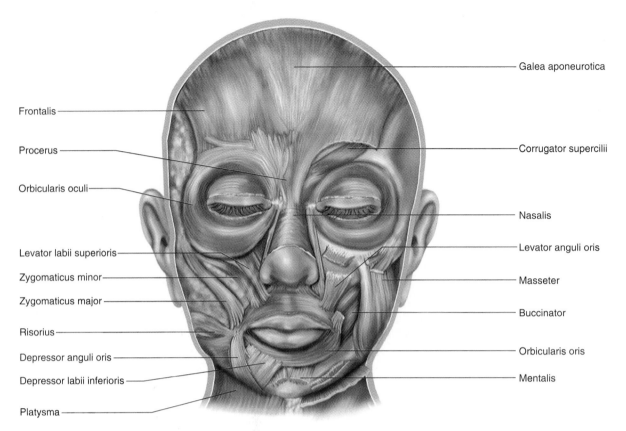

Galea aponeurotica

Frontalis

Procerus

Orbicularis oculi

Corrugator supercilii

Nasalis

Levator anguli oris

Levator labii superioris

Zygomaticus minor

Masseter

Zygomaticus major

Buccinator

Risorius

Depressor anguli oris

Orbicularis oris

Depressor labii inferioris

Mentalis

Platysma

Muscles of Facial Expression
Figure 11.3

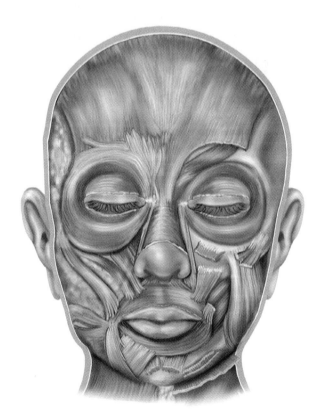

Muscles of Facial Expression
Figure 11.3

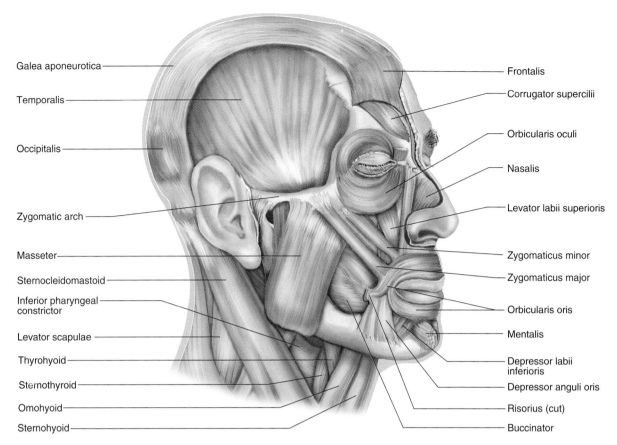

Galea aponeurotica

Temporalis

Occipitalis

Zygomatic arch

Masseter

Sternocleidomastoid

Inferior pharyngeal constrictor

Levator scapulae

Thyrohyoid

Sternothyroid

Omohyoid

Sternohyoid

Frontalis

Corrugator supercilii

Orbicularis oculi

Nasalis

Levator labii superioris

Zygomaticus minor

Zygomaticus major

Orbicularis oris

Mentalis

Depressor labii inferioris

Depressor anguli oris

Risorius (cut)

Buccinator

Muscles of Facial Expression (*Continued*)
Figure 11.3

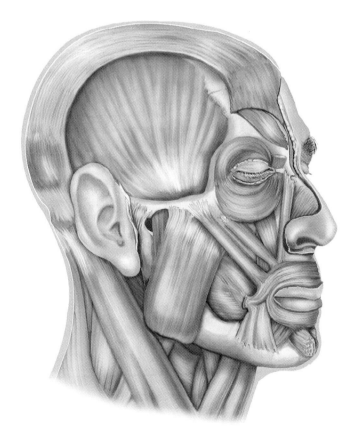

Muscles of Facial Expression
Figure 11.3

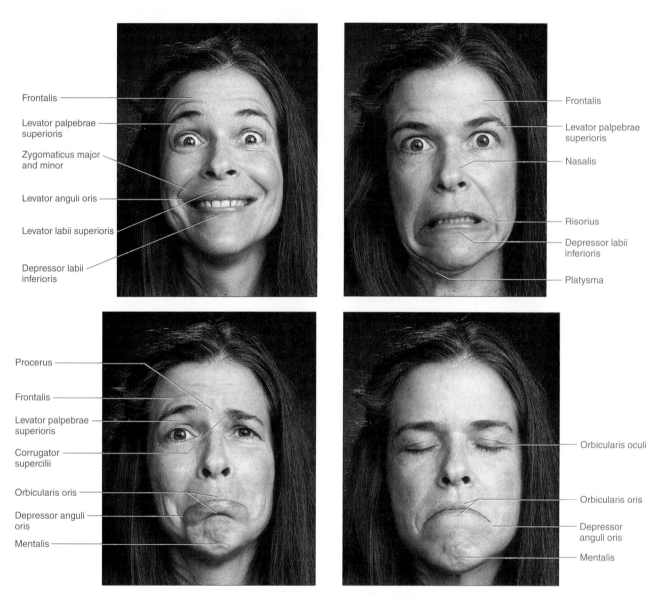

Expressions Produced by Several of the Facial Muscles

Figure 11.4

© The McGraw-Hill Companies, Inc./Joe DeGrandis, photographer

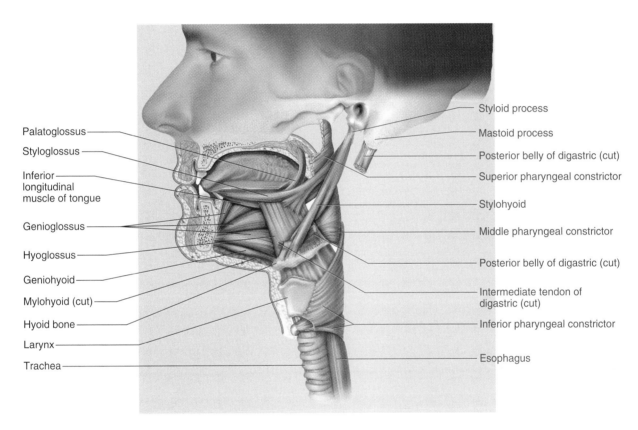

Palatoglossus

Styloglossus

Inferior longitudinal muscle of tongue

Genioglossus

Hyoglossus

Geniohyoid

Mylohyoid (cut)

Hyoid bone

Larynx

Trachea

Styloid process

Mastoid process

Posterior belly of digastric (cut)

Superior pharyngeal constrictor

Stylohyoid

Middle pharyngeal constrictor

Posterior belly of digastric (cut)

Intermediate tendon of digastric (cut)

Inferior pharyngeal constrictor

Esophagus

Muscles of the Tongue and Pharynx
Figure 11.5

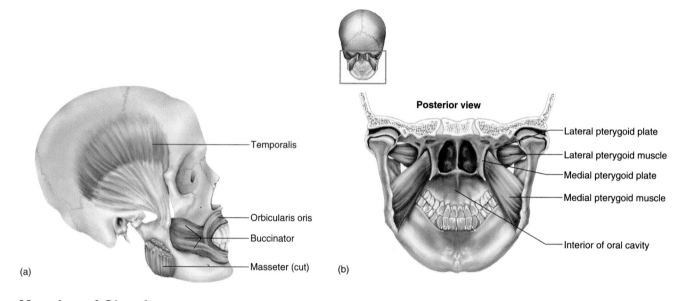

(a)

Temporalis

Orbicularis oris

Buccinator

Masseter (cut)

(b)

Posterior view

Lateral pterygoid plate

Lateral pterygoid muscle

Medial pterygoid plate

Medial pterygoid muscle

Interior of oral cavity

Muscles of Chewing
Figure 11.6

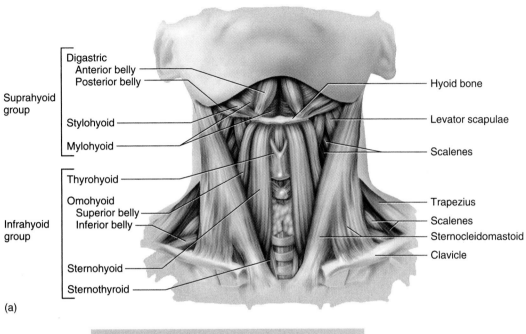

Suprahyoid group
- Digastric
  - Anterior belly
  - Posterior belly
- Stylohyoid
- Mylohyoid

Infrahyoid group
- Thyrohyoid
- Omohyoid
  - Superior belly
  - Inferior belly
- Sternohyoid
- Sternothyroid

Hyoid bone
Levator scapulae
Scalenes
Trapezius
Scalenes
Sternocleidomastoid
Clavicle

(a)

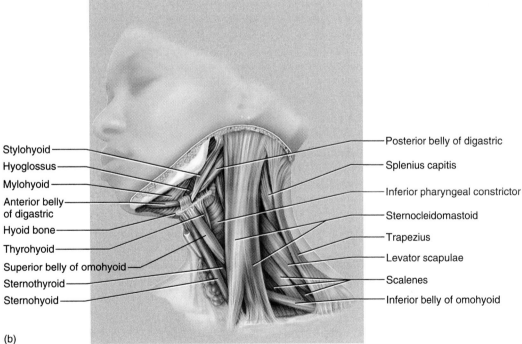

Stylohyoid
Hyoglossus
Mylohyoid
Anterior belly of digastric
Hyoid bone
Thyrohyoid
Superior belly of omohyoid
Sternothyroid
Sternohyoid

Posterior belly of digastric
Splenius capitis
Inferior pharyngeal constrictor
Sternocleidomastoid
Trapezius
Levator scapulae
Scalenes
Inferior belly of omohyoid

(b)

Muscles of the Neck

Figure 11.7

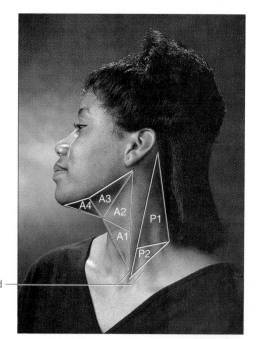

Anterior triangles
A1. Muscular
A2. Carotid
A3. Submandibular
A4. Suprahyoid

Posterior triangles
P1. Occipital
P2. Omoclavicular

Sternocleidomastoid

Triangles of the Neck
Figure 11.8

© The McGraw-Hill Companies, Inc./Joe DeGrandis, photographer

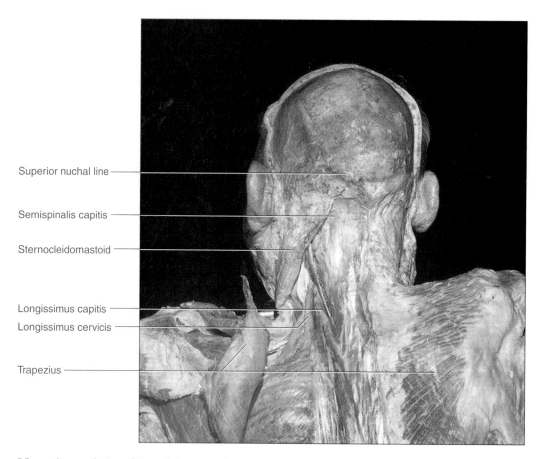

Superior nuchal line

Semispinalis capitis

Sternocleidomastoid

Longissimus capitis
Longissimus cervicis

Trapezius

Muscles of the Shoulder and Nuchal Regions
Figure 11.9

© The McGraw-Hill Companies, Inc./Rebecca Gray, photographer/Don Kincaid, dissections

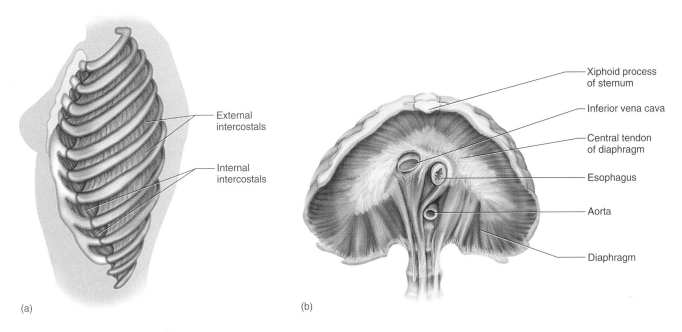

(a)

External
intercostals

Internal
intercostals

(b)

Xiphoid process
of sternum

Inferior vena cava

Central tendon
of diaphragm

Esophagus

Aorta

Diaphragm

Muscles of Respiration
Figure 11.10

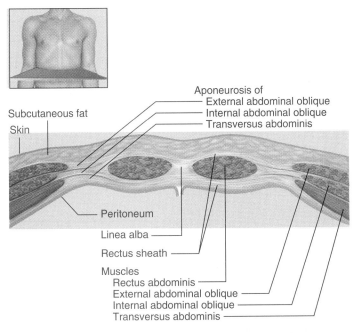

Subcutaneous fat

Skin

Aponeurosis of
External abdominal oblique
Internal abdominal oblique
Transversus abdominis

Peritoneum

Linea alba

Rectus sheath

Muscles
Rectus abdominis
External abdominal oblique
Internal abdominal oblique
Transversus abdominis

Cross Section of the Anterior Abdominal Wall
Figure 11.11

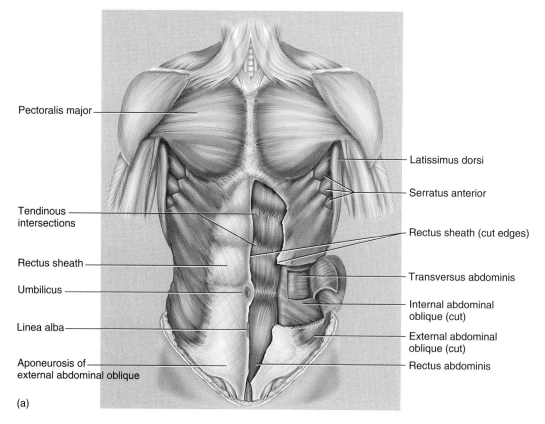

Pectoralis major

Latissimus dorsi

Serratus anterior

Tendinous intersections

Rectus sheath (cut edges)

Rectus sheath

Transversus abdominis

Umbilicus

Internal abdominal oblique (cut)

Linea alba

External abdominal oblique (cut)

Aponeurosis of external abdominal oblique

Rectus abdominis

(a)

Thoracic and Abdominal Muscles
Figure 11.13

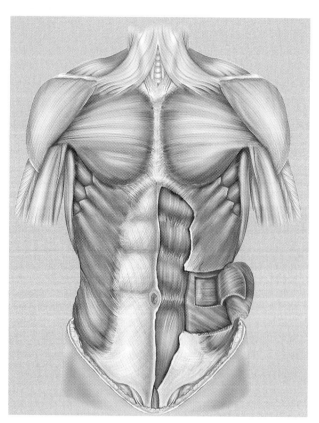

(a)

Thoracic and Abdominal Muscles
Figure 11.13

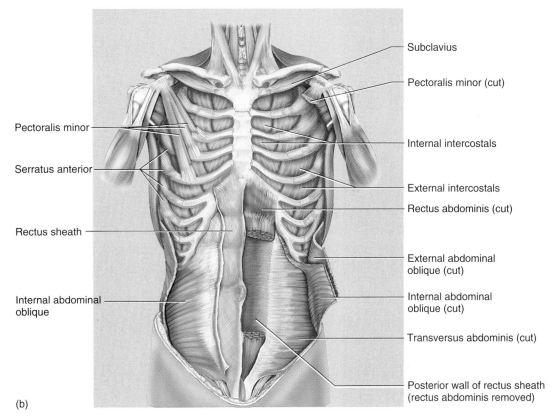

Subclavius

Pectoralis minor (cut)

Pectoralis minor

Serratus anterior

Internal intercostals

External intercostals

Rectus abdominis (cut)

Rectus sheath

External abdominal
oblique (cut)

Internal abdominal
oblique (cut)

Internal abdominal
oblique

Transversus abdominis (cut)

Posterior wall of rectus sheath
(rectus abdominis removed)

(b)

Thoracic and Abdominal Muscles (*Continued*)
Figure 11.13

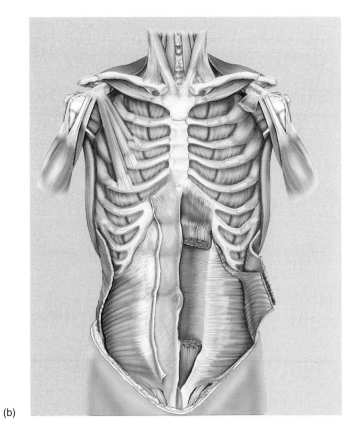

(b)

Thoracic and Abdominal Muscles
Figure 11.13

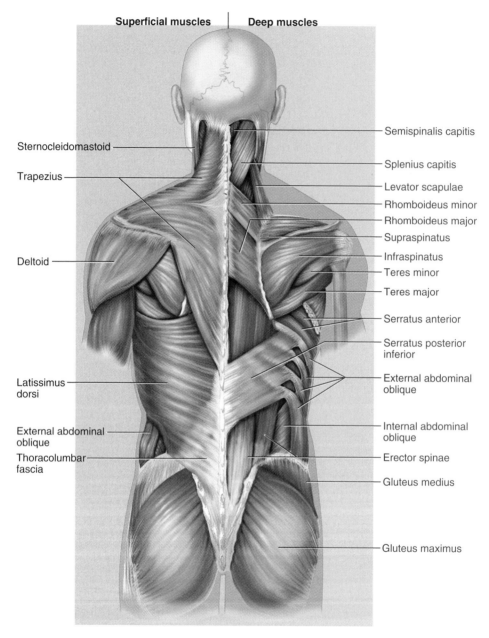

Superficial muscles | **Deep muscles**

Sternocleidomastoid

Trapezius

Deltoid

Latissimus dorsi

External abdominal oblique

Thoracolumbar fascia

Semispinalis capitis

Splenius capitis

Levator scapulae

Rhomboideus minor

Rhomboideus major

Supraspinatus

Infraspinatus

Teres minor

Teres major

Serratus anterior

Serratus posterior inferior

External abdominal oblique

Internal abdominal oblique

Erector spinae

Gluteus medius

Gluteus maximus

Neck, Back, and Gluteal Muscles
Figure 11.14

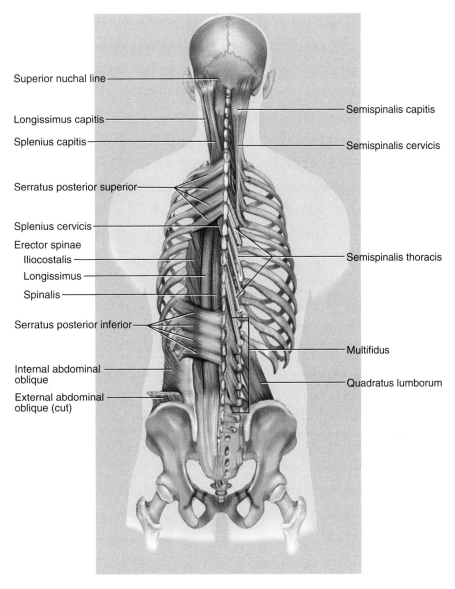

Superior nuchal line

Longissimus capitis

Splenius capitis

Serratus posterior superior

Splenius cervicis

Erector spinae
Iliocostalis
Longissimus
Spinalis

Serratus posterior inferior

Internal abdominal oblique

External abdominal oblique (cut)

Semispinalis capitis

Semispinalis cervicis

Semispinalis thoracis

Multifidus

Quadratus lumborum

Muscles Acting on the Vertebral Column
Figure 11.15

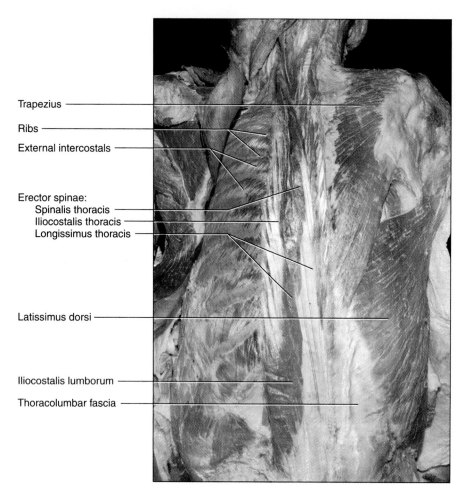

Trapezius

Ribs

External intercostals

Erector spinae:
    Spinalis thoracis
    Iliocostalis thoracis
    Longissimus thoracis

Latissimus dorsi

Iliocostalis lumborum

Thoracolumbar fascia

Some Deep Back Muscles of the Cadaver
Figure 11.16

© The McGraw-Hill Companies, Inc./Rebecca Gray, photographer/Don Kincaid, dissections

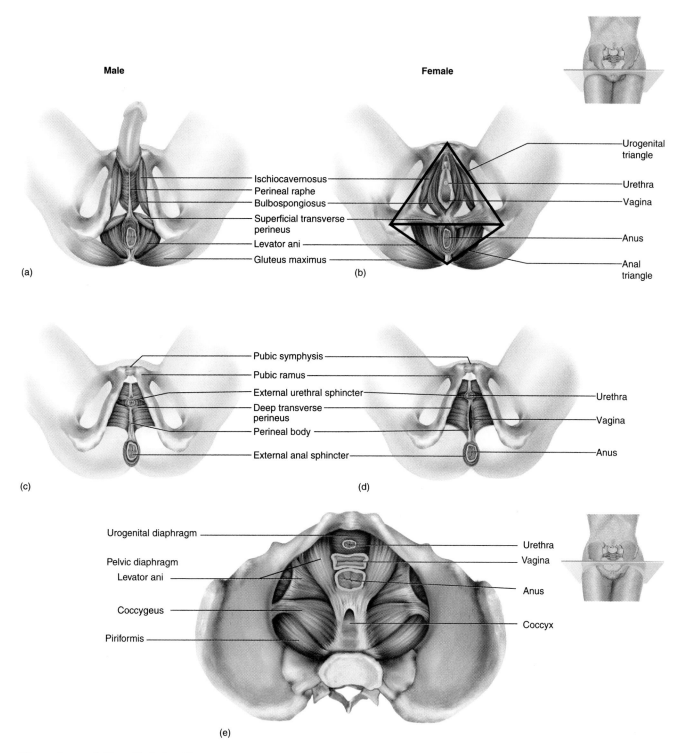

Male

Female

Ischiocavernosus
Perineal raphe
Bulbospongiosus
Superficial transverse
perineus
Levator ani
Gluteus maximus

(a)

(b)

Urogenital
triangle

Urethra

Vagina

Anus

Anal
triangle

Pubic symphysis
Pubic ramus
External urethral sphincter
Deep transverse
perineus
Perineal body

External anal sphincter

(c)

(d)

Urethra

Vagina

Anus

Urogenital diaphragm

Pelvic diaphragm
Levator ani

Coccygeus

Piriformis

Urethra
Vagina

Anus

Coccyx

(e)

Muscles of the Pelvic Floor
Figure 11.17

Lateral rotation
Trapezius (superior part)
Serratus anterior

Elevation
Levator scapulae
Trapezius (superior part)
Rhomboideus major
Rhomboideus minor

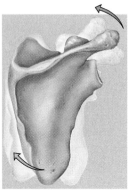

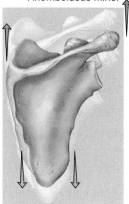

Medial rotation
Levator scapulae
Rhomboideus major
Rhomboideus minor

Depression
Trapezius (inferior part)
Serratus anterior

Retraction
Rhomboideus major
Rhomboideus minor
Trapezius

Protraction
Pectoralis minor
Serratus anterior

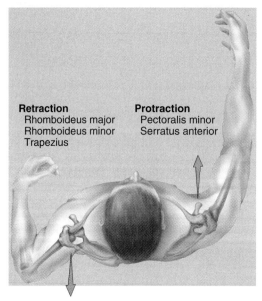

Actions of Some Thoracic Muscles on the Scapula
Figure 12.1

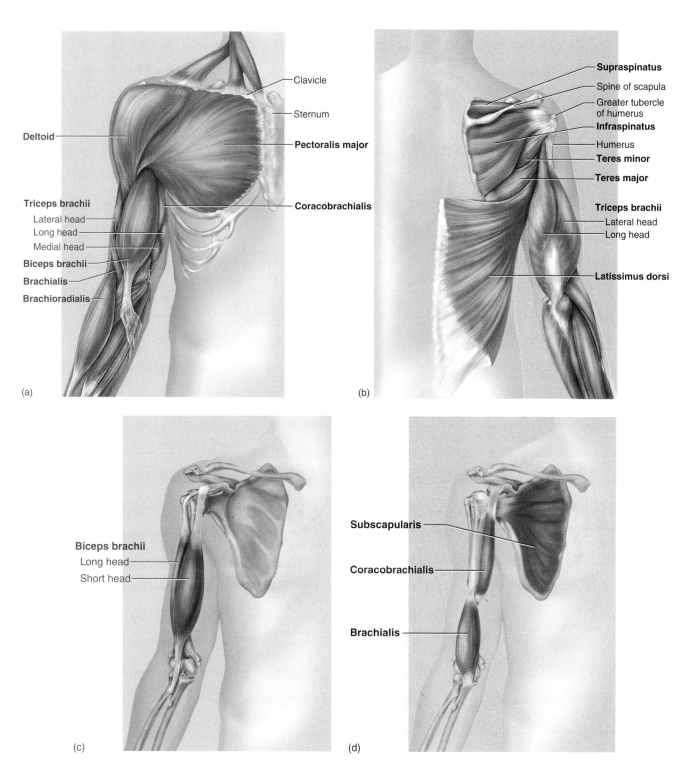

Pectoral and Brachial Muscles
Figure 12.2

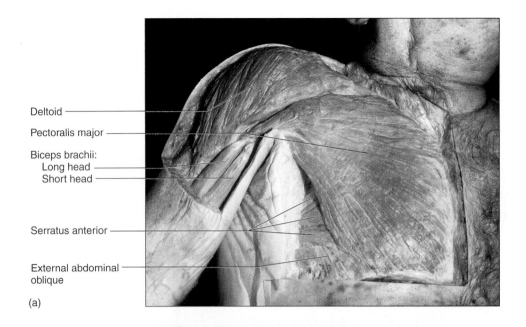

Deltoid

Pectoralis major

Biceps brachii:
    Long head
    Short head

Serratus anterior

External abdominal
oblique

(a)

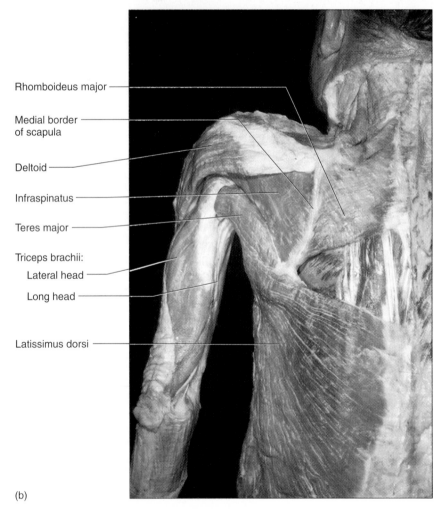

Rhomboideus major

Medial border
of scapula

Deltoid

Infraspinatus

Teres major

Triceps brachii:

    Lateral head

    Long head

Latissimus dorsi

(b)

Muscles of the Chest and Arm of the Cadaver
Figure 12.3

a,b: © The McGraw-Hill Companies, Inc./Rebecca Gray, photographer/Don Kincaid, dissections

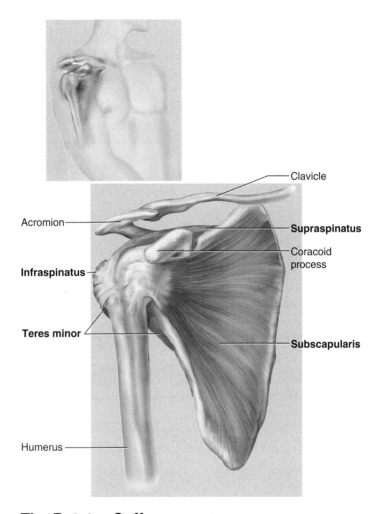

The Rotator Cuff
Figure 12.4

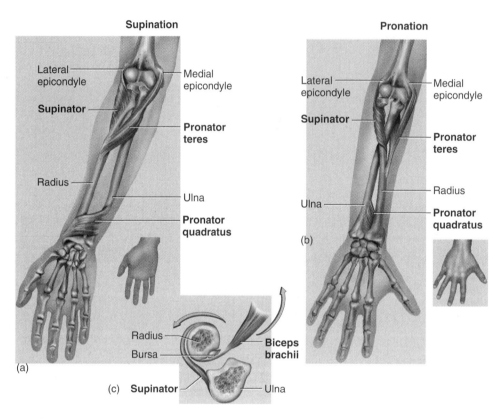

Actions of the Rotator Muscles on the Forearm
Figure 12.5

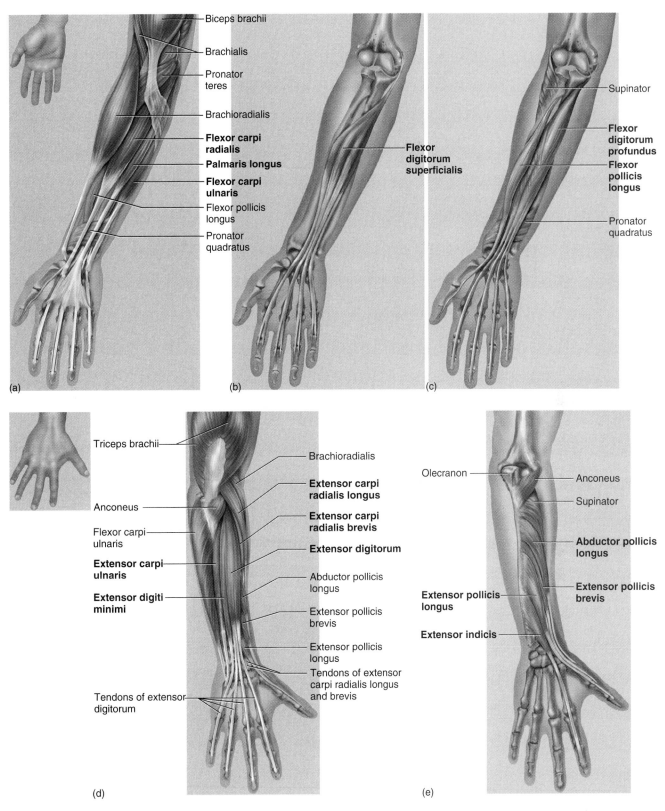

Muscles of the Forearm
Figure 12.6

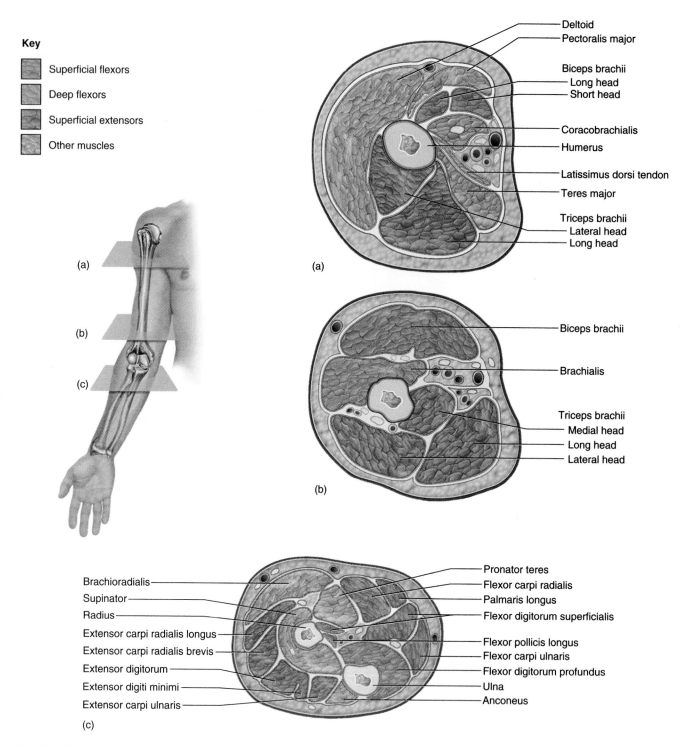

Key

- Superficial flexors
- Deep flexors
- Superficial extensors
- Other muscles

(a)

(b)

(c)

(a)

- Deltoid
- Pectoralis major
- Biceps brachii
  - Long head
  - Short head
- Coracobrachialis
- Humerus
- Latissimus dorsi tendon
- Teres major
- Triceps brachii
  - Lateral head
  - Long head

(b)

- Biceps brachii
- Brachialis
- Triceps brachii
  - Medial head
  - Long head
  - Lateral head

(c)

- Brachioradialis
- Supinator
- Radius
- Extensor carpi radialis longus
- Extensor carpi radialis brevis
- Extensor digitorum
- Extensor digiti minimi
- Extensor carpi ulnaris
- Pronator teres
- Flexor carpi radialis
- Palmaris longus
- Flexor digitorum superficialis
- Flexor pollicis longus
- Flexor carpi ulnaris
- Flexor digitorum profundus
- Ulna
- Anconeus

Serial Cross Sections Through the Upper Limb
Figure 12.7

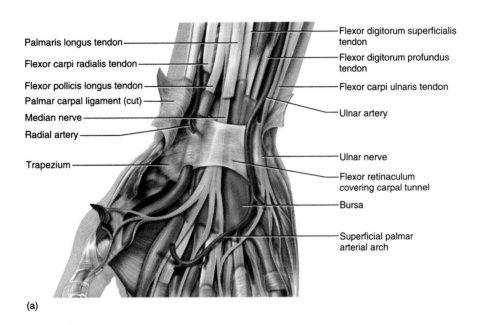

(a)

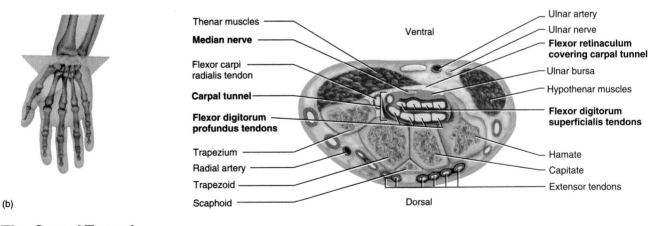

(b)

The Carpal Tunnel
Figure 12.8

Palmar aspect, superficial

Tendon sheath

Tendon of flexor
digitorum profundus

Tendon of flexor
digitorum superficialis

Lumbricals

**Opponens
digiti minimi**

**Flexor digiti
minimi brevis**

Abductor digiti minimi

Flexor retinaculum

Tendons of:
  Flexor carpi ulnaris

  Flexor digitorum
  superficialis

  Palmaris longus

**First dorsal
interosseous**

Tendon of flexor
pollicis longus

**Adductor
pollicis**

**Flexor pollicis
brevis**

**Abductor pollicis
brevis**

Opponens pollicis

Tendons of:
  Abductor pollicis
  longus
  Flexor carpi
  radialis
  Flexor pollicis
  longus

(a)

Palmar aspect, deep

**Palmar
interosseous**

**Opponens
digiti minimi**

Flexor
retinaculum (cut)

Carpal tunnel

Opponens pollicis

Tendons of:
  Abductor pollicis
  longus
  Flexor carpi radialis
  Flexor carpi
  ulnaris

(b)

Intrinsic Muscles of the Hand
Figure 12.9

Palmar aspect, superficial

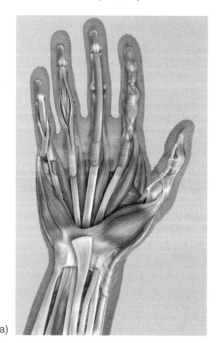

(a)

Palmar aspect, deep

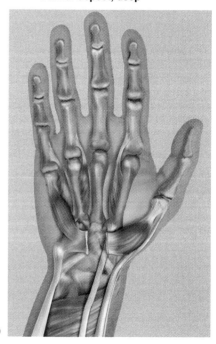

(b)

Intrinsic Muscles of the Hand
Figure 12.9

Dorsal aspect

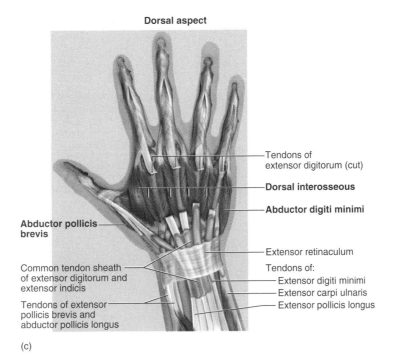

Tendons of
extensor digitorum (cut)

Dorsal interosseous

Abductor digiti minimi

Abductor pollicis brevis

Extensor retinaculum

Common tendon sheath
of extensor digitorum and
extensor indicis

Tendons of:
Extensor digiti minimi
Extensor carpi ulnaris
Extensor pollicis longus

Tendons of extensor
pollicis brevis and
abductor pollicis longus

(c)

Palmar aspect, cadaver

Tendon of
flexor
digitorum
superficialis

Lumbrical

Opponens digiti
minimi

Flexor digiti
minimi brevis

Abductor digiti
minimi

Pisiform bone

Flexor digitorum
superficialis

Adductor
pollicis

Flexor pollicis
brevis

Abductor
pollicis brevis

Tendon of
extensor
pollicis brevis

Tendon of flexor
carpi radialis

(d)

Intrinsic Muscles of the Hand (*Continued*)
Figure 12.9

d: © The McGraw-Hill Companies, Inc./Rebecca Gray,
photographer/Don Kincaid, dissections

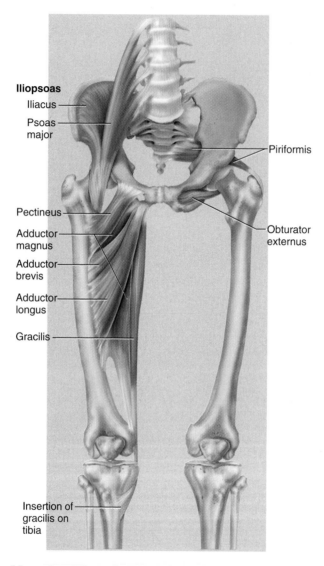

Iliopsoas
Iliacus
Psoas major
Pectineus
Adductor magnus
Adductor brevis
Adductor longus
Gracilis
Insertion of gracilis on tibia
Piriformis
Obturator externus

Muscles That Act on the Hip and Femur
Figure 12.10

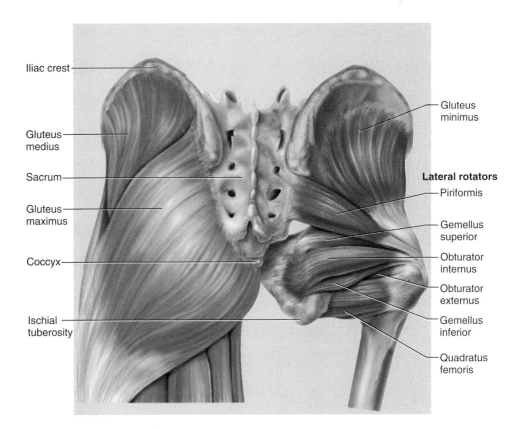

Gluteal Muscles
Figure 12.11

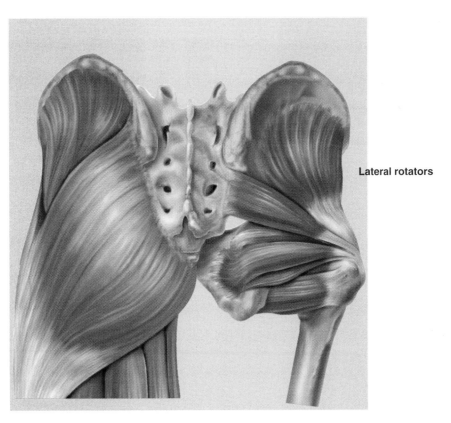

Gluteal Muscles
Figure 12.11

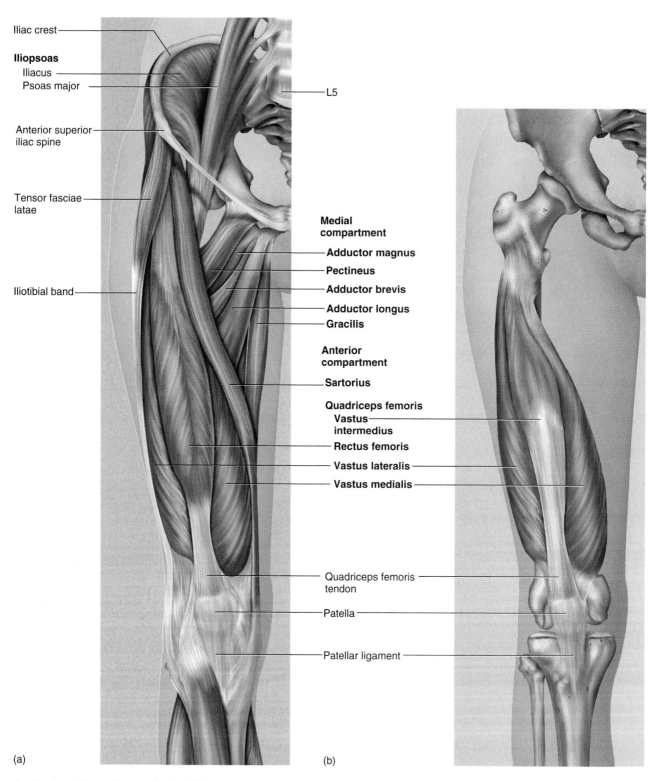

Iliac crest

Iliopsoas
  Iliacus
  Psoas major

Anterior superior
iliac spine

Tensor fasciae
latae

Iliotibial band

L5

**Medial
compartment**
  Adductor magnus
  Pectineus
  Adductor brevis
  Adductor longus
  Gracilis

**Anterior
compartment**
  Sartorius

Quadriceps femoris
  **Vastus
  intermedius**
  Rectus femoris
  Vastus lateralis
  Vastus medialis

Quadriceps femoris
tendon

Patella

Patellar ligament

(a)

(b)

Anterior Muscles of the Thigh
Figure 12.12

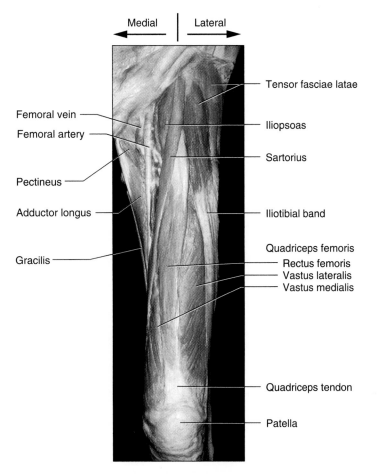

Medial | Lateral

Tensor fasciae latae

Femoral vein

Femoral artery

Iliopsoas

Sartorius

Pectineus

Adductor longus

Iliotibial band

Gracilis

Quadriceps femoris
Rectus femoris
Vastus lateralis
Vastus medialis

Quadriceps tendon

Patella

Anterior Superficial Thigh Muscles of the Cadaver
Figure 12.13

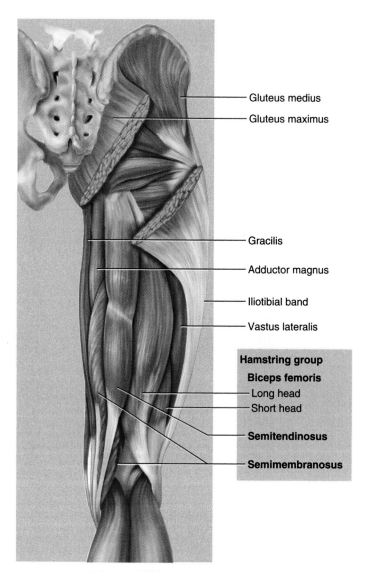

- Gluteus medius
- Gluteus maximus

- Gracilis
- Adductor magnus
- Iliotibial band
- Vastus lateralis

Hamstring group
Biceps femoris
- Long head
- Short head
Semitendinosus
Semimembranosus

Gluteal and Thigh Muscles
Figure 12.14

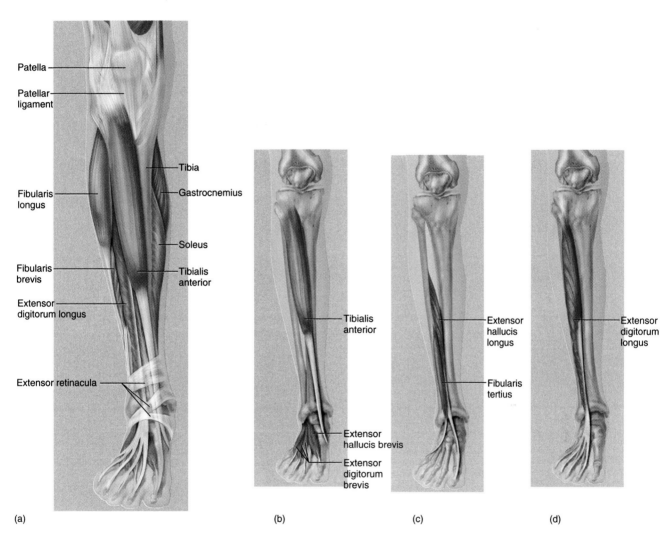

Anterior Muscles of the Leg
Figure 12.15

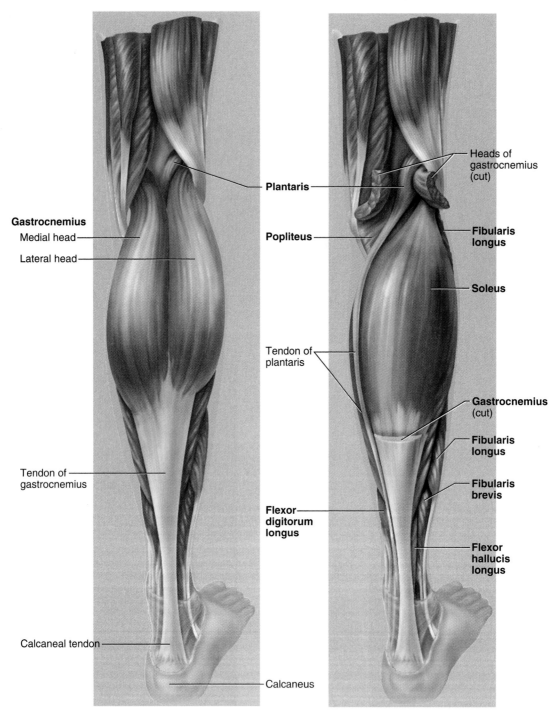

Gastrocnemius
Medial head
Lateral head

Plantaris

Popliteus

Tendon of plantaris

Tendon of gastrocnemius

Flexor digitorum longus

Calcaneal tendon

Calcaneus

Heads of gastrocnemius (cut)

Fibularis longus

Soleus

Gastrocnemius (cut)

Fibularis longus

Fibularis brevis

Flexor hallucis longus

Superficial Muscles of the Leg, Posterior Compartment
Figure 12.16

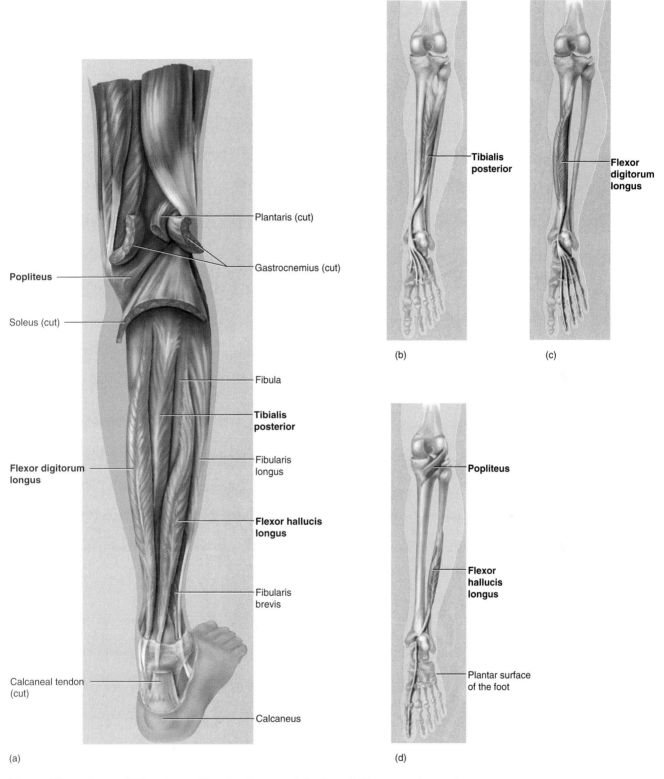

Deep Muscles of the Leg, Posterior and Lateral Compartments
Figure 12.17

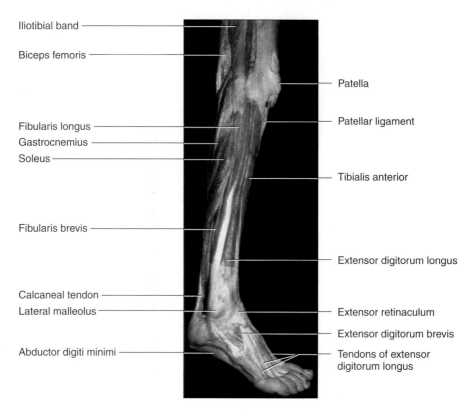

Iliotibial band

Biceps femoris

Fibularis longus
Gastrocnemius
Soleus

Fibularis brevis

Calcaneal tendon
Lateral malleolus

Abductor digiti minimi

Patella

Patellar ligament

Tibialis anterior

Extensor digitorum longus

Extensor retinaculum
Extensor digitorum brevis
Tendons of extensor
digitorum longus

Superficial Muscles of the Leg of the Cadaver
Figure 12.18

© The McGraw-Hill Companies, Inc./Rebecca Gray, photographer/Don Kincaid,
dissections

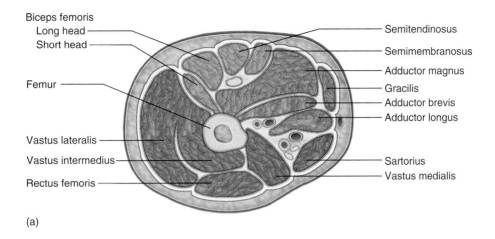

(a)

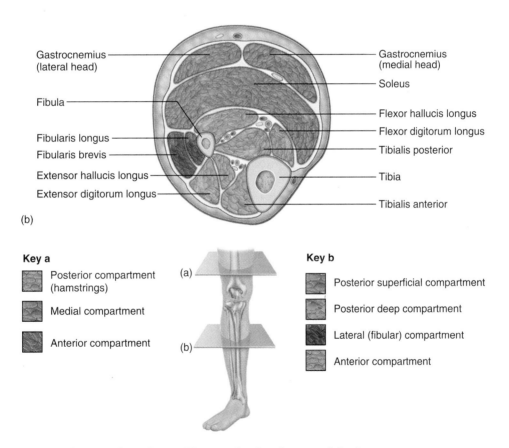

(b)

Key a

Posterior compartment (hamstrings)

Medial compartment

Anterior compartment

(a)

(b)

Key b

Posterior superficial compartment

Posterior deep compartment

Lateral (fibular) compartment

Anterior compartment

Serial Cross Sections Through the Lower Limb
Figure 12.19

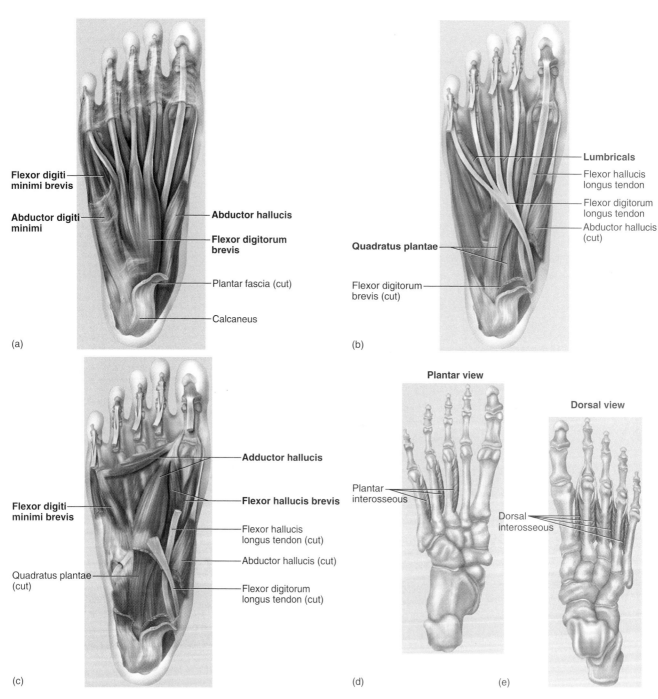

(a)

Flexor digiti minimi brevis

Abductor digiti minimi

Abductor hallucis

Flexor digitorum brevis

Plantar fascia (cut)

Calcaneus

(b)

Lumbricals

Flexor hallucis longus tendon

Flexor digitorum longus tendon

Abductor hallucis (cut)

Quadratus plantae

Flexor digitorum brevis (cut)

(c)

Adductor hallucis

Flexor hallucis brevis

Flexor digiti minimi brevis

Flexor hallucis longus tendon (cut)

Abductor hallucis (cut)

Quadratus plantae (cut)

Flexor digitorum longus tendon (cut)

(d)

Plantar view

Plantar interosseous

(e)

Dorsal view

Dorsal interosseous

Intrinsic Muscles of the Foot
Figure 12.20

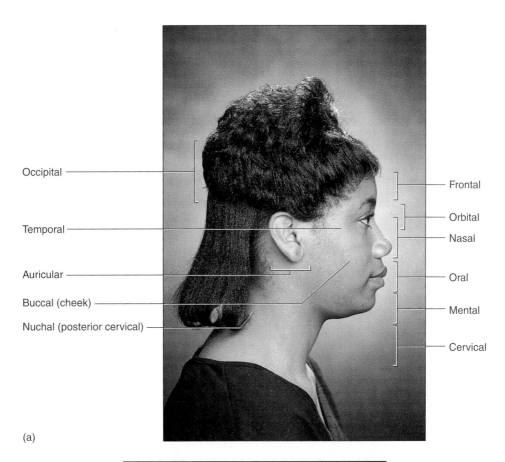

Occipital

Temporal

Auricular

Buccal (cheek)

Nuchal (posterior cervical)

Frontal

Orbital

Nasal

Oral

Mental

Cervical

(a)

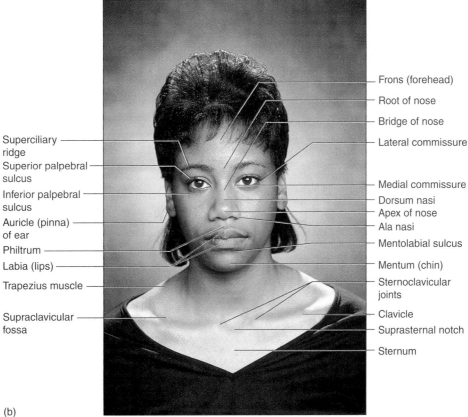

Superciliary ridge

Superior palpebral sulcus

Inferior palpebral sulcus

Auricle (pinna) of ear

Philtrum

Labia (lips)

Trapezius muscle

Supraclavicular fossa

Frons (forehead)

Root of nose

Bridge of nose

Lateral commissure

Medial commissure

Dorsum nasi

Apex of nose

Ala nasi

Mentolabial sulcus

Mentum (chin)

Sternoclavicular joints

Clavicle

Suprasternal notch

Sternum

(b)

The Head and Neck

Figure B.1

a,b: © The McGraw-Hill Companies, Inc./Joe DeGrandis, photographer

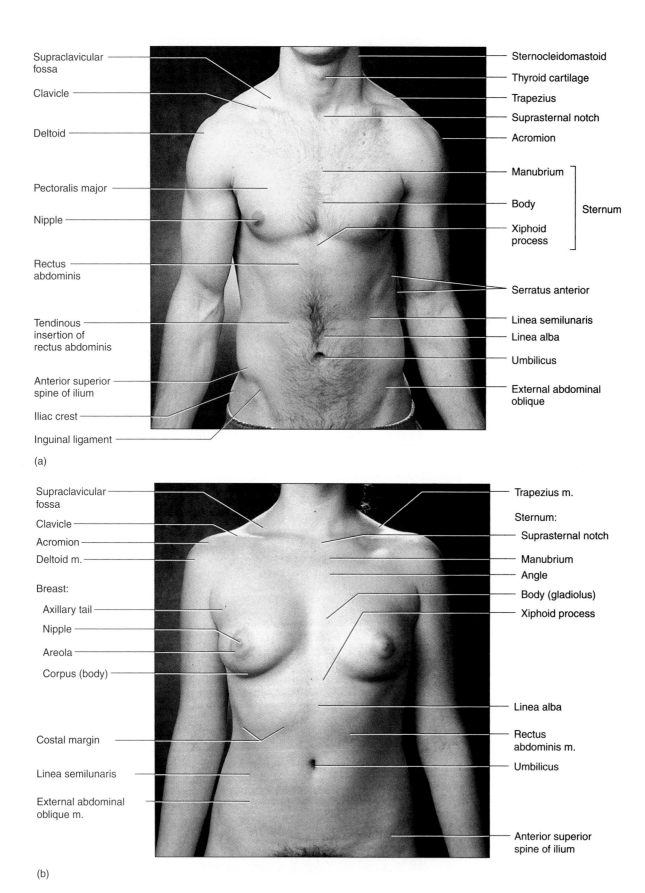

Supraclavicular fossa
Clavicle
Deltoid
Pectoralis major
Nipple
Rectus abdominis
Tendinous insertion of rectus abdominis
Anterior superior spine of ilium
Iliac crest
Inguinal ligament

Sternocleidomastoid
Thyroid cartilage
Trapezius
Suprasternal notch
Acromion
Manubrium
Body
Xiphoid process
Sternum
Serratus anterior
Linea semilunaris
Linea alba
Umbilicus
External abdominal oblique

(a)

Supraclavicular fossa
Clavicle
Acromion
Deltoid m.
Breast:
Axillary tail
Nipple
Areola
Corpus (body)
Costal margin
Linea semilunaris
External abdominal oblique m.

Trapezius m.
Sternum:
Suprasternal notch
Manubrium
Angle
Body (gladiolus)
Xiphoid process
Linea alba
Rectus abdominis m.
Umbilicus
Anterior superior spine of ilium

(b)

The Thorax and Abdomen, Ventral Aspect
Figure B.2

a,b: © The McGraw-Hill Companies, Inc./Joe DeGrandis, photographer

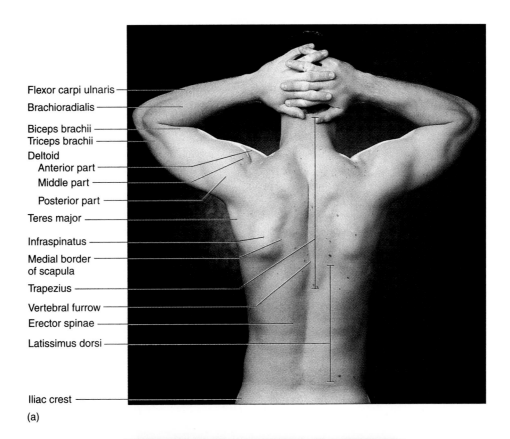

Flexor carpi ulnaris
Brachioradialis
Biceps brachii
Triceps brachii
Deltoid
    Anterior part
    Middle part
    Posterior part
Teres major
Infraspinatus
Medial border
of scapula
Trapezius
Vertebral furrow
Erector spinae
Latissimus dorsi

Iliac crest
(a)

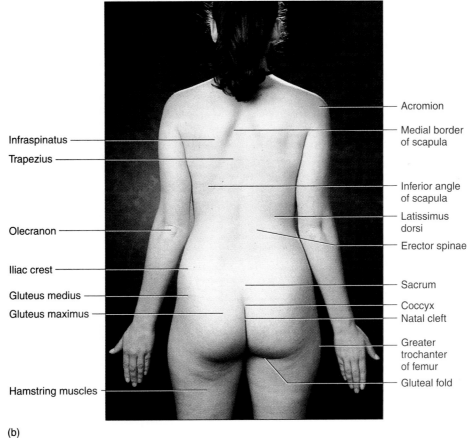

Infraspinatus
Trapezius

Olecranon

Iliac crest

Gluteus medius
Gluteus maximus

Hamstring muscles

Acromion
Medial border
of scapula

Inferior angle
of scapula
Latissimus
dorsi
Erector spinae

Sacrum

Coccyx
Natal cleft

Greater
trochanter
of femur
Gluteal fold

(b)

The Back and Gluteal Region
Figure B.3

a,b: © The McGraw-Hill Companies, Inc./Joe DeGrandis, photographer

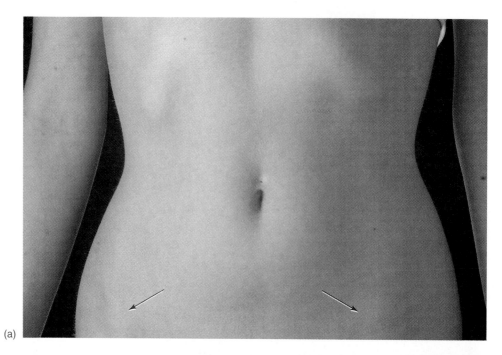

(a)

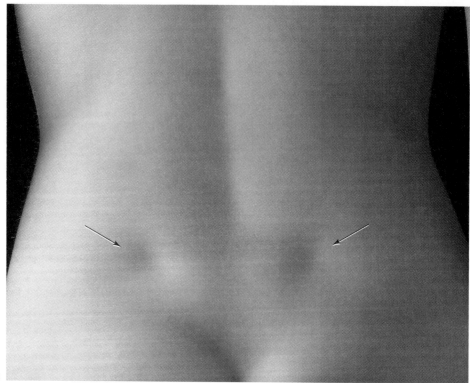

(b)

The Pelvic Region
Figure B.4

a,b: © The McGraw-Hill Companies, Inc./Joe DeGrandis, photographer

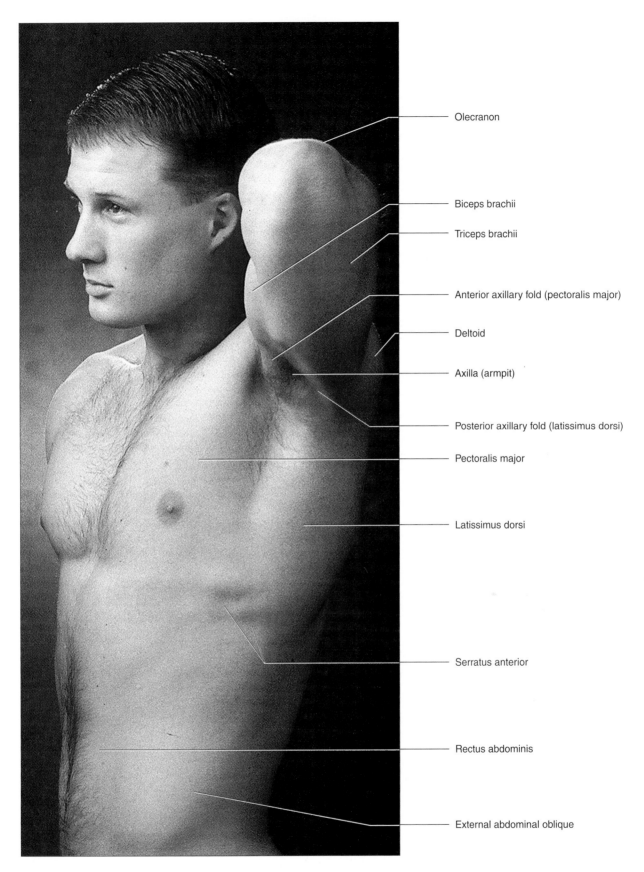

Olecranon

Biceps brachii

Triceps brachii

Anterior axillary fold (pectoralis major)

Deltoid

Axilla (armpit)

Posterior axillary fold (latissimus dorsi)

Pectoralis major

Latissimus dorsi

Serratus anterior

Rectus abdominis

External abdominal oblique

The Axillary Region
Figure B.5
© The McGraw-Hill Companies, Inc./Joe DeGrandis, photographer

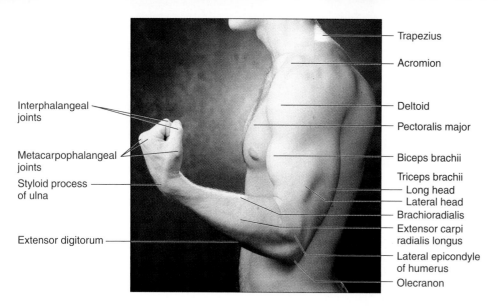

The Upper Limb, Lateral aspect
Figure B.6

© The McGraw-Hill Companies, Inc./Joe DeGrandis, photographer

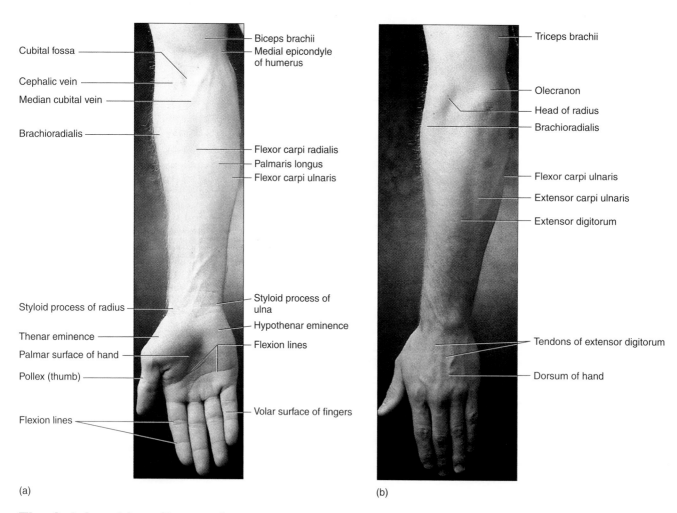

(a)

(b)

The Antebrachium (forearm)
Figure B.7

a,b: © The McGraw-Hill Companies, Inc./Joe DeGrandis, photographer

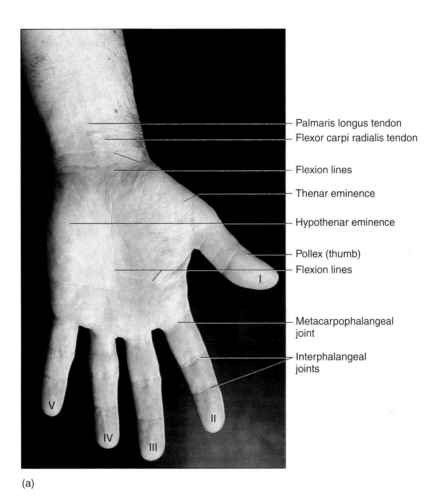

Palmaris longus tendon
Flexor carpi radialis tendon

Flexion lines

Thenar eminence

Hypothenar eminence

Pollex (thumb)
Flexion lines

Metacarpophalangeal joint

Interphalangeal joints

I

V

IV

III

II

(a)

The Wrist and Hand
Figure B.8

a: © The McGraw-Hill Companies, Inc./Joe DeGrandis, photographer

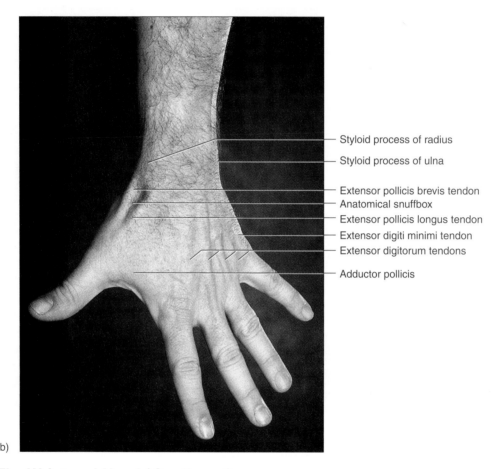

(b)

Styloid process of radius
Styloid process of ulna

Extensor pollicis brevis tendon
Anatomical snuffbox
Extensor pollicis longus tendon
Extensor digiti minimi tendon
Extensor digitorum tendons

Adductor pollicis

The Wrist and Hand (*Continued*)
Figure B.8

b: © The McGraw-Hill Companies, Inc./Joe DeGrandis, photographer

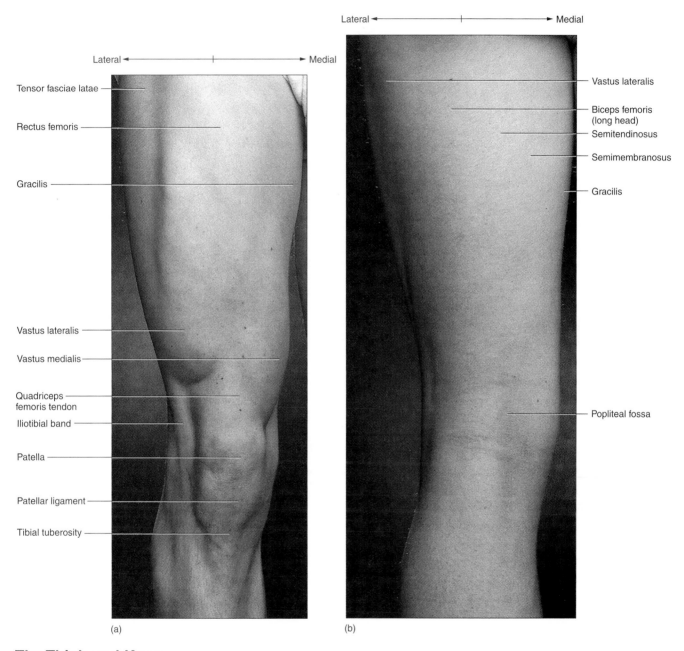

Lateral ◄———————|———————► Medial

Tensor fasciae latae

Rectus femoris

Gracilis

Vastus lateralis

Vastus medialis

Quadriceps
femoris tendon

Iliotibial band

Patella

Patellar ligament

Tibial tuberosity

(a)

Lateral ◄———————|———————► Medial

Vastus lateralis

Biceps femoris
(long head)

Semitendinosus

Semimembranosus

Gracilis

Popliteal fossa

(b)

The Thigh and Knee
Figure B.9

a,b: © The McGraw-Hill Companies, Inc./Joe DeGrandis, photographer

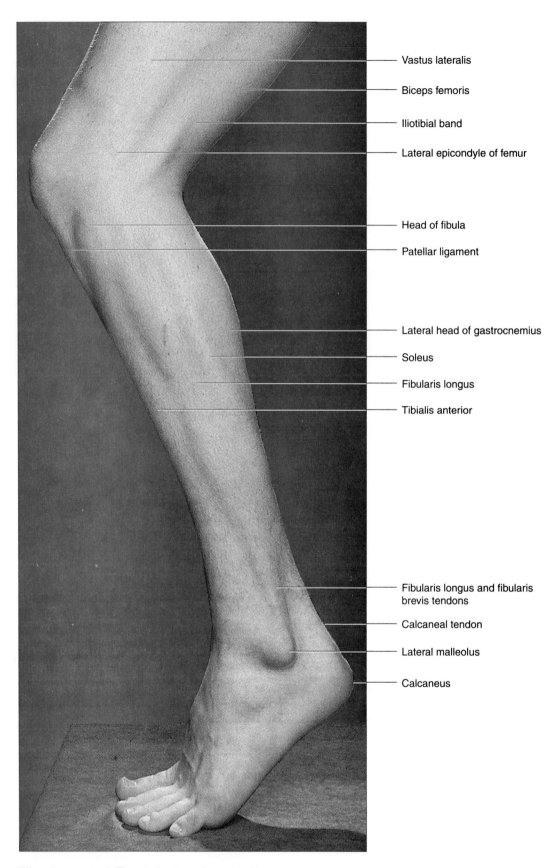

Vastus lateralis

Biceps femoris

Iliotibial band

Lateral epicondyle of femur

Head of fibula

Patellar ligament

Lateral head of gastrocnemius

Soleus

Fibularis longus

Tibialis anterior

Fibularis longus and fibularis
brevis tendons

Calcaneal tendon

Lateral malleolus

Calcaneus

The Leg and Foot, Lateral aspect
Figure B.10
© The McGraw-Hill Companies, Inc./Joe DeGrandis, photographer

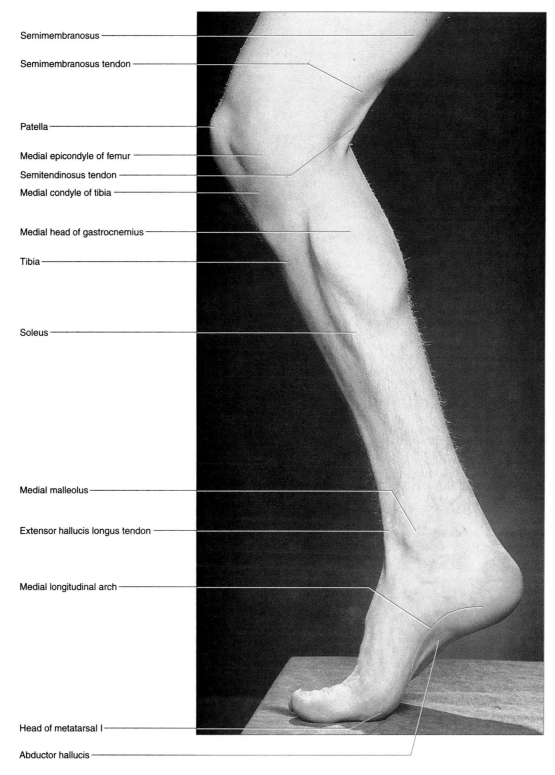

Semimembranosus

Semimembranosus tendon

Patella

Medial epicondyle of femur

Semitendinosus tendon

Medial condyle of tibia

Medial head of gastrocnemius

Tibia

Soleus

Medial malleolus

Extensor hallucis longus tendon

Medial longitudinal arch

Head of metatarsal I

Abductor hallucis

The Leg and Foot, Medial Aspect
Figure B.11
© The McGraw-Hill Companies, Inc./Joe DeGrandis, photographer

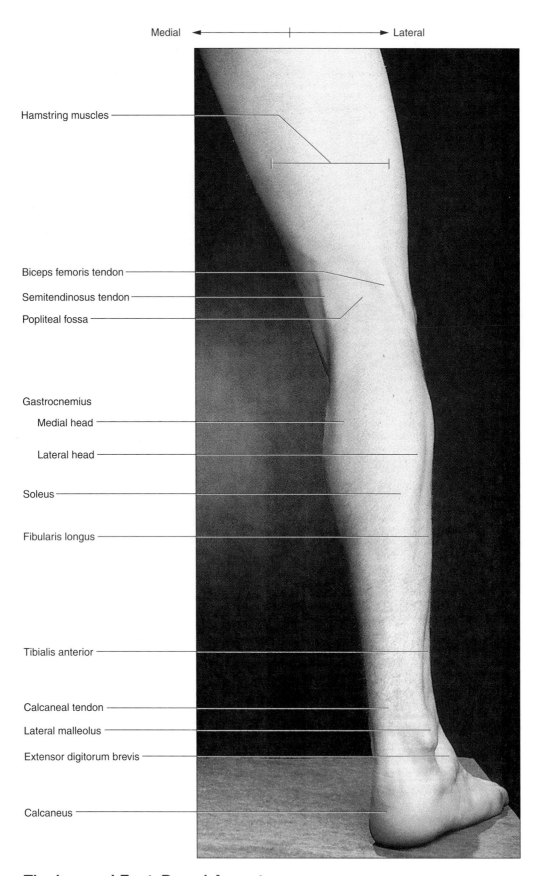

Medial ◄——————|——————► Lateral

Hamstring muscles

Biceps femoris tendon

Semitendinosus tendon

Popliteal fossa

Gastrocnemius

    Medial head

    Lateral head

Soleus

Fibularis longus

Tibialis anterior

Calcaneal tendon

Lateral malleolus

Extensor digitorum brevis

Calcaneus

The Leg and Foot, Dorsal Aspect
Figure B.12

© The McGraw-Hill Companies, Inc./Joe DeGrandis, photographer

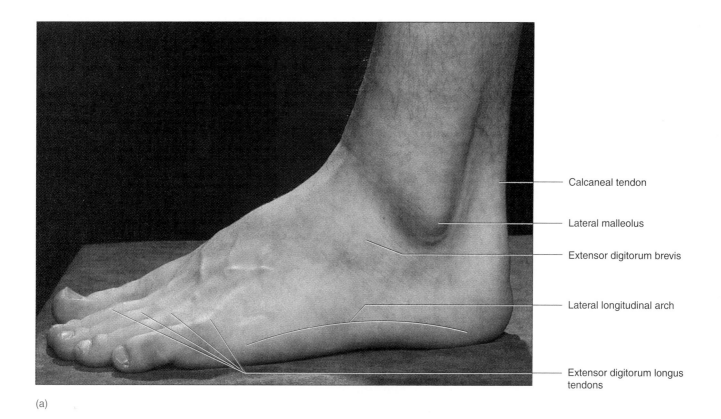

— Calcaneal tendon

— Lateral malleolus

— Extensor digitorum brevis

— Lateral longitudinal arch

— Extensor digitorum longus tendons

(a)

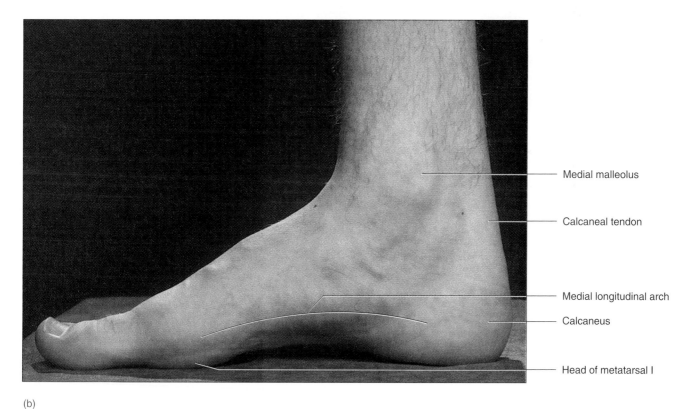

— Medial malleolus

— Calcaneal tendon

— Medial longitudinal arch

— Calcaneus

— Head of metatarsal I

(b)

The Foot
Figure B.13

a,b: © The McGraw-Hill Companies, Inc./Joe DeGrandis, photographer

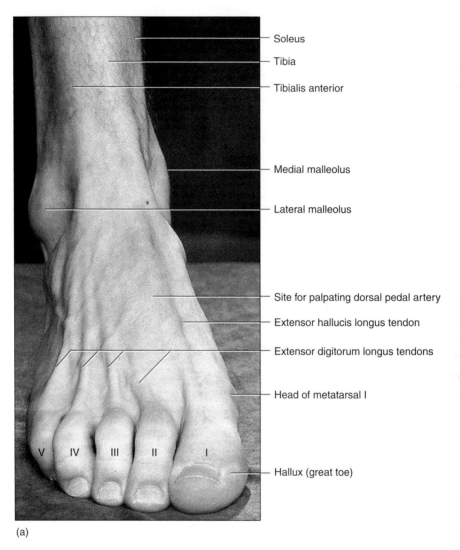

Soleus

Tibia

Tibialis anterior

Medial malleolus

Lateral malleolus

Site for palpating dorsal pedal artery

Extensor hallucis longus tendon

Extensor digitorum longus tendons

Head of metatarsal I

V IV III II I

Hallux (great toe)

(a)

The Foot
Figure B.14

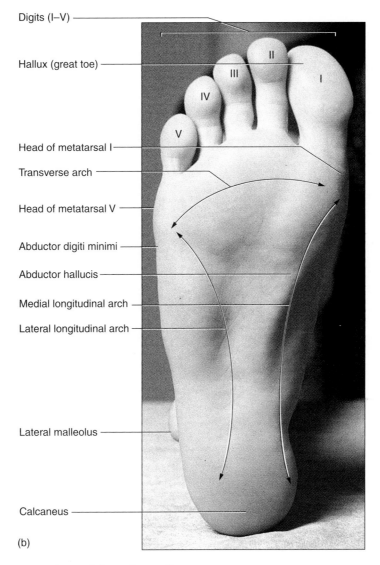

Digits (I–V)

Hallux (great toe)

II
III
IV
V
I

Head of metatarsal I

Transverse arch

Head of metatarsal V

Abductor digiti minimi

Abductor hallucis

Medial longitudinal arch

Lateral longitudinal arch

Lateral malleolus

Calcaneus

(b)

The Foot (*Continued*)
Figure B.14
b: © The McGraw-Hill Companies, Inc./Joe DeGrandis, photographer

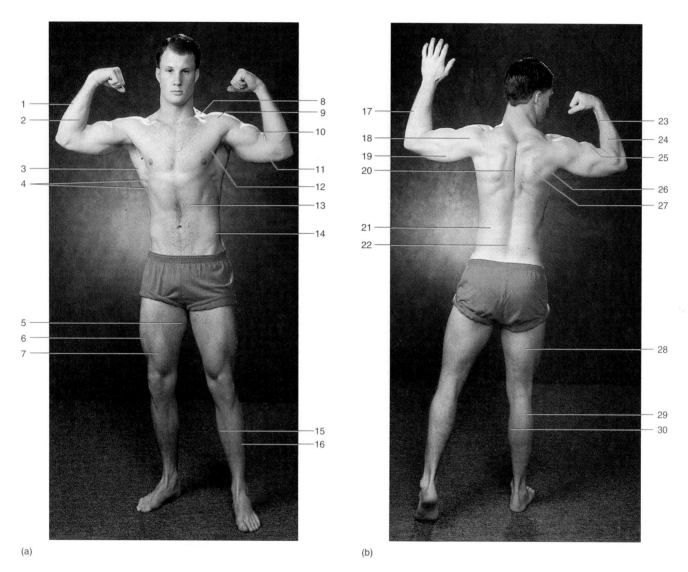

Muscle Self-Test
Figure B.15

a,b: © The McGraw-Hill Companies, Inc./Joe DeGrandis, photographer

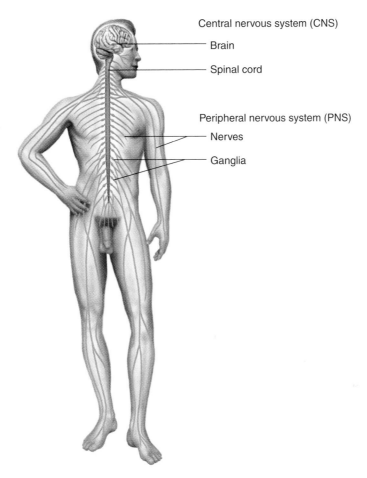

Central nervous system (CNS)

Brain

Spinal cord

Peripheral nervous system (PNS)

Nerves

Ganglia

The Nervous System
Figure 13.1

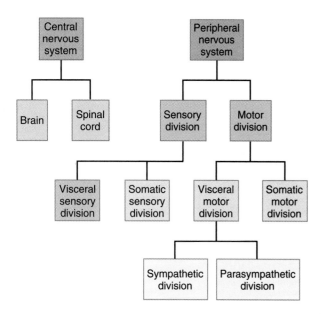

Subdivisions of the Nervous System
Figure 13.2

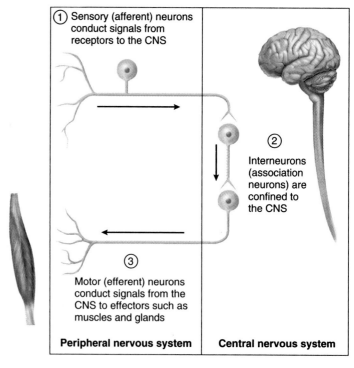

① Sensory (afferent) neurons conduct signals from receptors to the CNS

② Interneurons (association neurons) are confined to the CNS

③ Motor (efferent) neurons conduct signals from the CNS to effectors such as muscles and glands

Peripheral nervous system | **Central nervous system**

Functional Classes of Neurons
Figure 13.3

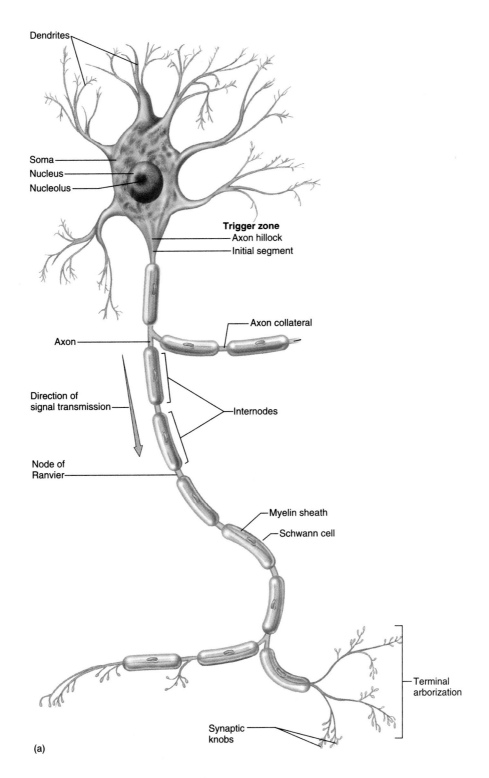

Dendrites

Soma

Nucleus

Nucleolus

Trigger zone

Axon hillock

Initial segment

Axon collateral

Axon

Direction of
signal transmission

Internodes

Node of
Ranvier

Myelin sheath

Schwann cell

Terminal
arborization

Synaptic
knobs

(a)

A Representative Neuron
Figure 13.4

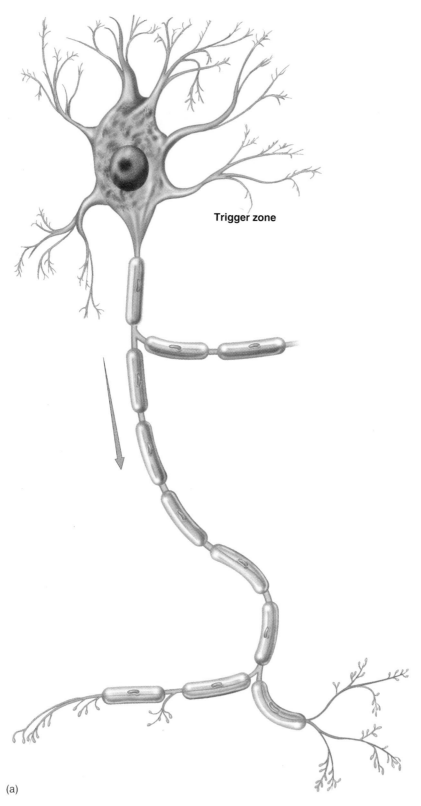

Trigger zone

(a)

A Representative Neuron
Figure 13.4

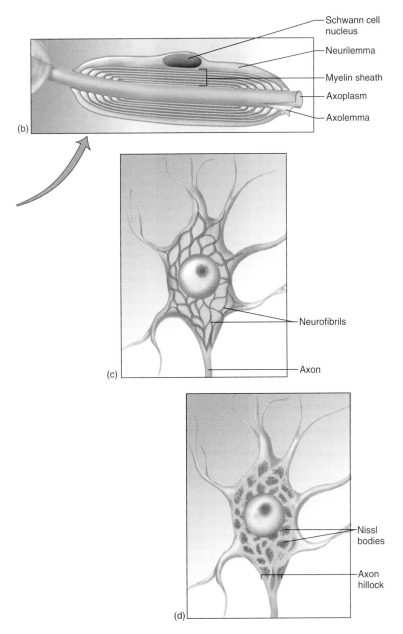

(b)

Schwann cell nucleus

Neurilemma

Myelin sheath

Axoplasm

Axolemma

(c)

Neurofibrils

Axon

(d)

Nissl bodies

Axon hillock

A Representative Neuron (*Continued*)
Figure 13.4

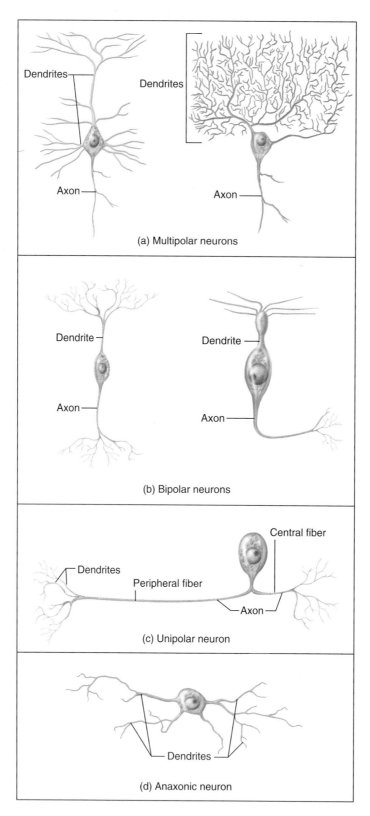

Variation in Neuronal Structure

Figure 13.5

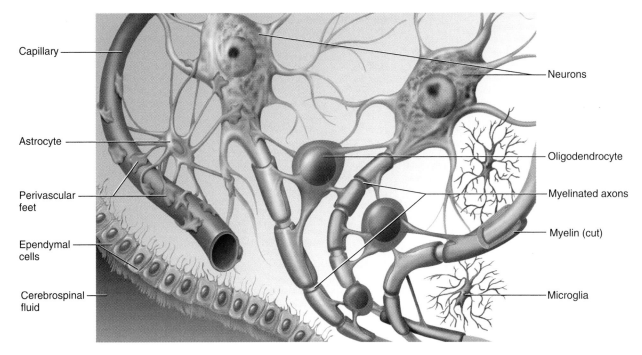

Neuroglia of the Central Nervous System
Figure 13.6

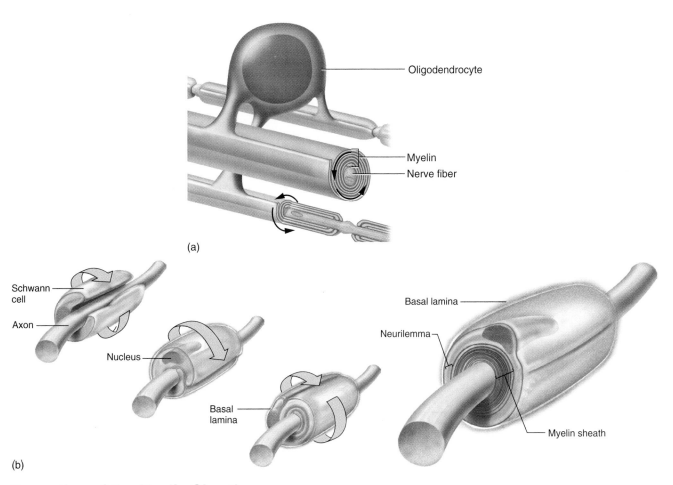

Formation of the Myelin Sheath
Figure 13.7

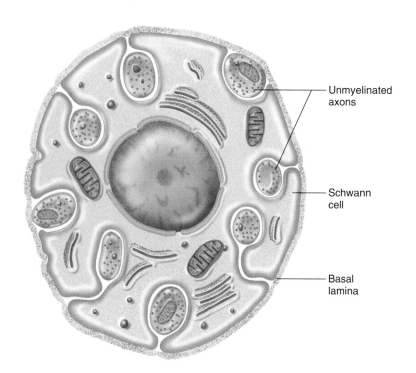

Unmyelinated
axons

Schwann
cell

Basal
lamina

Unmyelinated Nerve Fibers
Figure 13.8

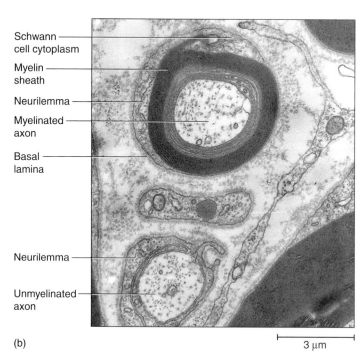

Schwann
cell cytoplasm

Myelin
sheath

Neurilemma

Myelinated
axon

Basal
lamina

Neurilemma

Unmyelinated
axon

(b)

3 μm

Myelinated and Unmyelinated Axons (TEM)
Figure 13.9

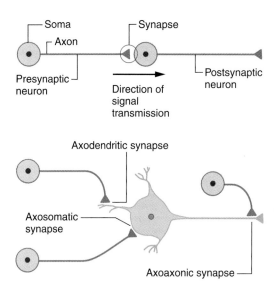

Synaptic Relationships Between Neurons
Figure 13.11

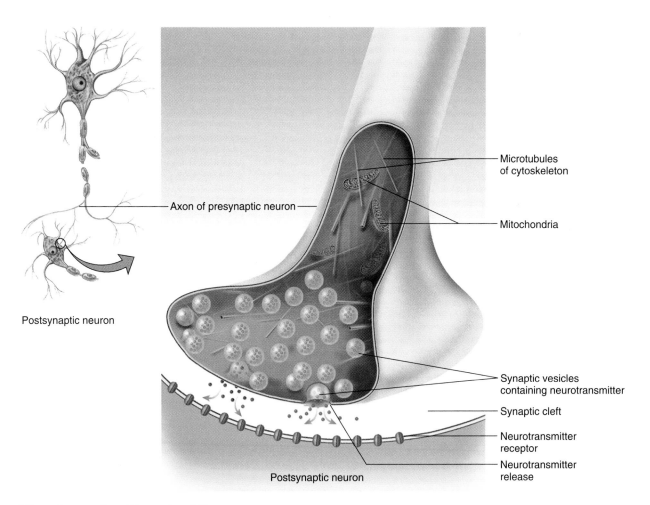

Structure of a Chemical Synapse
Figure 13.12

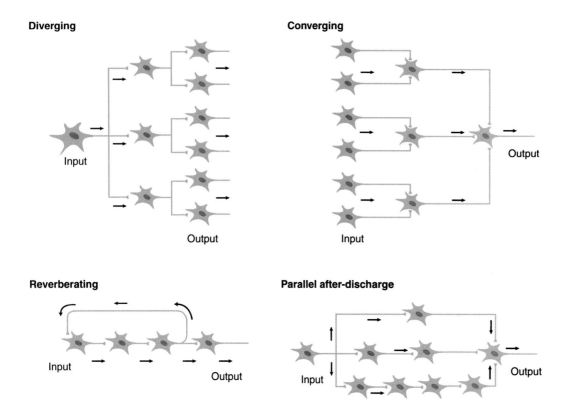

Four Types of Neuronal Circuits
Figure 13.13

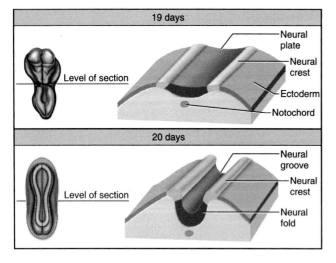

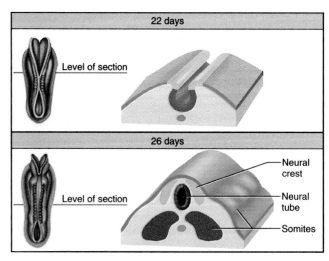

Formation of the Neural Tube
Figure 13.14

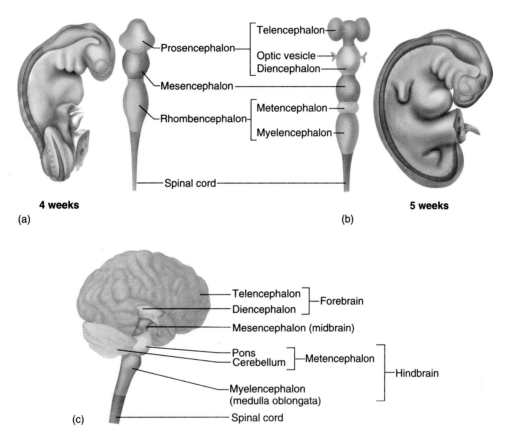

4 weeks

(a)

Prosencephalon
Mesencephalon
Rhombencephalon

Telencephalon
Optic vesicle
Diencephalon
Metencephalon
Myelencephalon

Spinal cord

5 weeks

(b)

Telencephalon
Diencephalon
— Forebrain
Mesencephalon (midbrain)
Pons
Cerebellum
— Metencephalon
Myelencephalon
(medulla oblongata)
— Hindbrain
Spinal cord

(c)

Primary and Secondary Vesicles of the Embryonic Brain
Figure 13.15

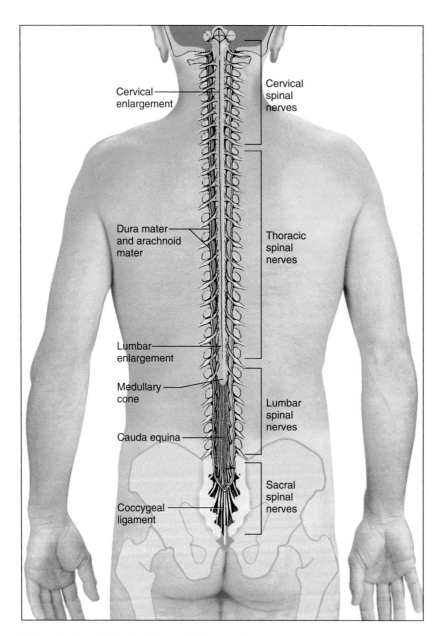

The Spinal Cord, Dorsal Aspect
Figure 14.1

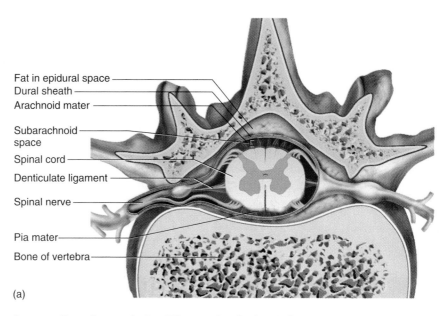

Fat in epidural space
Dural sheath
Arachnoid mater
Subarachnoid space
Spinal cord
Denticulate ligament
Spinal nerve
Pia mater
Bone of vertebra

(a)

Cross Section of the Thoracic Spinal Cord
Figure 14.2

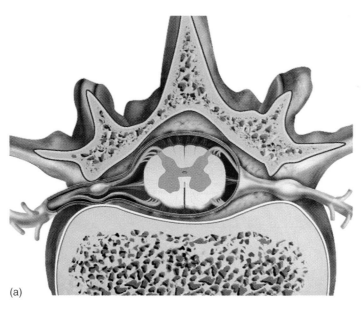

(a)

Cross Section of the Thoracic Spinal Cord
Figure 14.2

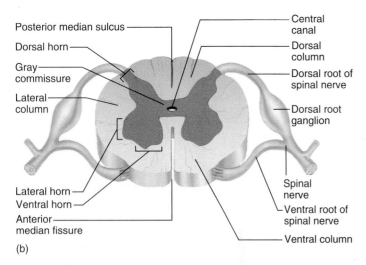

Cross Section of the Thoracic Spinal Cord
Figure 14.2

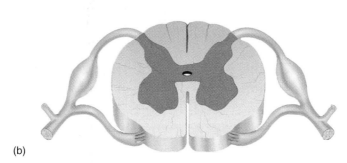

(b)

Cross Section of the Thoracic Spinal Cord
Figure 14.2

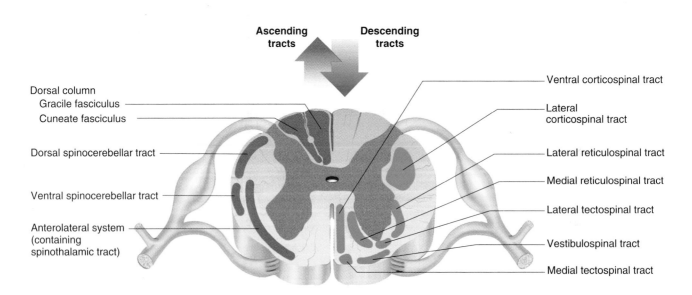

Tracts of the Spinal Cord
Figure 14.3

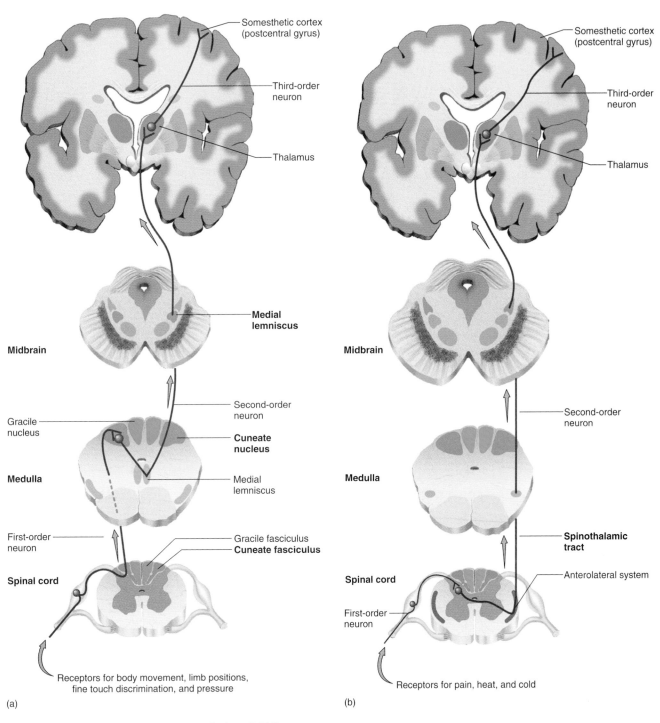

Two Ascending Pathways of the CNS
Figure 14.4

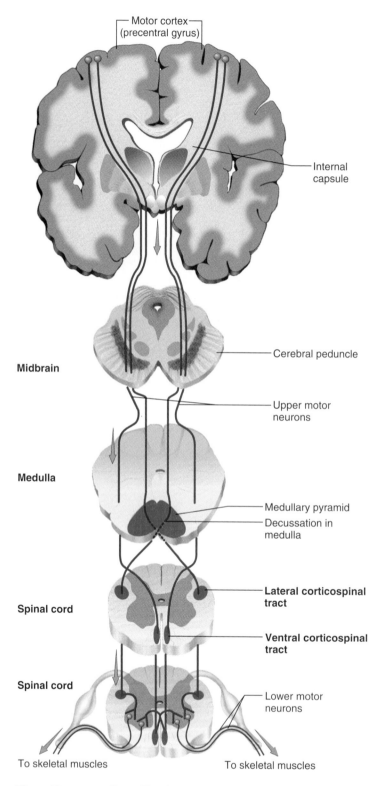

Motor cortex
(precentral gyrus)

Internal
capsule

Midbrain

Cerebral peduncle

Upper motor
neurons

Medulla

Medullary pyramid

Decussation in
medulla

Spinal cord

**Lateral corticospinal
tract**

**Ventral corticospinal
tract**

Spinal cord

Lower motor
neurons

To skeletal muscles

To skeletal muscles

Two Descending Pathways of the CNS
Figure 14.5

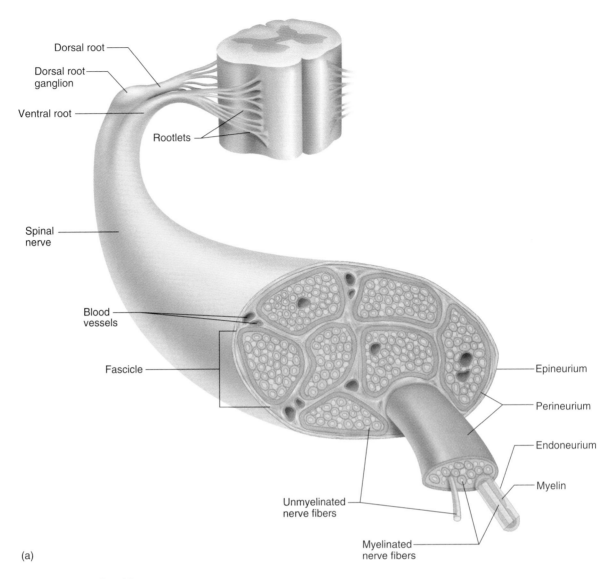

(a)

Anatomy of a Nerve
Figure 14.7

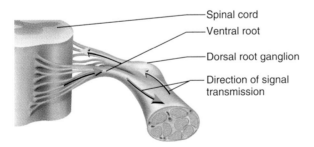

Anatomy of a Ganglion
Figure 14.8

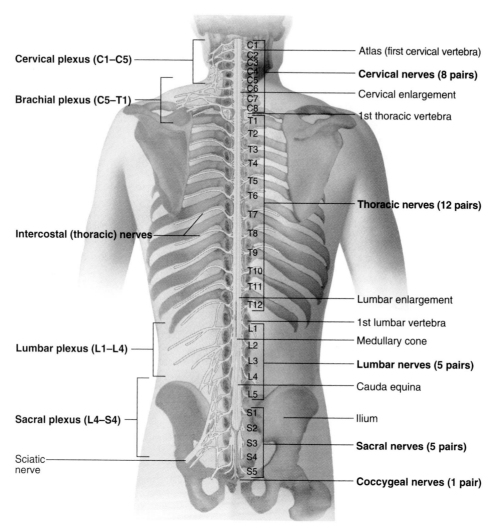

Cervical plexus (C1–C5)

Brachial plexus (C5–T1)

Intercostal (thoracic) nerves

Lumbar plexus (L1–L4)

Sacral plexus (L4–S4)

Sciatic
nerve

C1
C2
C3
C4
C5
C6
C7
C8
T1
T2
T3
T4
T5
T6
T7
T8
T9
T10
T11
T12
L1
L2
L3
L4
L5
S1
S2
S3
S4
S5

Atlas (first cervical vertebra)

Cervical nerves (8 pairs)

Cervical enlargement

1st thoracic vertebra

Thoracic nerves (12 pairs)

Lumbar enlargement

1st lumbar vertebra

Medullary cone

Lumbar nerves (5 pairs)

Cauda equina

Ilium

Sacral nerves (5 pairs)

Coccygeal nerves (1 pair)

The Spinal Nerve Roots and Plexuses, Dorsal View
Figure 14.9

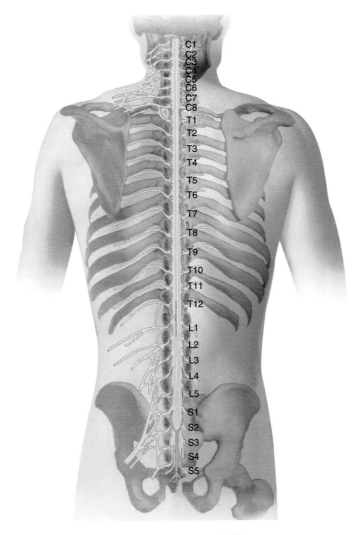

C1
C2
C3
C4
C5
C6
C7
C8
T1
T2
T3
T4
T5
T6
T7
T8
T9
T10
T11
T12
L1
L2
L3
L4
L5
S1
S2
S3
S4
S5

**The Spinal Nerve Roots and Plexuses,
Dorsal View**
Figure 14.9

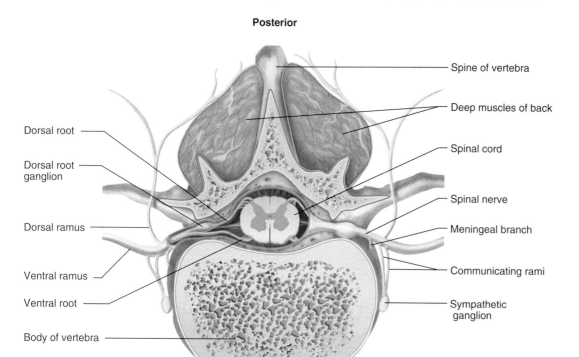

Posterior

Spine of vertebra

Deep muscles of back

Spinal cord

Spinal nerve

Meningeal branch

Communicating rami

Sympathetic ganglion

Dorsal root

Dorsal root ganglion

Dorsal ramus

Ventral ramus

Ventral root

Body of vertebra

Anterior

Branches of a Spinal Nerve in Relation to the Spinal Cord and Vertebra (cross section)
Figure 14.10

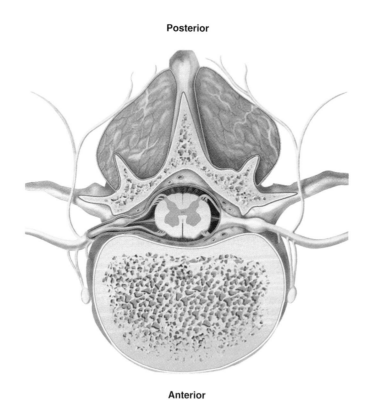

Posterior

Anterior

Branches of a Spinal Nerve in Relation to the Spinal Cord and Vertebra (cross section)
Figure 14.10

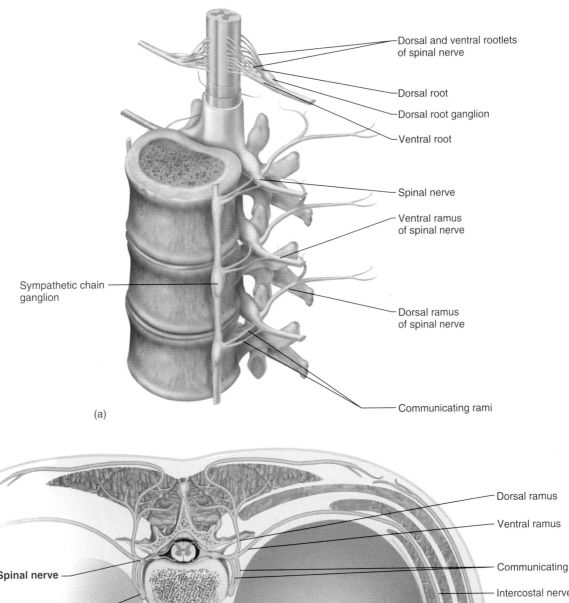

Dorsal and ventral rootlets
of spinal nerve

Dorsal root

Dorsal root ganglion

Ventral root

Spinal nerve

Ventral ramus
of spinal nerve

Dorsal ramus
of spinal nerve

Sympathetic chain
ganglion

Communicating rami

(a)

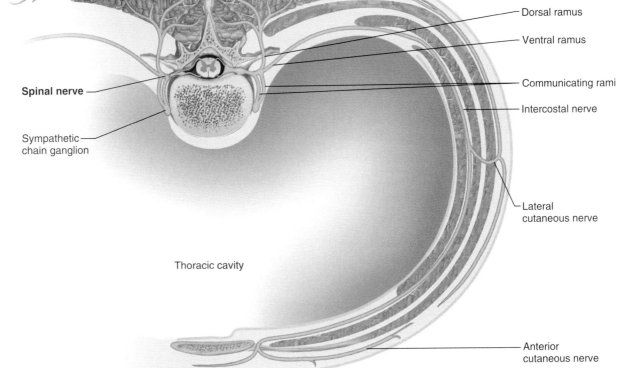

Dorsal ramus

Ventral ramus

Communicating rami

Intercostal nerve

Spinal nerve

Sympathetic
chain ganglion

Lateral
cutaneous nerve

Thoracic cavity

Anterior
cutaneous nerve

(b)

Rami of the Spinal Nerves
Figure 14.12

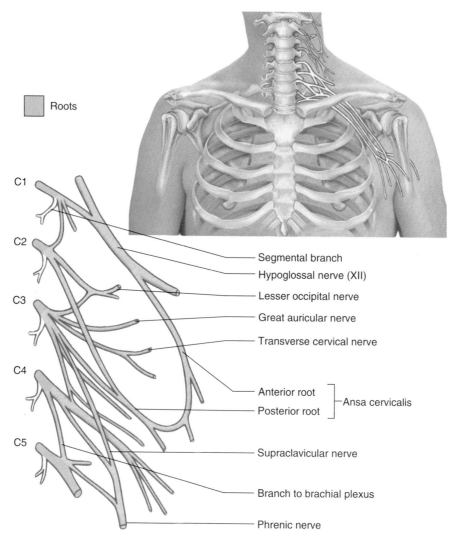

Roots

C1

C2

C3

C4

C5

Segmental branch
Hypoglossal nerve (XII)
Lesser occipital nerve
Great auricular nerve
Transverse cervical nerve

Anterior root ⎤
                ⎦ Ansa cervicalis
Posterior root

Supraclavicular nerve

Branch to brachial plexus

Phrenic nerve

The Cervical Plexus
Figure 14.13

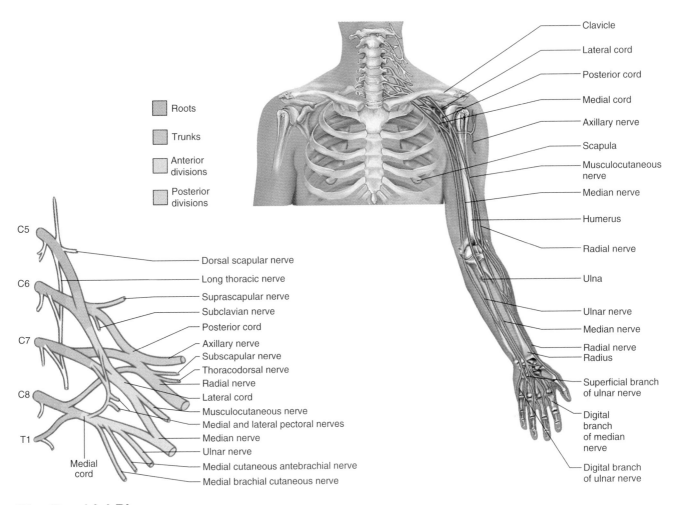

Roots

Trunks

Anterior
divisions

Posterior
divisions

C5

C6

C7

C8

T1

Medial
cord

Dorsal scapular nerve
Long thoracic nerve
Suprascapular nerve
Subclavian nerve
Posterior cord
Axillary nerve
Subscapular nerve
Thoracodorsal nerve
Radial nerve
Lateral cord
Musculocutaneous nerve
Medial and lateral pectoral nerves
Median nerve
Ulnar nerve
Medial cutaneous antebrachial nerve
Medial brachial cutaneous nerve

Clavicle
Lateral cord
Posterior cord
Medial cord
Axillary nerve
Scapula
Musculocutaneous
nerve
Median nerve
Humerus
Radial nerve
Ulna
Ulnar nerve
Median nerve
Radial nerve
Radius
Superficial branch
of ulnar nerve
Digital
branch
of median
nerve
Digital branch
of ulnar nerve

The Brachial Plexus
Figure 14.14

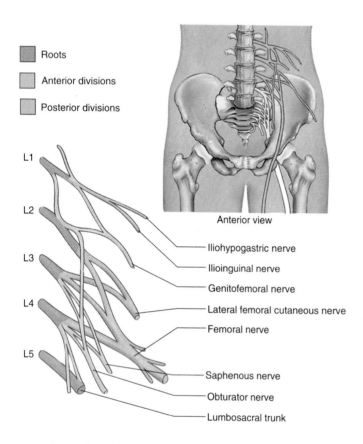

Roots

Anterior divisions

Posterior divisions

L1

L2

L3

L4

L5

Anterior view

Iliohypogastric nerve

Ilioinguinal nerve

Genitofemoral nerve

Lateral femoral cutaneous nerve

Femoral nerve

Saphenous nerve

Obturator nerve

Lumbosacral trunk

The Lumbar Plexus
Figure 14.16

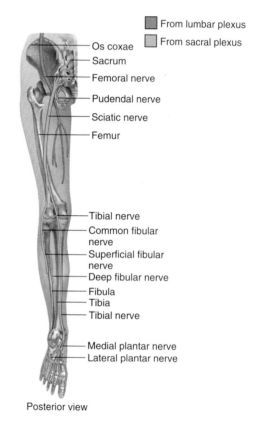

From lumbar plexus

From sacral plexus

Os coxae

Sacrum

Femoral nerve

Pudendal nerve

Sciatic nerve

Femur

Tibial nerve

Common fibular nerve

Superficial fibular nerve

Deep fibular nerve

Fibula

Tibia

Tibial nerve

Medial plantar nerve

Lateral plantar nerve

Posterior view

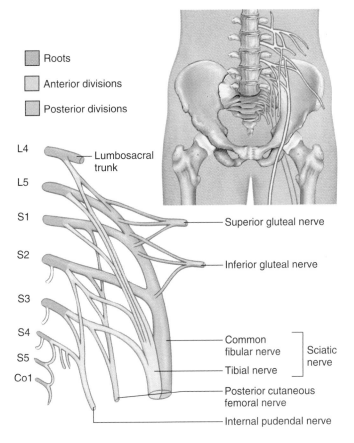

Roots

Anterior divisions

Posterior divisions

L4 — Lumbosacral trunk

L5

S1 — Superior gluteal nerve

S2 — Inferior gluteal nerve

S3

S4 — Common fibular nerve ⎤ Sciatic

S5 — Tibial nerve ⎦ nerve

Co1 — Posterior cutaneous femoral nerve

Internal pudendal nerve

The Sacral and Coccygeal Plexuses
Figure 14.17

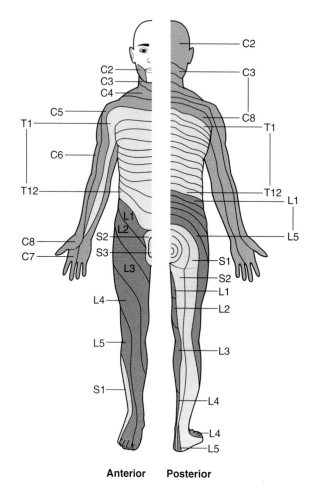

Anterior   **Posterior**

A Dermatome Map of the Body
Figure 14.18

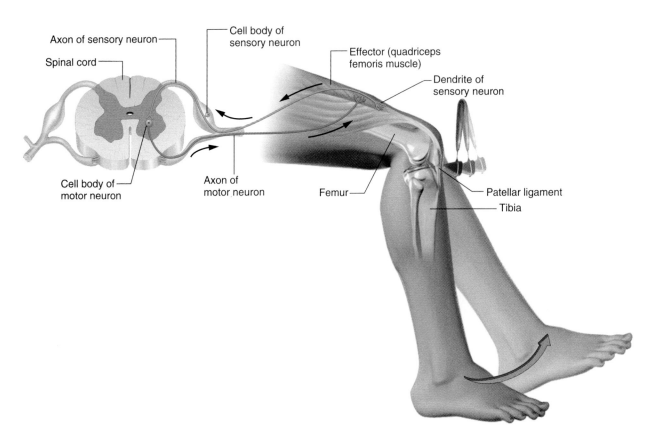

A Representative Reflex Arc
Figure 14.19

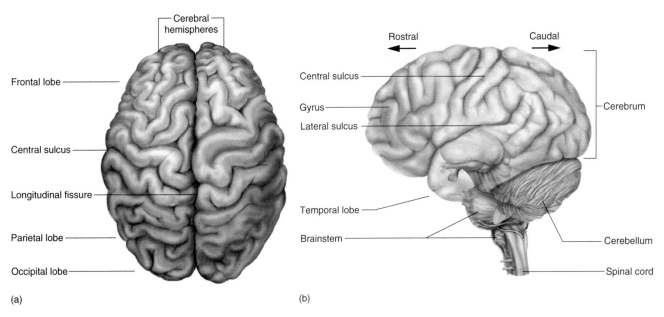

Surface Anatomy of the Brain
Figure 15.1

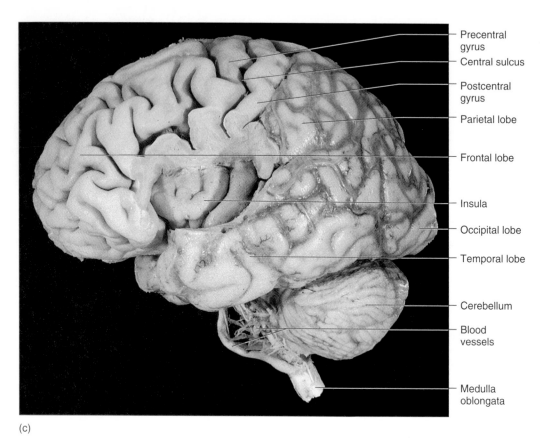

(c)

Surface Anatomy of the Brain (*Continued*)
Figure 15.1

c: © The McGraw-Hill Companies, Inc./Rebecca Gray, photographer/Don Kincaid, dissections

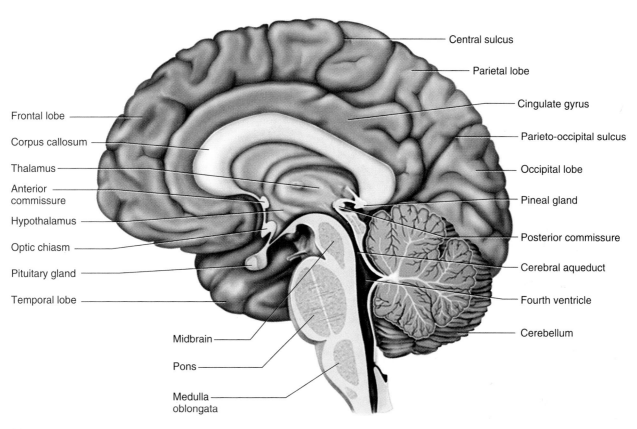

Central sulcus

Parietal lobe

Cingulate gyrus

Parieto-occipital sulcus

Occipital lobe

Pineal gland

Posterior commissure

Cerebral aqueduct

Fourth ventricle

Cerebellum

Frontal lobe

Corpus callosum

Thalamus

Anterior commissure

Hypothalamus

Optic chiasm

Pituitary gland

Temporal lobe

Midbrain

Pons

Medulla oblongata

(a)

Medial Aspect of the Brain
Figure 15.2

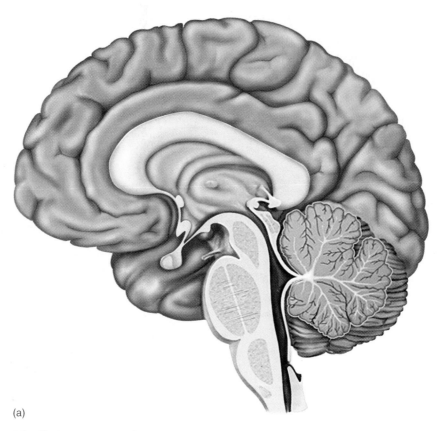

(a)

Medial Aspect of the Brain
Figure 15.2

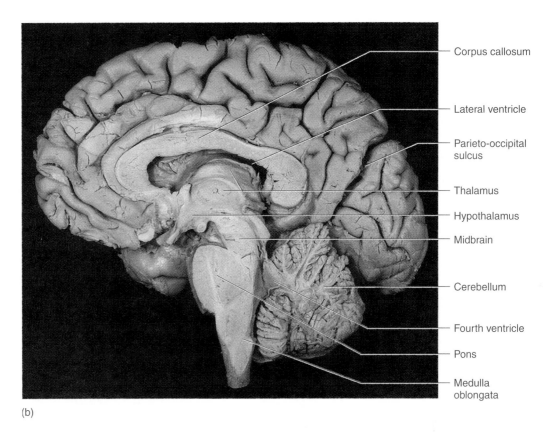

(b)

Medial Aspect of the Brain (*Continued*)
Figure 15.2

b: © The McGraw-Hill Companies, Inc./Dennis Strete, photographer

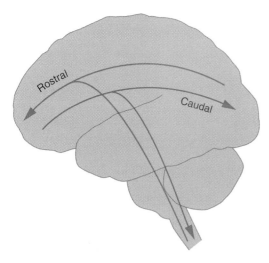

Directional Terms in CNS Anatomy
Figure 15.3

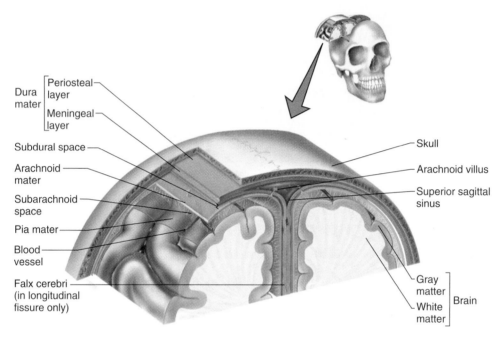

Dura mater — Periosteal layer

Meningeal layer

Subdural space

Arachnoid mater

Subarachnoid space

Pia mater

Blood vessel

Falx cerebri (in longitudinal fissure only)

Skull

Arachnoid villus

Superior sagittal sinus

Gray matter

White matter

Brain

The Meninges of the Brain
Figure 15.4

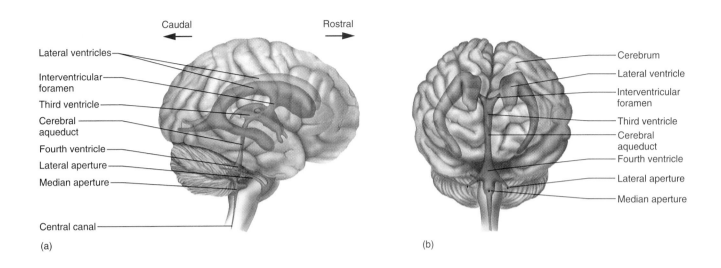

Caudal ←   **Rostral** →

(a)
- Lateral ventricles
- Interventricular foramen
- Third ventricle
- Cerebral aqueduct
- Fourth ventricle
- Lateral aperture
- Median aperture
- Central canal

(b)
- Cerebrum
- Lateral ventricle
- Interventricular foramen
- Third ventricle
- Cerebral aqueduct
- Fourth ventricle
- Lateral aperture
- Median aperture

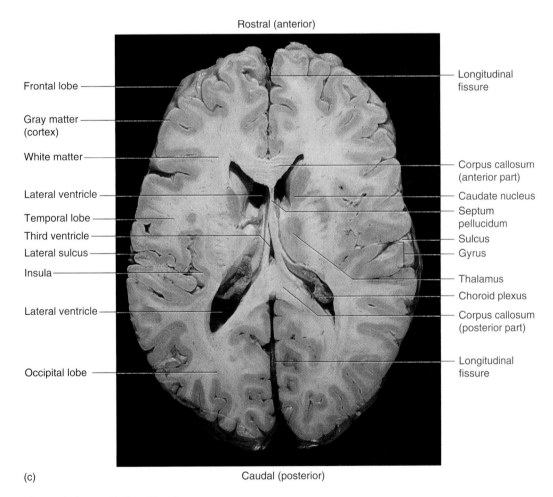

Rostral (anterior)

- Frontal lobe
- Gray matter (cortex)
- White matter
- Lateral ventricle
- Temporal lobe
- Third ventricle
- Lateral sulcus
- Insula
- Lateral ventricle
- Occipital lobe

- Longitudinal fissure
- Corpus callosum (anterior part)
- Caudate nucleus
- Septum pellucidum
- Sulcus
- Gyrus
- Thalamus
- Choroid plexus
- Corpus callosum (posterior part)
- Longitudinal fissure

Caudal (posterior)

(c)

Ventricles of the Brain
Figure 15.5

c: © The McGraw-Hill Companies, Inc./Rebecca Gray, photographer/Don Kincaid, dissections

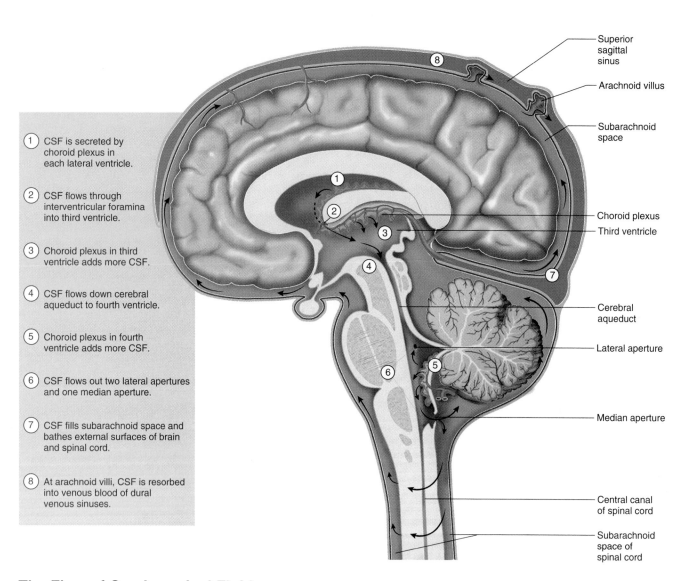

① CSF is secreted by choroid plexus in each lateral ventricle.

② CSF flows through interventricular foramina into third ventricle.

③ Choroid plexus in third ventricle adds more CSF.

④ CSF flows down cerebral aqueduct to fourth ventricle.

⑤ Choroid plexus in fourth ventricle adds more CSF.

⑥ CSF flows out two lateral apertures and one median aperture.

⑦ CSF fills subarachnoid space and bathes external surfaces of brain and spinal cord.

⑧ At arachnoid villi, CSF is resorbed into venous blood of dural venous sinuses.

Superior sagittal sinus

Arachnoid villus

Subarachnoid space

Choroid plexus

Third ventricle

Cerebral aqueduct

Lateral aperture

Median aperture

Central canal of spinal cord

Subarachnoid space of spinal cord

The Flow of Cerebrospinal Fluid
Figure 15.6

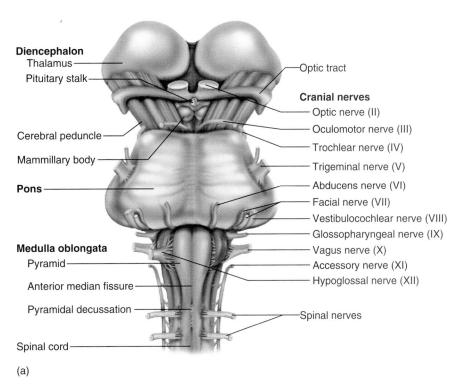

Diencephalon
Thalamus
Pituitary stalk

Cerebral peduncle

Mammillary body

Pons

Medulla oblongata
Pyramid

Anterior median fissure

Pyramidal decussation

Spinal cord

Optic tract

Cranial nerves
Optic nerve (II)
Oculomotor nerve (III)
Trochlear nerve (IV)
Trigeminal nerve (V)
Abducens nerve (VI)
Facial nerve (VII)
Vestibulocochlear nerve (VIII)
Glossopharyngeal nerve (IX)
Vagus nerve (X)
Accessory nerve (XI)
Hypoglossal nerve (XII)

Spinal nerves

(a)

The Brainstem
Figure 15.7

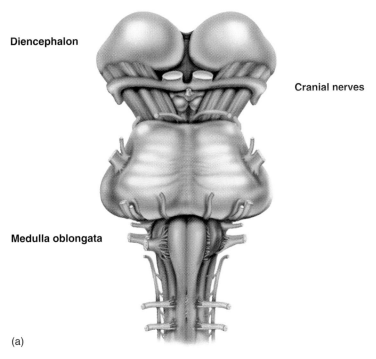

Diencephalon

Cranial nerves

Medulla oblongata

(a)

The Brainstem
Figure 15.7

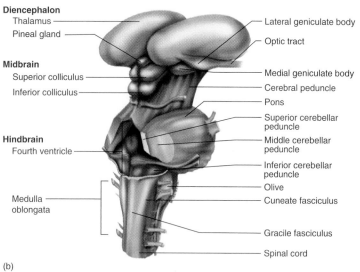

Diencephalon
  Thalamus
  Pineal gland

Midbrain
  Superior colliculus
  Inferior colliculus

Hindbrain
  Fourth ventricle

Medulla
oblongata

Lateral geniculate body
Optic tract
Medial geniculate body
Cerebral peduncle
Pons
Superior cerebellar peduncle
Middle cerebellar peduncle
Inferior cerebellar peduncle
Olive
Cuneate fasciculus
Gracile fasciculus
Spinal cord

(b)

The Brainstem (*Continued*)
Figure 15.7

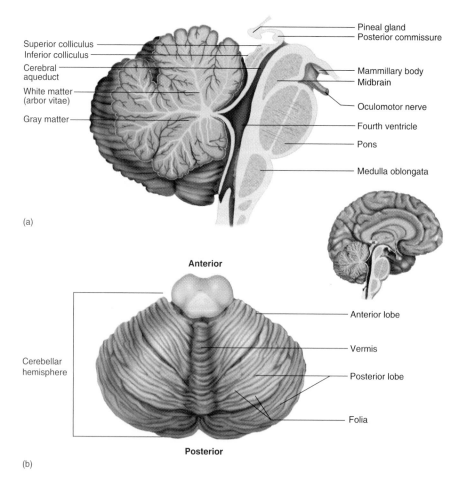

Superior colliculus
Inferior colliculus
Cerebral aqueduct
White matter (arbor vitae)
Gray matter

Pineal gland
Posterior commissure
Mammillary body
Midbrain
Oculomotor nerve
Fourth ventricle
Pons
Medulla oblongata

(a)

Anterior

Cerebellar hemisphere

Anterior lobe
Vermis
Posterior lobe
Folia

Posterior

(b)

The Cerebellum
Figure 15.8

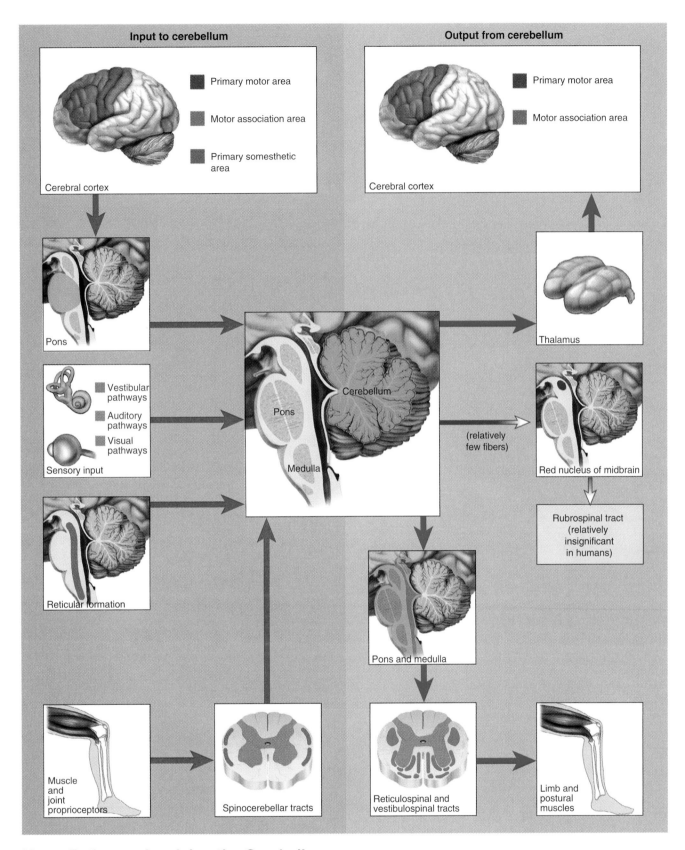

Motor Pathways Involving the Cerebellum
Figure 15.9

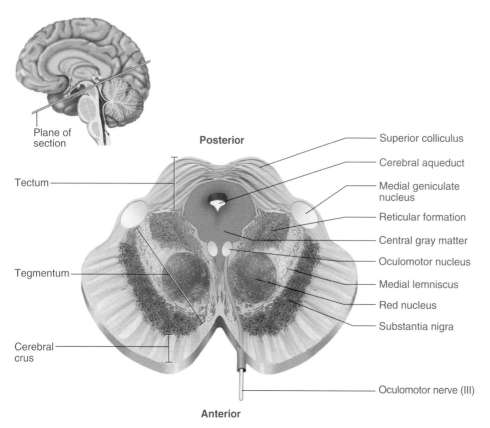

Plane of section

Posterior

Tectum

Tegmentum

Cerebral crus

Anterior

Superior colliculus

Cerebral aqueduct

Medial geniculate nucleus

Reticular formation

Central gray matter

Oculomotor nucleus

Medial lemniscus

Red nucleus

Substantia nigra

Oculomotor nerve (III)

Cross Section of the Midbrain
Figure 15.10

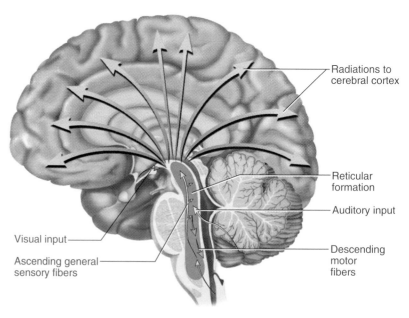

Radiations to cerebral cortex

Reticular formation

Auditory input

Descending motor fibers

Visual input

Ascending general sensory fibers

The Reticular Formation
Figure 15.11

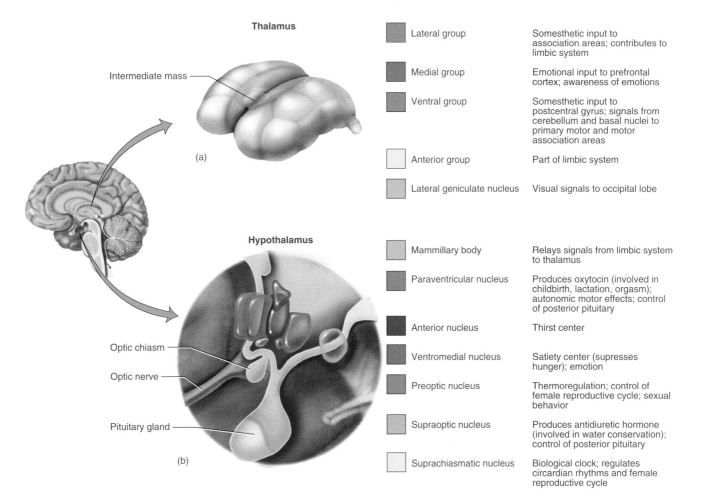

Thalamus

Intermediate mass

(a)

Color	Label	Description
	Lateral group	Somesthetic input to association areas; contributes to limbic system
	Medial group	Emotional input to prefrontal cortex; awareness of emotions
	Ventral group	Somesthetic input to postcentral gyrus; signals from cerebellum and basal nuclei to primary motor and motor association areas
	Anterior group	Part of limbic system
	Lateral geniculate nucleus	Visual signals to occipital lobe

Hypothalamus

Optic chiasm
Optic nerve
Pituitary gland
(b)

Color	Label	Description
	Mammillary body	Relays signals from limbic system to thalamus
	Paraventricular nucleus	Produces oxytocin (involved in childbirth, lactation, orgasm); autonomic motor effects; control of posterior pituitary
	Anterior nucleus	Thirst center
	Ventromedial nucleus	Satiety center (supresses hunger); emotion
	Preoptic nucleus	Thermoregulation; control of female reproductive cycle; sexual behavior
	Supraoptic nucleus	Produces antidiuretic hormone (involved in water conservation); control of posterior pituitary
	Suprachiasmatic nucleus	Biological clock; regulates circardian rhythms and female reproductive cycle

The Diencephalon
Figure 15.12

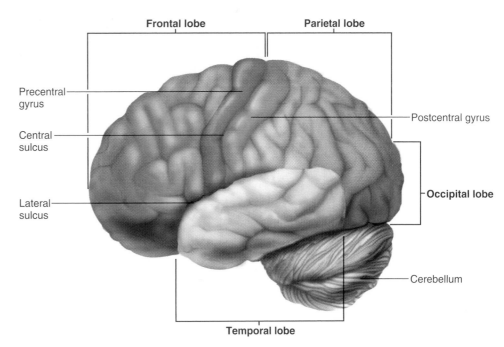

Frontal lobe   Parietal lobe

Precentral gyrus

Central sulcus

Lateral sulcus

Postcentral gyrus

Occipital lobe

Cerebellum

Temporal lobe

Lobes of the Cerebrum
Figure 15.13

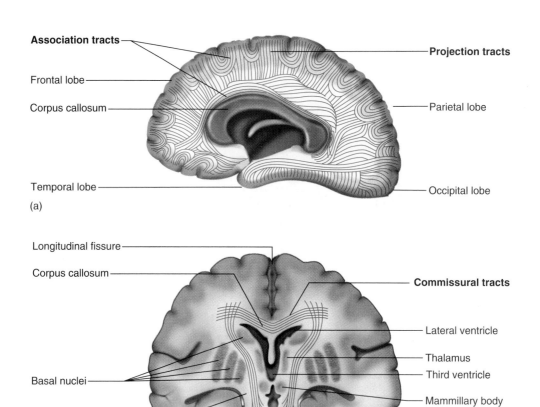

Association tracts

Projection tracts

Frontal lobe

Corpus callosum

Parietal lobe

Temporal lobe

Occipital lobe

(a)

Longitudinal fissure

Corpus callosum

Commissural tracts

Lateral ventricle

Thalamus

Third ventricle

Basal nuclei

Mammillary body

Cerebral peduncle

Projection tracts

Pons

Pyramid

Decussation in pyramids

Medulla oblongata

(b)

Tracts of Cerebral White Matter
Figure 15.14

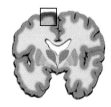

Cortical surface

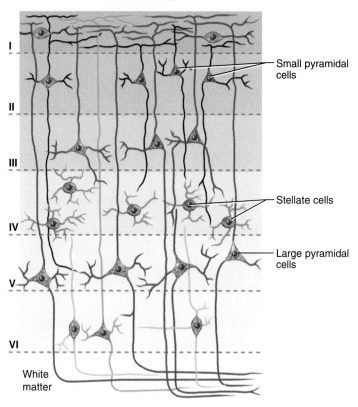

I

Small pyramidal cells

II

III

Stellate cells

IV

Large pyramidal cells

V

VI

White matter

Histology of the Neocortex
Figure 15.15

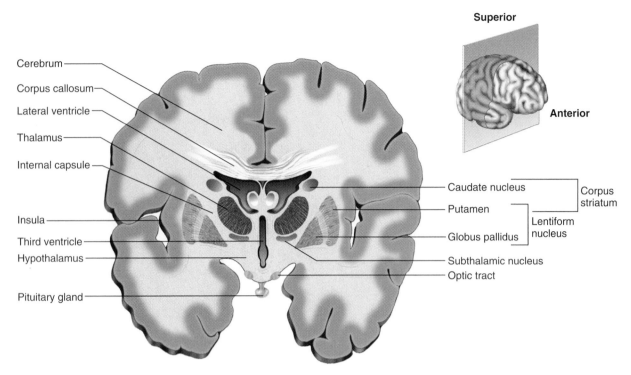

Cerebrum
Corpus callosum
Lateral ventricle
Thalamus
Internal capsule
Insula
Third ventricle
Hypothalamus
Pituitary gland

Superior
Anterior

Caudate nucleus
Putamen
Globus pallidus
Subthalamic nucleus
Optic tract

Corpus striatum
Lentiform nucleus

The Basal Nuclei
Figure 15.16

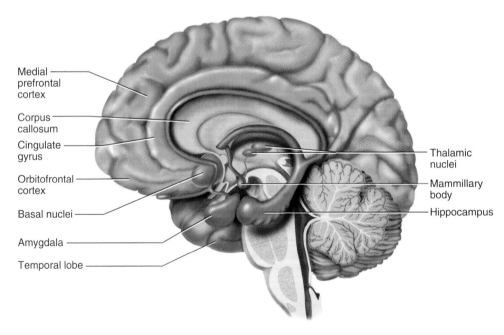

Medial prefrontal cortex
Corpus callosum
Cingulate gyrus
Orbitofrontal cortex
Basal nuclei
Amygdala
Temporal lobe

Thalamic nuclei
Mammillary body
Hippocampus

The Limbic System
Figure 15.17

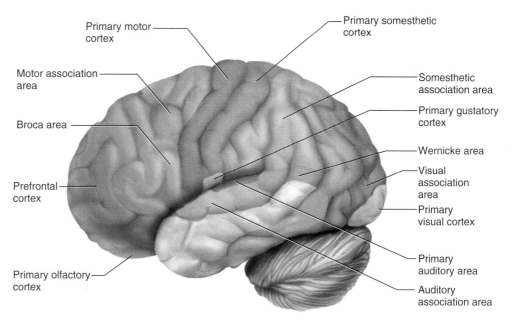

Some Functional Regions of the Cerebral Cortex
Figure 15.18

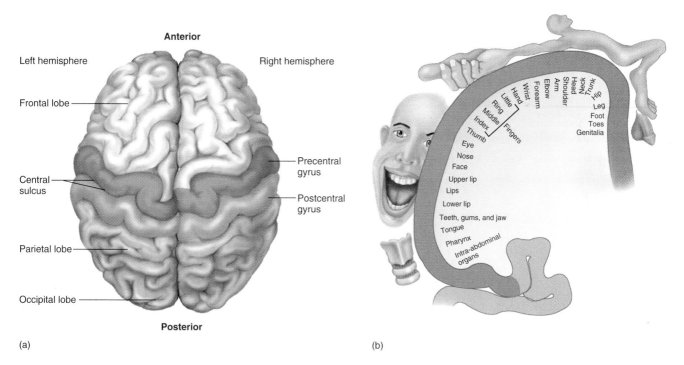

(a)

(b)

The Primary Somesthetic Area (postcentral gyrus)
Figure 15.19

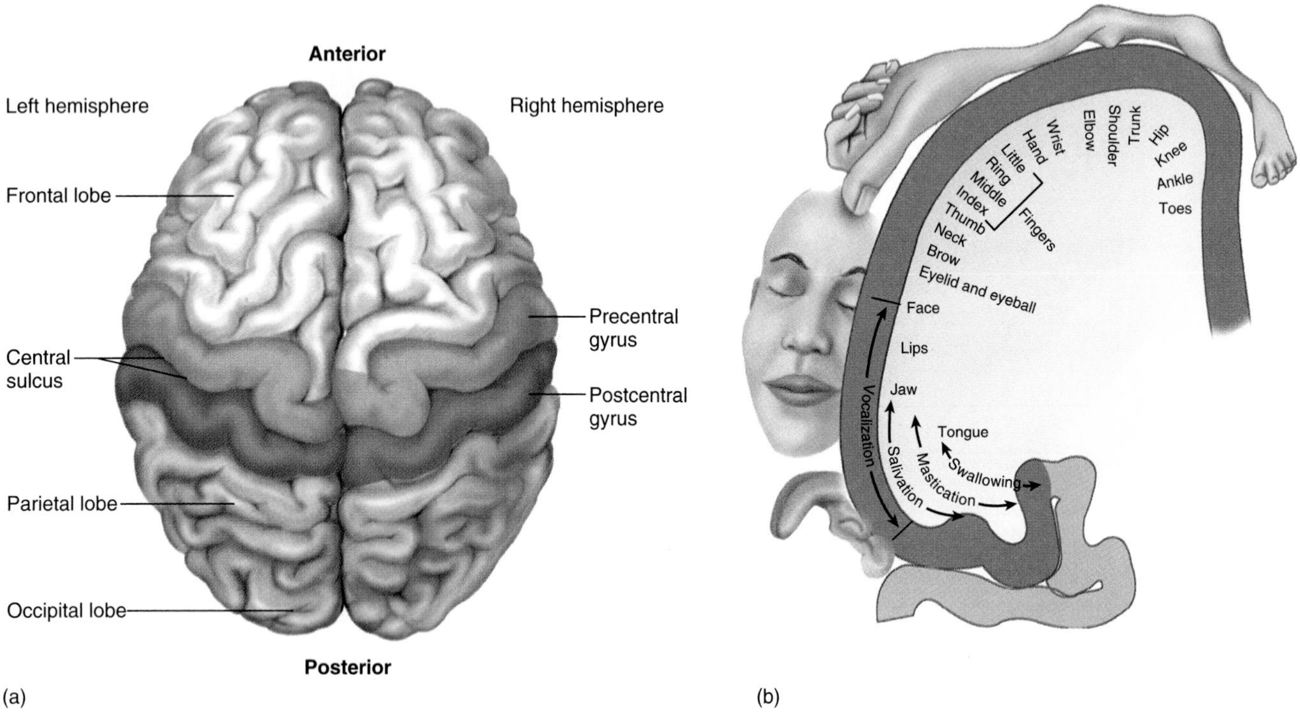
Anterior
Left hemisphere
Right hemisphere
Frontal lobe
Precentral gyrus
Central sulcus
Postcentral gyrus
Parietal lobe
Occipital lobe
Posterior
(a)
Trunk
Shoulder
Elbow
Wrist
Hand
Hip
Knee
Ankle
Toes
Little
Ring
Middle
Index
Thumb
Fingers
Neck
Brow
Eyelid and eyeball
Face
Lips
Jaw
Tongue
Swallowing
Mastication
Salivation
Vocalization
(b)

The Primary Motor Area (precentral gyrus)
Figure 15.20

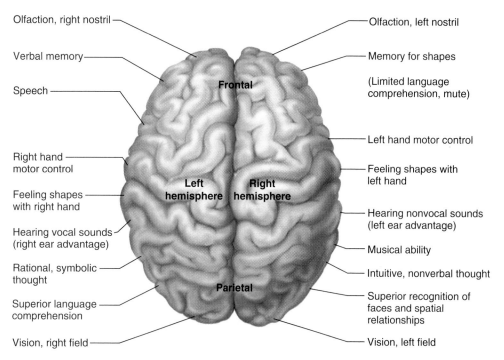

Olfaction, right nostril

Verbal memory

Speech

Right hand motor control

Feeling shapes with right hand

Hearing vocal sounds (right ear advantage)

Rational, symbolic thought

Superior language comprehension

Vision, right field

Frontal

Left hemisphere

Right hemisphere

Parietal

Olfaction, left nostril

Memory for shapes

(Limited language comprehension, mute)

Left hand motor control

Feeling shapes with left hand

Hearing nonvocal sounds (left ear advantage)

Musical ability

Intuitive, nonverbal thought

Superior recognition of faces and spatial relationships

Vision, left field

Lateralization of Cerebral Functions
Figure 15.22

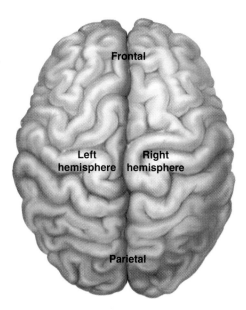

Frontal

Left hemisphere

Right hemisphere

Parietal

Lateralization of Cerebral Functions
Figure 15.22

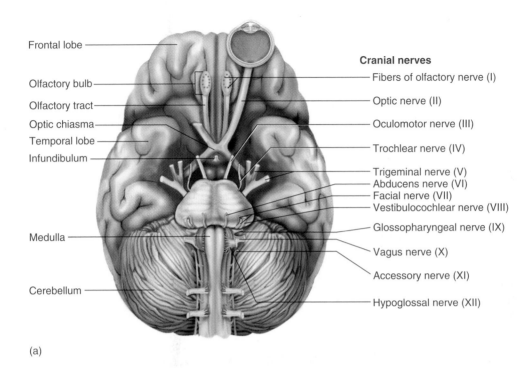

(a)

Frontal lobe

Olfactory bulb

Olfactory tract

Optic chiasma

Temporal lobe

Infundibulum

Medulla

Cerebellum

Cranial nerves

Fibers of olfactory nerve (I)

Optic nerve (II)

Oculomotor nerve (III)

Trochlear nerve (IV)

Trigeminal nerve (V)

Abducens nerve (VI)

Facial nerve (VII)

Vestibulocochlear nerve (VIII)

Glossopharyngeal nerve (IX)

Vagus nerve (X)

Accessory nerve (XI)

Hypoglossal nerve (XII)

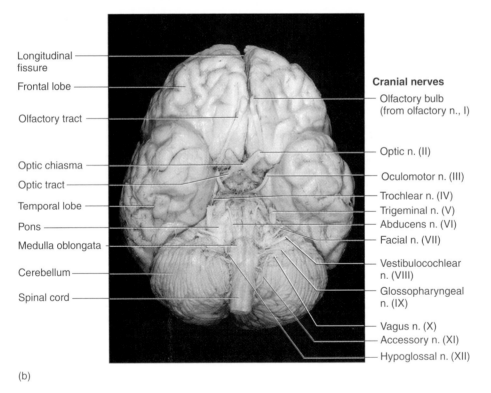

(b)

Longitudinal fissure

Frontal lobe

Olfactory tract

Optic chiasma

Optic tract

Temporal lobe

Pons

Medulla oblongata

Cerebellum

Spinal cord

Cranial nerves

Olfactory bulb (from olfactory n., I)

Optic n. (II)

Oculomotor n. (III)

Trochlear n. (IV)

Trigeminal n. (V)

Abducens n. (VI)

Facial n. (VII)

Vestibulocochlear n. (VIII)

Glossopharyngeal n. (IX)

Vagus n. (X)

Accessory n. (XI)

Hypoglossal n. (XII)

The Cranial Nerves
Figure 15.23

b: © The McGraw-Hill Companies, Inc./Rebecca Gray, photographer/Don Kincaid, dissections

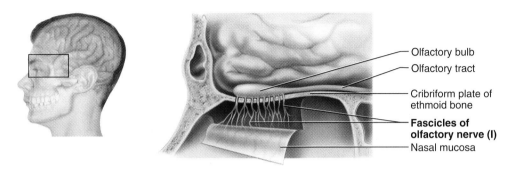

- Olfactory bulb
- Olfactory tract
- Cribriform plate of ethmoid bone
- **Fascicles of olfactory nerve (I)**
- Nasal mucosa

The Olfactory Nerve
Figure 15.24

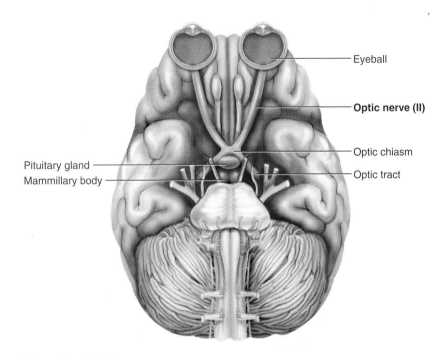

- Eyeball
- **Optic nerve (II)**
- Optic chiasm
- Pituitary gland
- Mammillary body
- Optic tract

The Optic Nerve
Figure 15.25

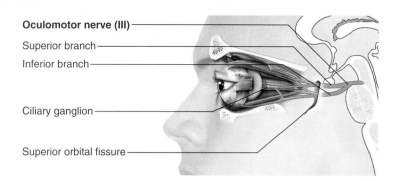

The Oculomotor Nerve
Figure 15.26

Oculomotor nerve (III)
Superior branch
Inferior branch
Ciliary ganglion
Superior orbital fissure

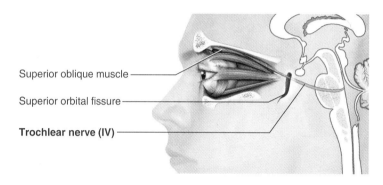

The Trochlear Nerve
Figure 15.27

Superior oblique muscle
Superior orbital fissure
Trochlear nerve (IV)

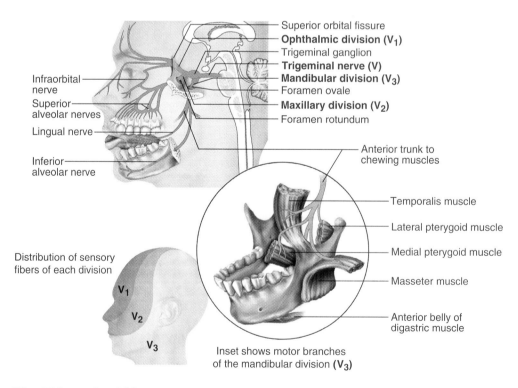

The Trigeminal Nerve
Figure 15.28

Superior orbital fissure
Ophthalmic division (V₁)
Trigeminal ganglion
Trigeminal nerve (V)
Mandibular division (V₃)
Foramen ovale
Maxillary division (V₂)
Foramen rotundum

Infraorbital nerve
Superior alveolar nerves
Lingual nerve
Inferior alveolar nerve

Distribution of sensory fibers of each division

V₁
V₂
V₃

Anterior trunk to chewing muscles
Temporalis muscle
Lateral pterygoid muscle
Medial pterygoid muscle
Masseter muscle
Anterior belly of digastric muscle

Inset shows motor branches of the mandibular division (V₃)

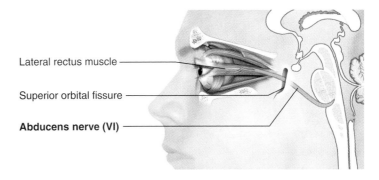

Lateral rectus muscle

Superior orbital fissure

Abducens nerve (VI)

The Abducens Nerve
Figure 15.29

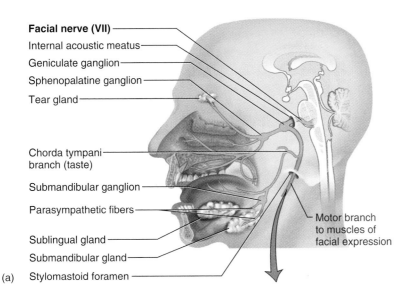

Facial nerve (VII)
Internal acoustic meatus
Geniculate ganglion
Sphenopalatine ganglion
Tear gland

Chorda tympani
branch (taste)

Submandibular ganglion

Parasympathetic fibers

Sublingual gland

Submandibular gland

(a) Stylomastoid foramen

Motor branch
to muscles of
facial expression

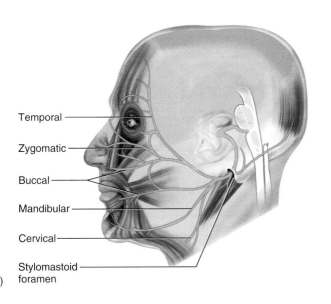

Temporal

Zygomatic

Buccal

Mandibular

Cervical

Stylomastoid
(b) foramen

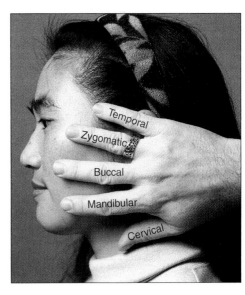

(c)

The Facial Nerve
Figure 15.30

c: © The McGraw-Hill Companies, Inc./Joe DeGrandis, photographer

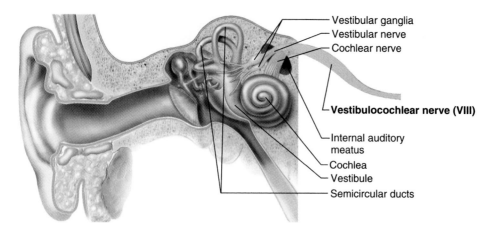

Vestibular ganglia
Vestibular nerve
Cochlear nerve

Vestibulocochlear nerve (VIII)

Internal auditory meatus
Cochlea
Vestibule
Semicircular ducts

The Vestibulocochlear Nerve
Figure 15.31

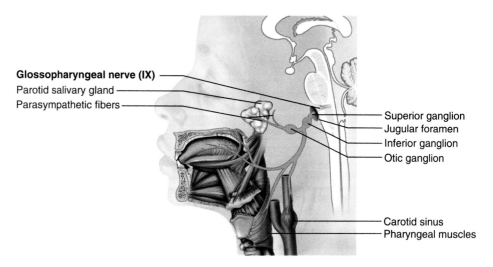

Glossopharyngeal nerve (IX)
Parotid salivary gland
Parasympathetic fibers

Superior ganglion
Jugular foramen
Inferior ganglion
Otic ganglion

Carotid sinus
Pharyngeal muscles

The Glossopharyngeal Nerve
Figure 15.32

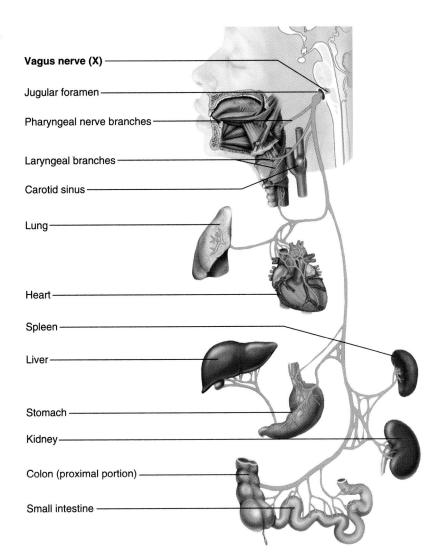

Vagus nerve (X)

Jugular foramen

Pharyngeal nerve branches

Laryngeal branches

Carotid sinus

Lung

Heart

Spleen

Liver

Stomach

Kidney

Colon (proximal portion)

Small intestine

The Vagus Nerve
Figure 15.33

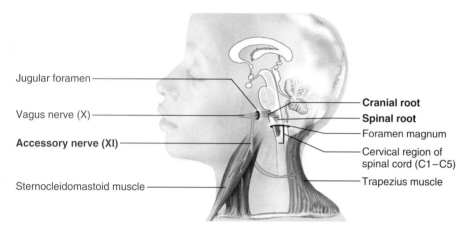

Jugular foramen

Vagus nerve (X)

Accessory nerve (XI)

Sternocleidomastoid muscle

Cranial root

Spinal root

Foramen magnum

Cervical region of spinal cord (C1–C5)

Trapezius muscle

The Accessory Nerve
Figure 15.34

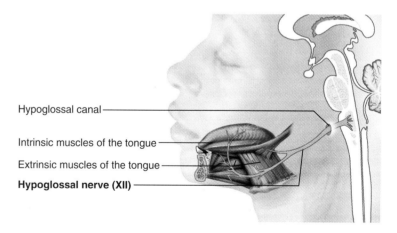

Hypoglossal canal

Intrinsic muscles of the tongue

Extrinsic muscles of the tongue

Hypoglossal nerve (XII)

The Hypoglossal Nerve
Figure 15.35

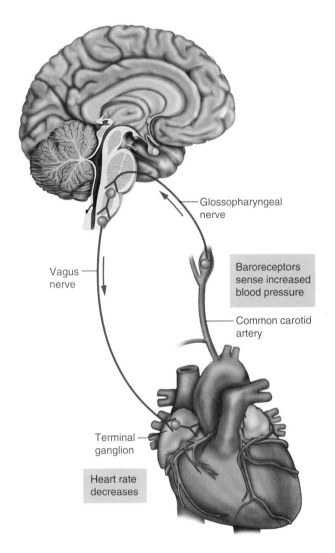

Glossopharyngeal
nerve

Vagus
nerve

Baroreceptors
sense increased
blood pressure

Common carotid
artery

Terminal
ganglion

Heart rate
decreases

**An Autonomic Reflex Arc in the
Regulation of Blood Pressure**
Figure 16.1

Somatic efferent innervation

Autonomic efferent innervation

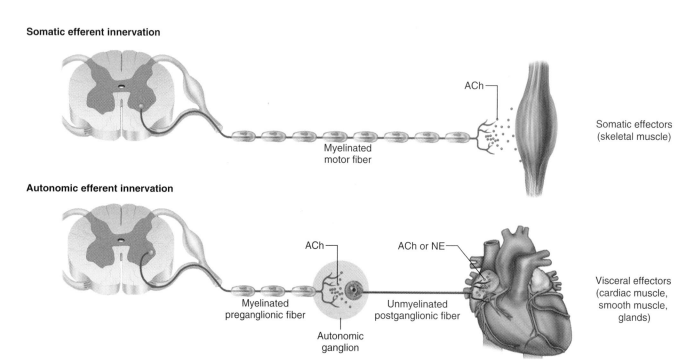

ACh

Myelinated
motor fiber

Somatic effectors
(skeletal muscle)

ACh

ACh or NE

Myelinated
preganglionic fiber

Unmyelinated
postganglionic fiber

Autonomic
ganglion

Visceral effectors
(cardiac muscle,
smooth muscle,
glands)

Comparison of Somatic and Autonomic Efferent Pathways
Figure 16.2

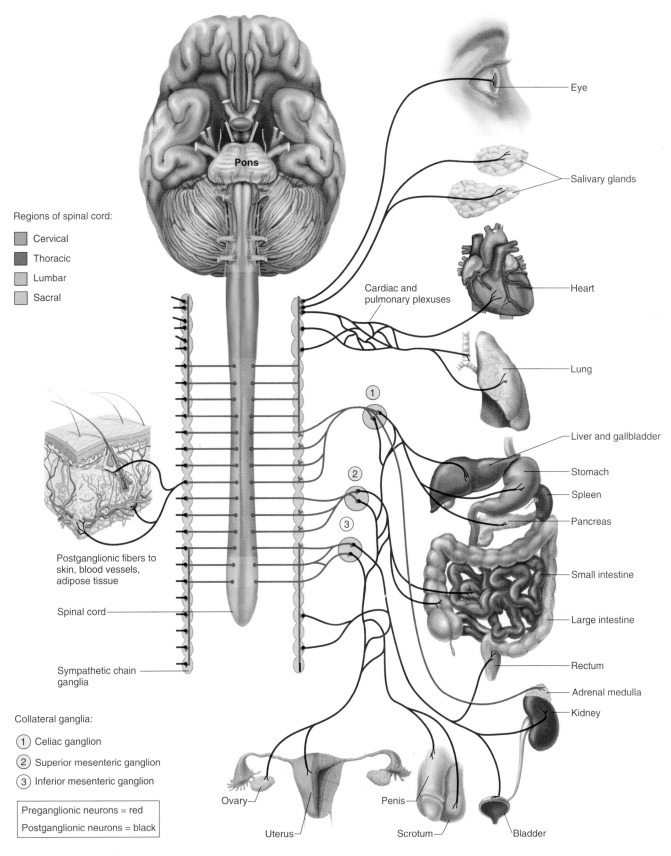

Regions of spinal cord:

- Cervical
- Thoracic
- Lumbar
- Sacral

Eye

Salivary glands

Cardiac and
pulmonary plexuses

Heart

Lung

Liver and gallbladder

Stomach

Spleen

Pancreas

Small intestine

Large intestine

Rectum

Adrenal medulla

Kidney

Pons

① Celiac ganglion

② Superior mesenteric ganglion

③ Inferior mesenteric ganglion

Postganglionic fibers to
skin, blood vessels,
adipose tissue

Spinal cord

Sympathetic chain
ganglia

Collateral ganglia:

① Celiac ganglion

② Superior mesenteric ganglion

③ Inferior mesenteric ganglion

Preganglionic neurons = red
Postganglionic neurons = black

Ovary

Uterus

Penis

Scrotum

Bladder

Sympathetic Pathways
Figure 16.4

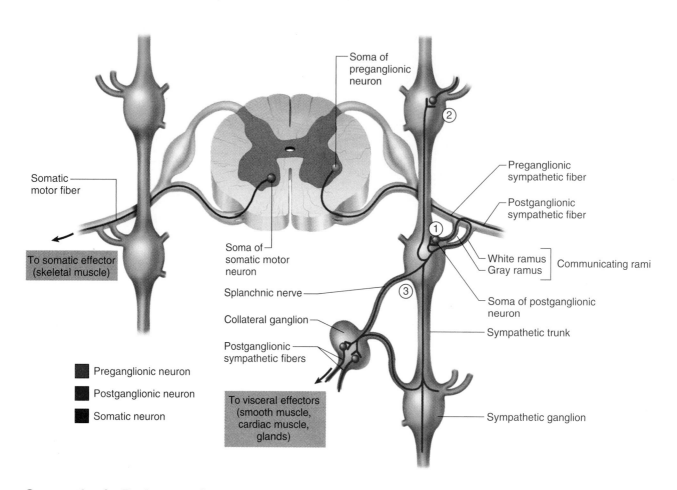

Sympathetic Pathways Compared to Somatic Efferent Pathways
Figure 16.5

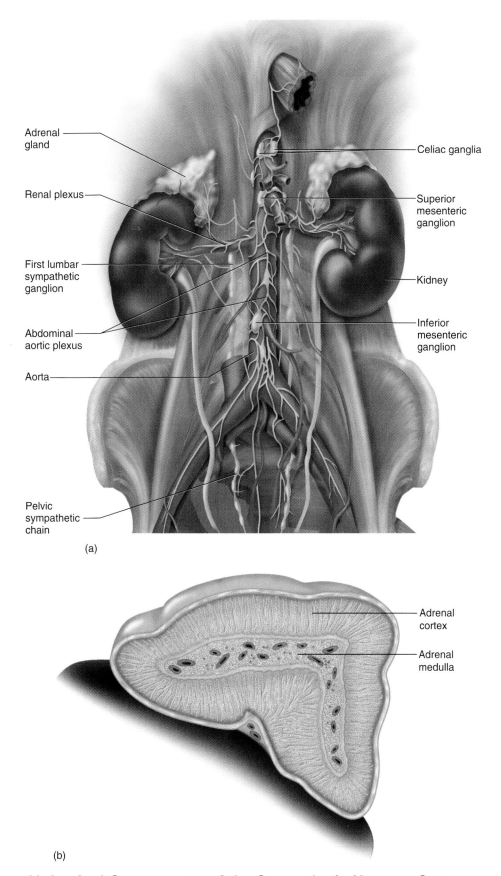

Adrenal gland

Renal plexus

First lumbar sympathetic ganglion

Abdominal aortic plexus

Aorta

Celiac ganglia

Superior mesenteric ganglion

Kidney

Inferior mesenteric ganglion

Pelvic sympathetic chain

(a)

Adrenal cortex

Adrenal medulla

(b)

Abdominal Components of the Sympathetic Nervous System
Figure 16.6

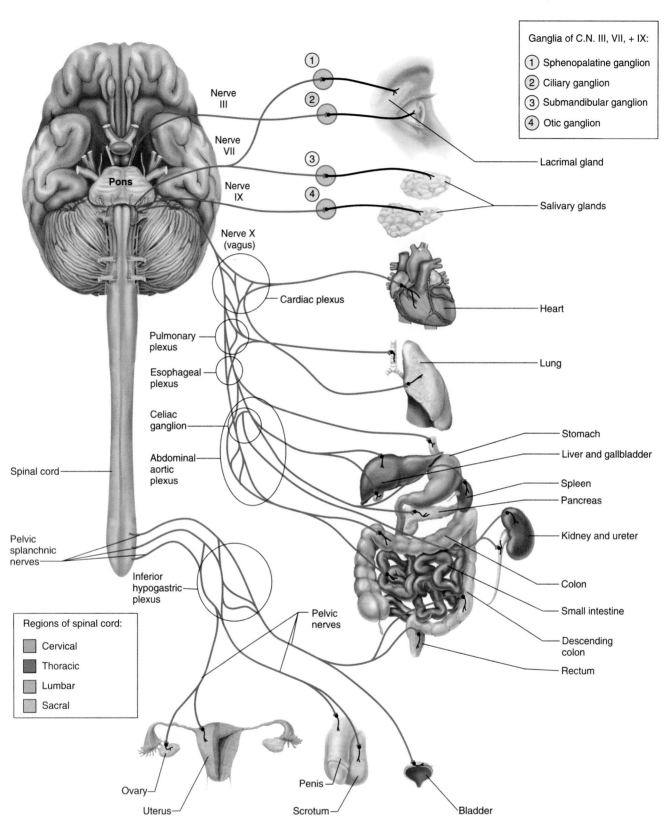

Parasympathetic Pathways
Figure 16.7

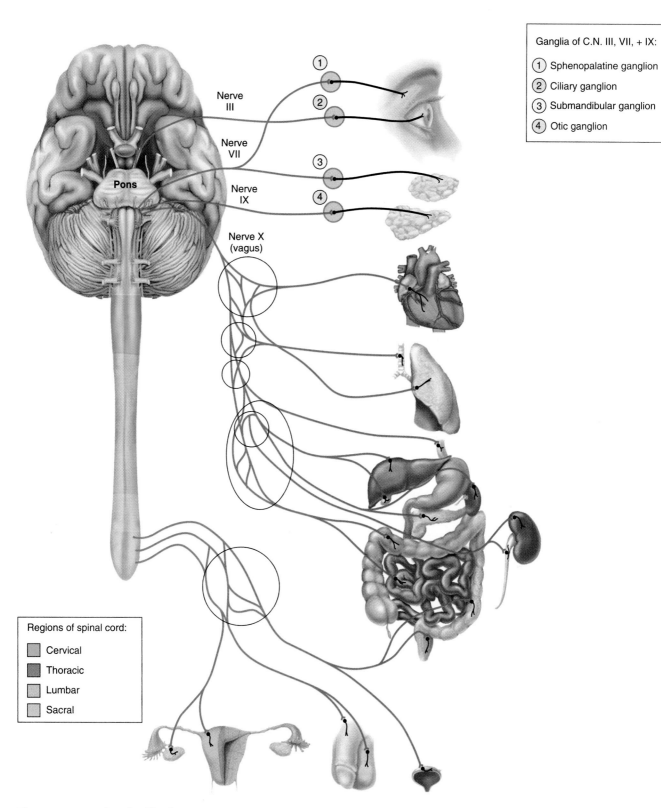

Ganglia of C.N. III, VII, + IX:

(1) Sphenopalatine ganglion
(2) Ciliary ganglion
(3) Submandibular ganglion
(4) Otic ganglion

Nerve III

Nerve VII

Pons

Nerve IX

Nerve X (vagus)

Regions of spinal cord:

- Cervical
- Thoracic
- Lumbar
- Sacral

Parasympathetic Pathways
Figure 16.7

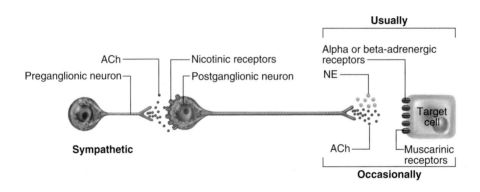

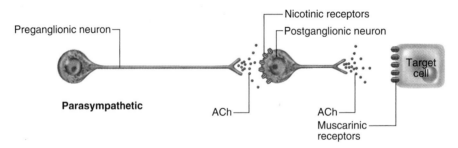

Neurotransmitters and Receptors of the Autonomic Nervous System

Figure 16.8

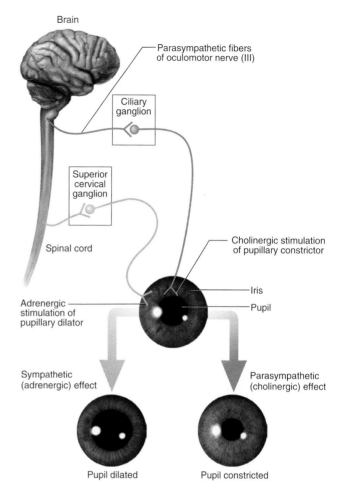

Dual Innervation of the Iris
Figure 16.9

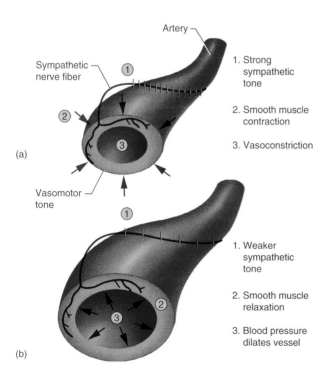

(a)

1. Strong sympathetic tone

2. Smooth muscle contraction

3. Vasoconstriction

1. Weaker sympathetic tone

2. Smooth muscle relaxation

3. Blood pressure dilates vessel

(b)

Sympathetic and Vasomotor Tone
Figure 16.10

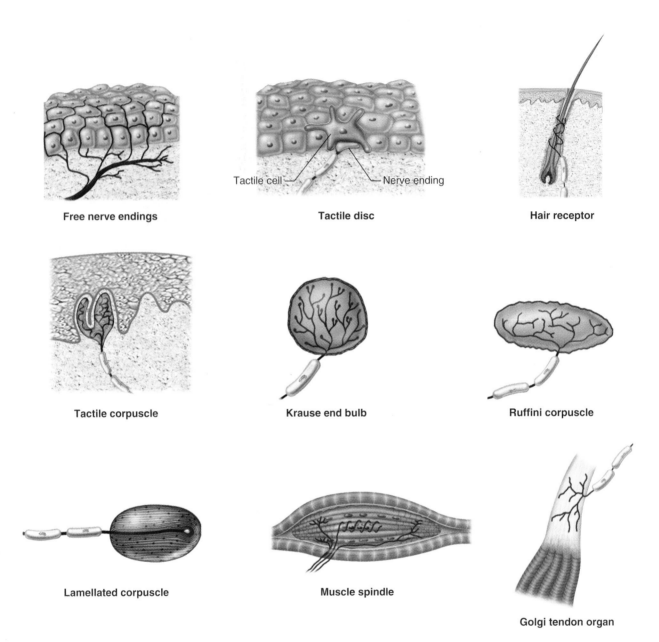

Free nerve endings

Tactile disc

Tactile cell — Nerve ending

Hair receptor

Tactile corpuscle

Krause end bulb

Ruffini corpuscle

Lamellated corpuscle

Muscle spindle

Golgi tendon organ

Receptors of the General (somesthetic) Senses
Figure 17.1

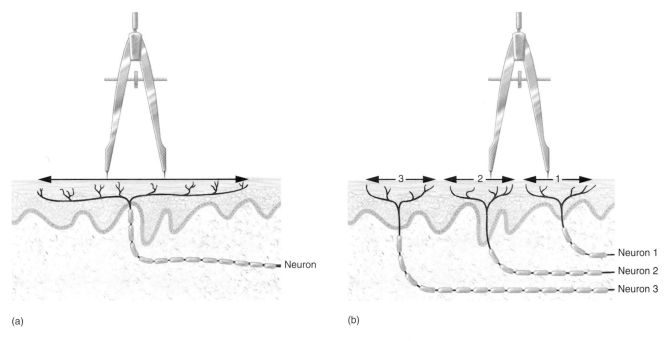

(a)

(b)

Receptive Fields of Sensory Neurons
Figure 17.2

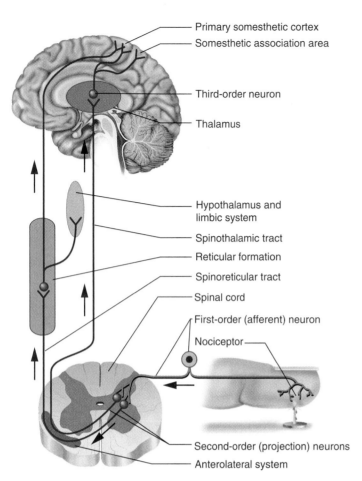

Projection Pathways for Pain
Figure 17.3

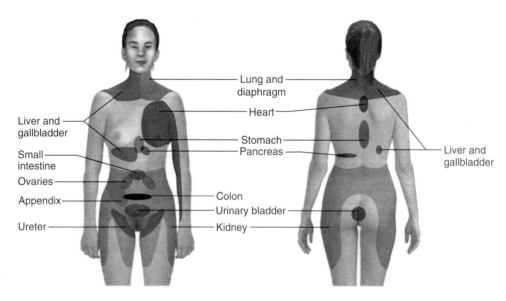

Referred Pain.
Figure 17.4

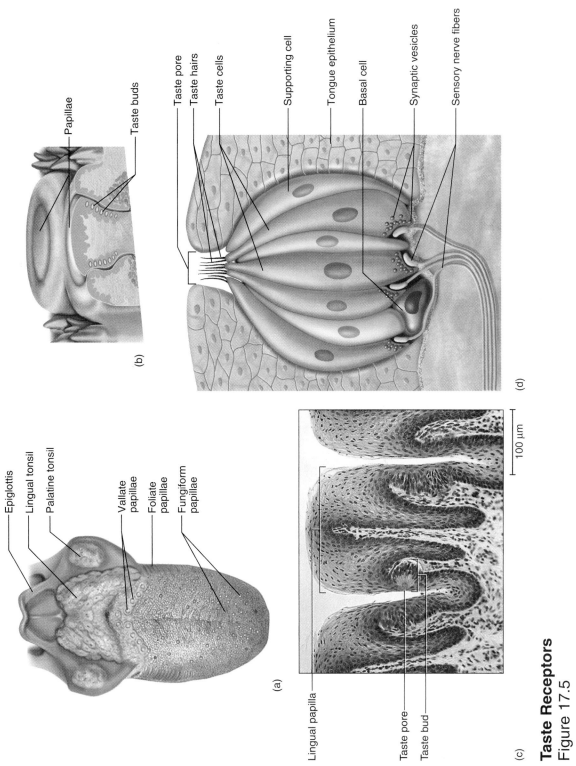

Taste Receptors
Figure 17.5

(a)

- Epiglottis
- Lingual tonsil
- Palatine tonsil
- Vallate papillae
- Foliate papillae
- Fungiform papillae

(b)

- Papillae
- Taste buds

(c)

- Lingual papilla
- Taste pore
- Taste bud
- 100 μm

(d)

- Taste pore
- Taste hairs
- Taste cells
- Supporting cell
- Tongue epithelium
- Basal cell
- Synaptic vesicles
- Sensory nerve fibers

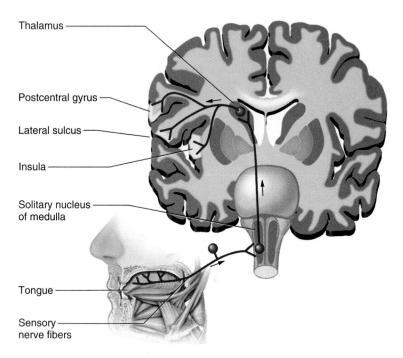

Gustatory Projection Pathways to the Cerebral Cortex
Figure 17.6

Labels:
Thalamus
Postcentral gyrus
Lateral sulcus
Insula
Solitary nucleus of medulla
Tongue
Sensory nerve fibers

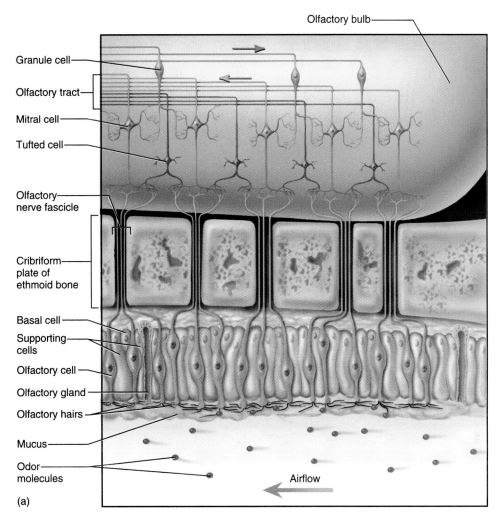

Olfactory bulb

Granule cell

Olfactory tract

Mitral cell

Tufted cell

Olfactory nerve fascicle

Cribriform plate of ethmoid bone

Basal cell

Supporting cells

Olfactory cell

Olfactory gland

Olfactory hairs

Mucus

Odor molecules

Airflow

(a)

Olfactory Receptors
Figure 17.7

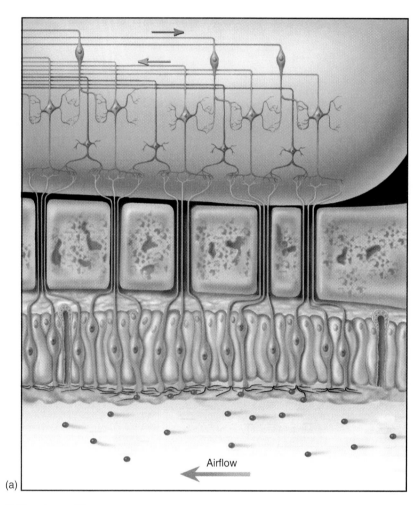

(a)

Olfactory Receptors
Figure 17.7

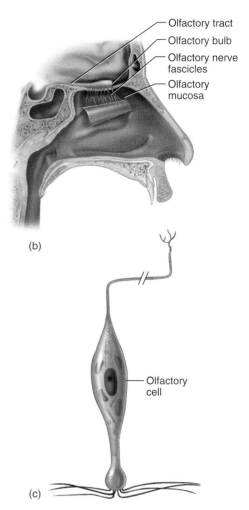

(b)

(c)

Olfactory tract
Olfactory bulb
Olfactory nerve
fascicles
Olfactory
mucosa

Olfactory
cell

**Olfactory Receptors
(*Continued*)**
Figure 17.7

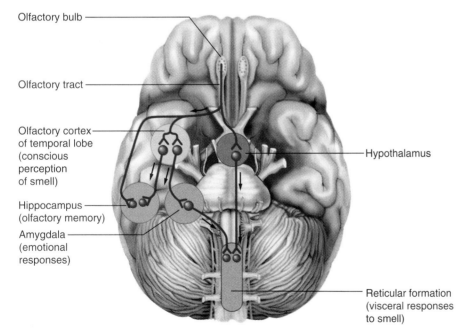

Olfactory bulb

Olfactory tract

Olfactory cortex
of temporal lobe
(conscious
perception
of smell)

Hippocampus
(olfactory memory)

Amygdala
(emotional
responses)

Hypothalamus

Reticular formation
(visceral responses
to smell)

Olfactory Projection Pathways in the Brain
Figure 17.8

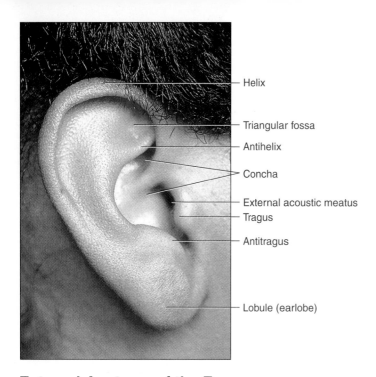

Helix

Triangular fossa

Antihelix

Concha

External acoustic meatus

Tragus

Antitragus

Lobule (earlobe)

External Anatomy of the Ear
Figure 17.9

© The McGraw-Hill Companies, Inc./Joe DeGrandis, photographer

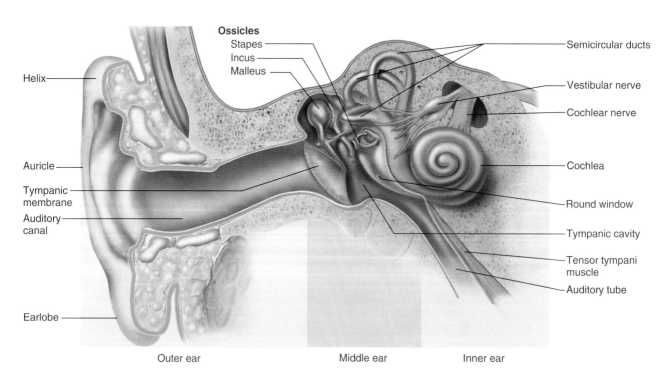

Ossicles

Stapes

Incus

Malleus

Semicircular ducts

Vestibular nerve

Cochlear nerve

Helix

Auricle

Tympanic membrane

Auditory canal

Cochlea

Round window

Tympanic cavity

Tensor tympani muscle

Auditory tube

Earlobe

Outer ear   Middle ear   Inner ear

Internal Anatomy of the Ear
Figure 17.10

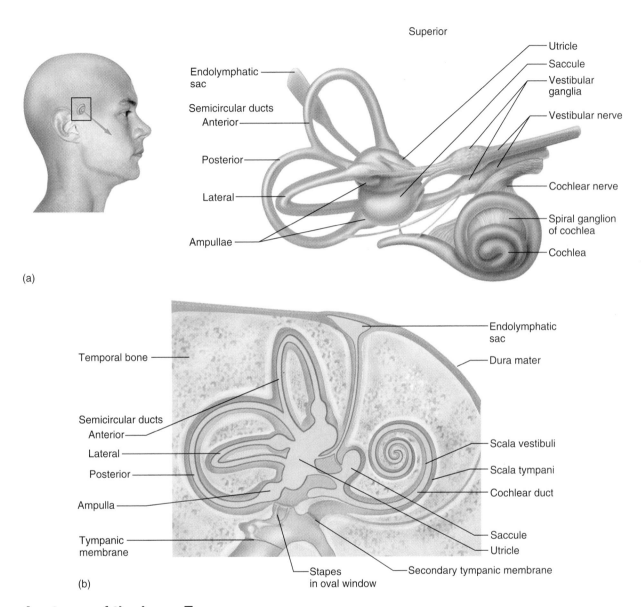

(a)

(b)

Anatomy of the Inner Ear
Figure 17.11

317

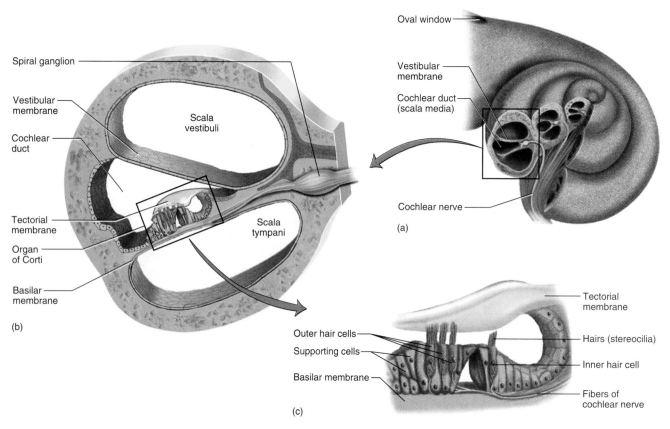

Spiral ganglion

Vestibular
membrane

Cochlear
duct

Scala
vestibuli

Tectorial
membrane

Organ
of Corti

Basilar
membrane

Scala
tympani

(b)

Oval window

Vestibular
membrane

Cochlear duct
(scala media)

Cochlear nerve

(a)

Tectorial
membrane

Outer hair cells

Supporting cells

Basilar membrane

Hairs (stereocilia)

Inner hair cell

Fibers of
cochlear nerve

(c)

Anatomy of the Cochlea
Figure 17.12

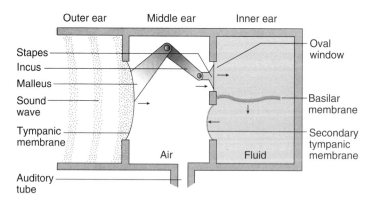

Outer ear   Middle ear   Inner ear

Stapes

Incus

Malleus

Sound
wave

Tympanic
membrane

Auditory
tube

Air

Fluid

Oval
window

Basilar
membrane

Secondary
tympanic
membrane

Mechanical Model of Auditory Function
Figure 17.14

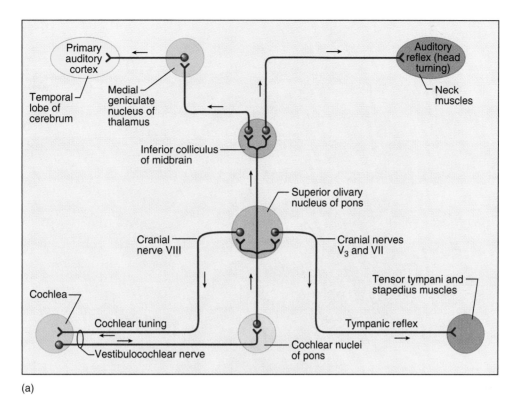

(a)

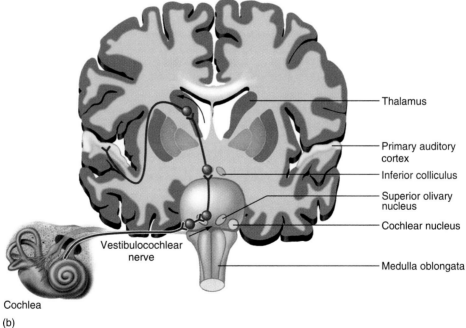

(b)

Auditory Pathways in the Brain
Figure 17.15

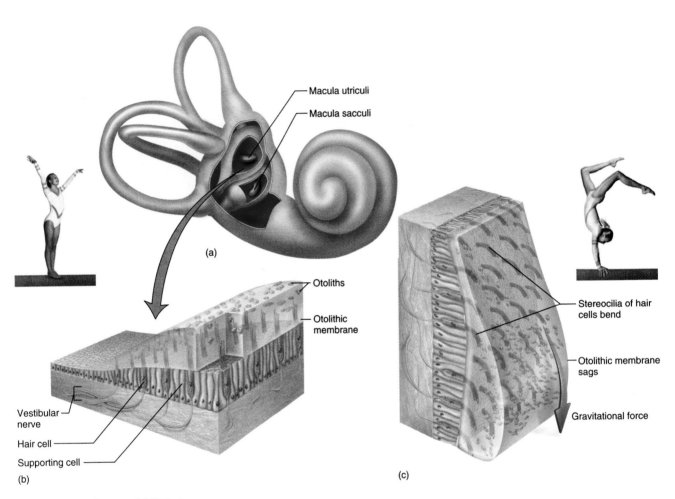

Macula utriculi

Macula sacculi

(a)

Otoliths

Otolithic membrane

Vestibular nerve

Hair cell

Supporting cell

(b)

Stereocilia of hair cells bend

Otolithic membrane sags

Gravitational force

(c)

The Saccule and Utricle
Figure 17.16

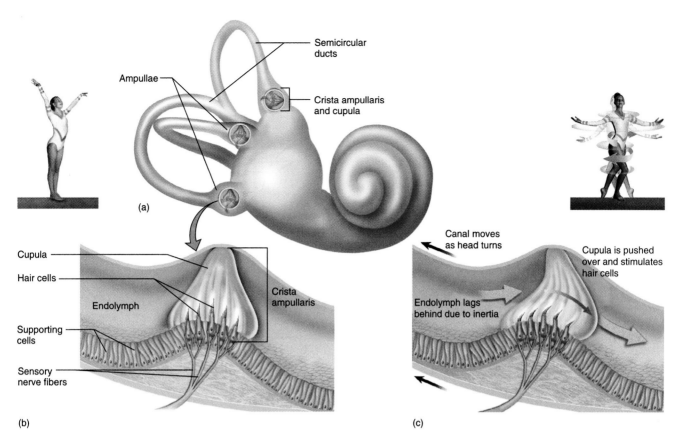

Structure and Function of the Semicircular Ducts
Figure 17.17

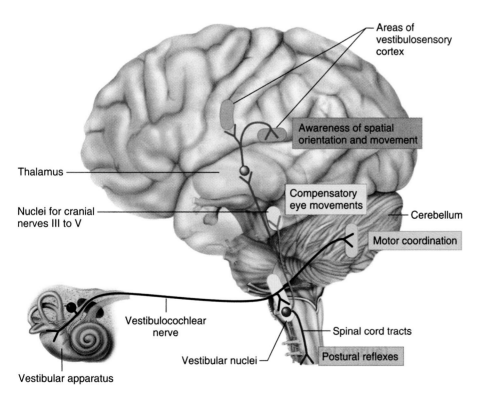

Vestibular Projection Pathways in the Brain
Figure 17.18

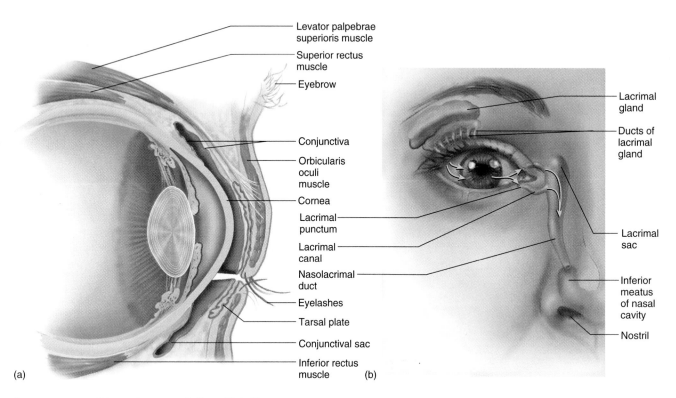

Levator palpebrae
superioris muscle

Superior rectus
muscle

Eyebrow

Conjunctiva

Orbicularis
oculi
muscle

Cornea

Lacrimal
punctum

Lacrimal
canal

Nasolacrimal
duct

Eyelashes

Tarsal plate

Conjunctival sac

Inferior rectus
muscle

Lacrimal
gland

Ducts of
lacrimal
gland

Lacrimal
sac

Inferior
meatus
of nasal
cavity

Nostril

(a)

(b)

Accessory Structures of the Orbit
Figure 17.20

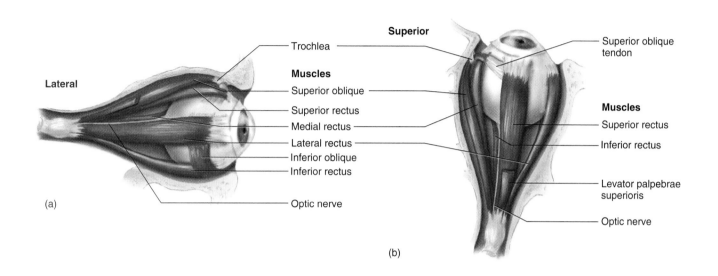

Lateral

Superior

Trochlea

Muscles
Superior oblique
Superior rectus
Medial rectus
Lateral rectus
Inferior oblique
Inferior rectus

Optic nerve

(a)

Superior oblique
tendon

Muscles
Superior rectus

Inferior rectus

Levator palpebrae
superioris

Optic nerve

(b)

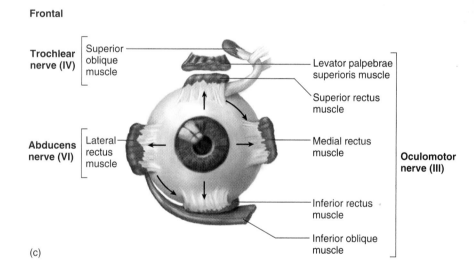

Frontal

Trochlear
nerve (IV)
Superior
oblique
muscle

Levator palpebrae
superioris muscle

Superior rectus
muscle

Abducens
nerve (VI)
Lateral
rectus
muscle

Medial rectus
muscle

Oculomotor
nerve (III)

Inferior rectus
muscle

Inferior oblique
muscle

(c)

Extrinsic Muscles of the Eye
Figure 17.21

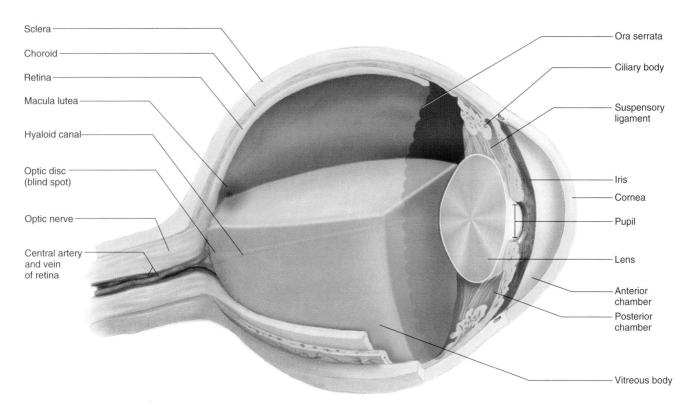

Sclera

Choroid

Retina

Macula lutea

Hyaloid canal

Optic disc
(blind spot)

Optic nerve

Central artery
and vein
of retina

Ora serrata

Ciliary body

Suspensory
ligament

Iris

Cornea

Pupil

Lens

Anterior
chamber

Posterior
chamber

Vitreous body

Anatomy of the Eye
Figure 17.22

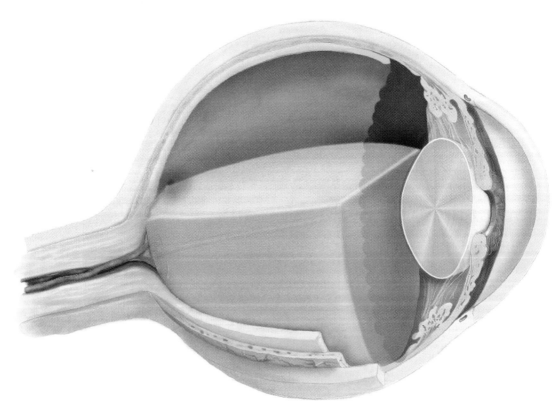

Anatomy of the Eye
Figure 17.22

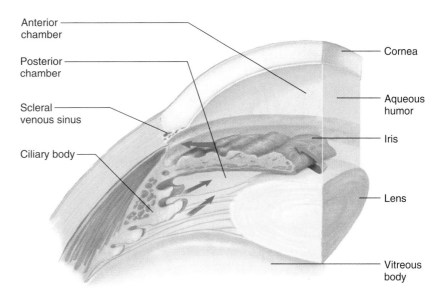

Production and Reabsorption of Aqueous Humor
Figure 17.23

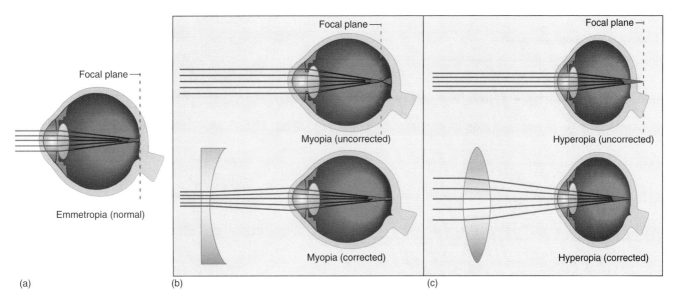

(a) Emmetropia (normal)

Focal plane

Myopia (uncorrected)

Myopia (corrected)

Hyperopia (uncorrected)

Hyperopia (corrected)

(b)

(c)

Two Common Visual Defects and the Effects of Corrective Lenses
Figure 17.26

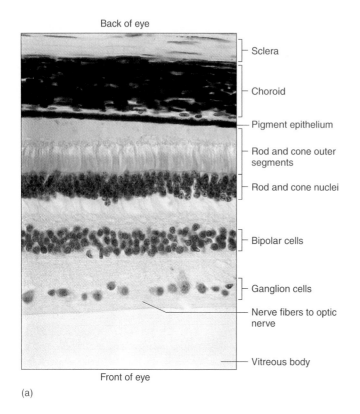

Back of eye

Sclera

Choroid

Pigment epithelium

Rod and cone outer segments

Rod and cone nuclei

Bipolar cells

Ganglion cells

Nerve fibers to optic nerve

Vitreous body

Front of eye

(a)

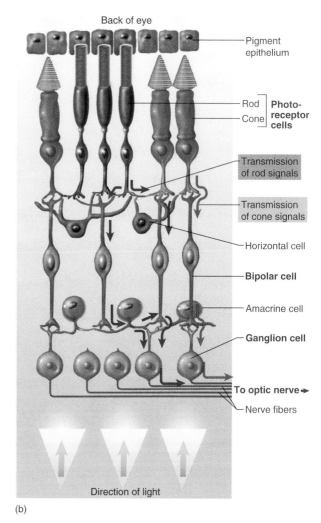

Back of eye

Pigment epithelium

Rod | **Photo-receptor cells**
Cone |

Transmission of rod signals

Transmission of cone signals

Horizontal cell

Bipolar cell

Amacrine cell

Ganglion cell

To optic nerve ➡

Nerve fibers

Direction of light

(b)

Histology of the Retina
Figure 17.27

a: © The McGraw-Hill Companies, Inc./Joe DeGrandis, photographer

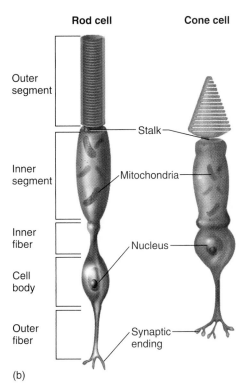

Rod and Cone Cells
Figure 17.28

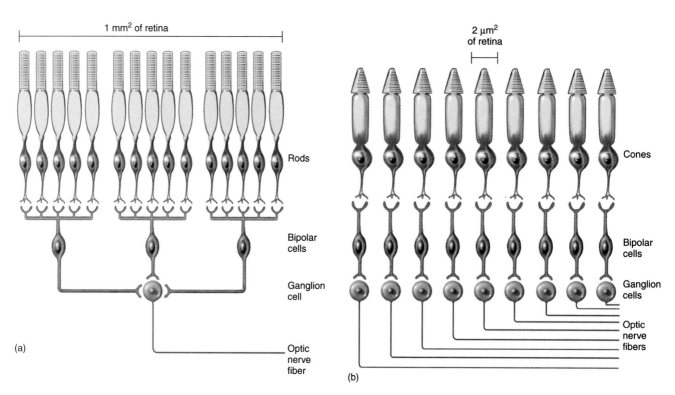

The Duplicity Theory of Vision
Figure 17.29

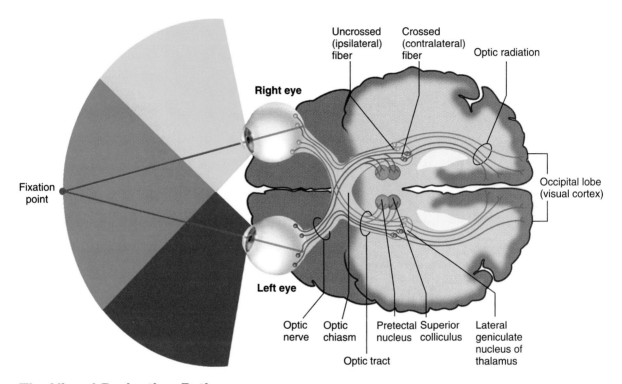

The Visual Projection Pathway
Figure 17.30

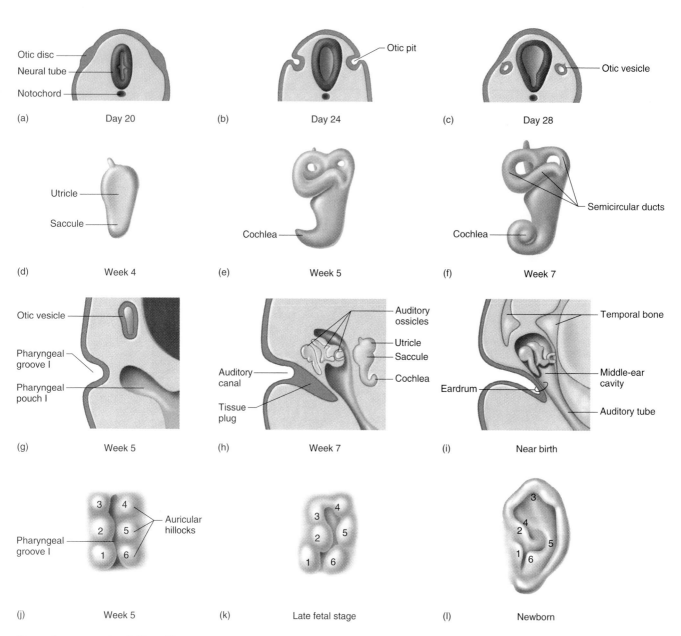

(a) Day 20
- Otic disc
- Neural tube
- Notochord

(b) Day 24
- Otic pit

(c) Day 28
- Otic vesicle

(d) Week 4
- Utricle
- Saccule

(e) Week 5
- Cochlea

(f) Week 7
- Semicircular ducts
- Cochlea

(g) Week 5
- Otic vesicle
- Pharyngeal groove I
- Pharyngeal pouch I

(h) Week 7
- Auditory ossicles
- Utricle
- Saccule
- Cochlea
- Auditory canal
- Tissue plug

(i) Near birth
- Temporal bone
- Middle-ear cavity
- Auditory tube
- Eardrum

(j) Week 5
- Pharyngeal groove I
- Auricular hillocks

(k) Late fetal stage

(l) Newborn

Development of the Ear
Figure 17.31

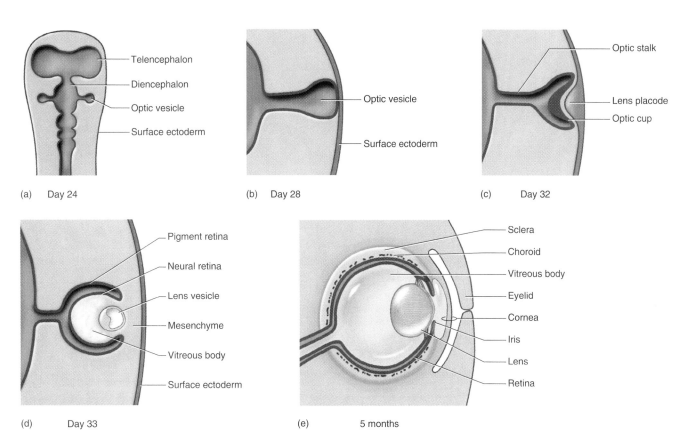

(a) Day 24

- Telencephalon
- Diencephalon
- Optic vesicle
- Surface ectoderm

(b) Day 28

- Optic vesicle
- Surface ectoderm

(c) Day 32

- Optic stalk
- Lens placode
- Optic cup

(d) Day 33

- Pigment retina
- Neural retina
- Lens vesicle
- Mesenchyme
- Vitreous body
- Surface ectoderm

(e) 5 months

- Sclera
- Choroid
- Vitreous body
- Eyelid
- Cornea
- Iris
- Lens
- Retina

Development of the Eye
Figure 17.32

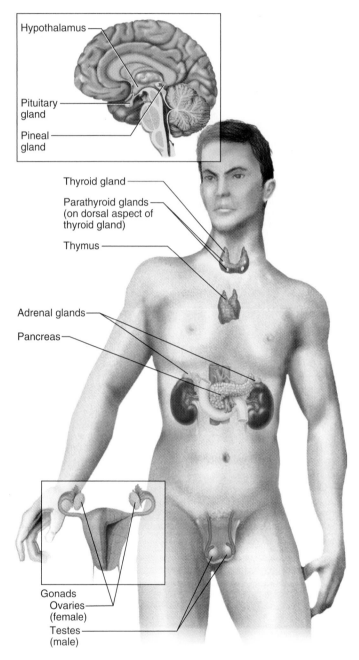

Major Organs of the Endocrine System
Figure 18.1

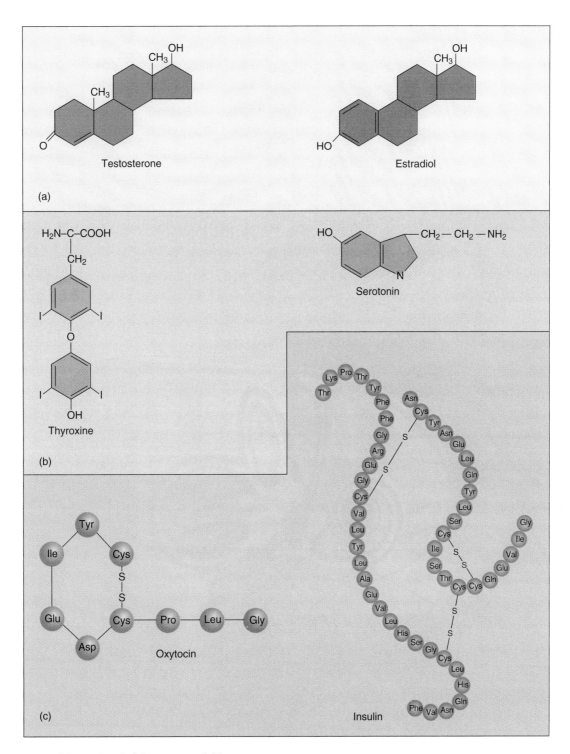

The Chemical Classes of Hormones
Figure 18.2

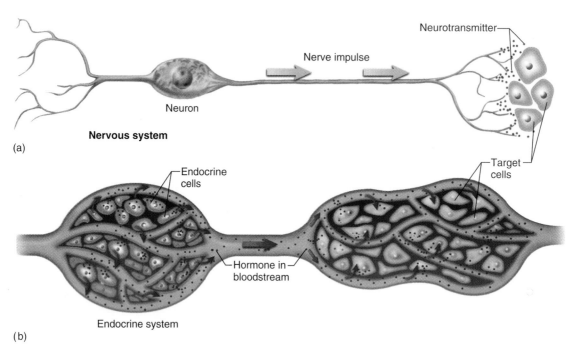

(a)

Nervous system

Neuron

Nerve impulse

Neurotransmitter

Target cells

(b)

Endocrine cells

Hormone in bloodstream

Endocrine system

Communication by the Nervous and Endocrine Systems
Figure 18.3

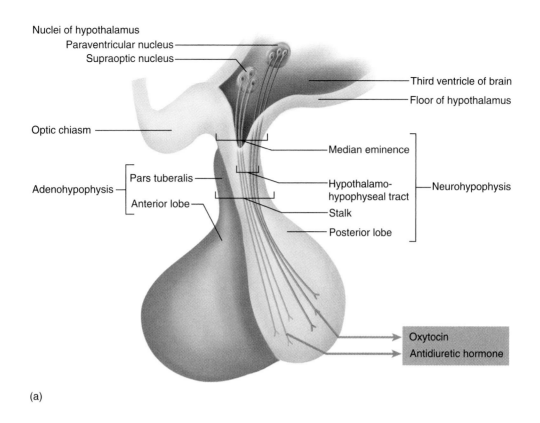

(a)

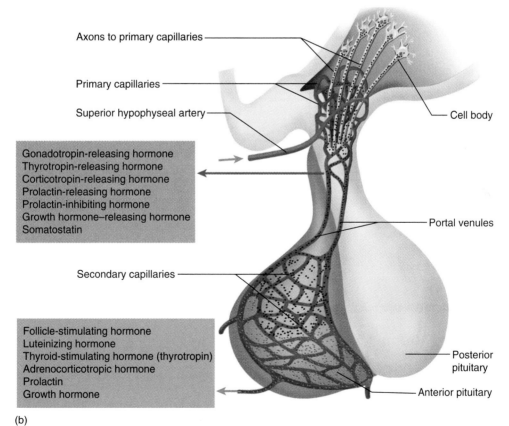

(b)

Gross Anatomy of the Pituitary Gland
Figure 18.4

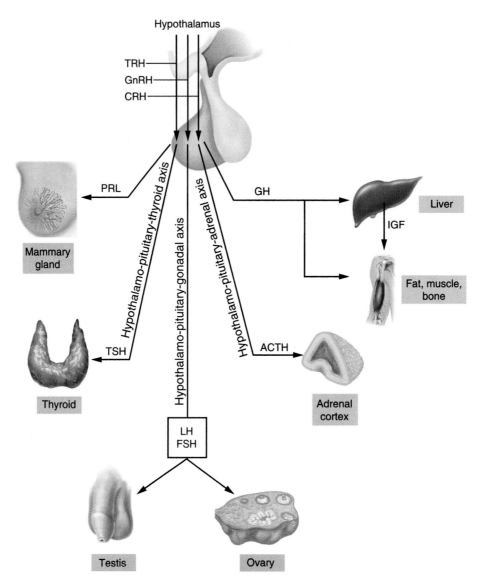

Hormones and Target Organs of the Anterior Pituitary Gland
Figure 18.6

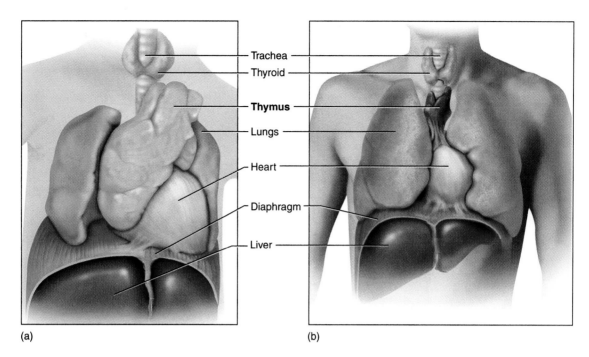

The Thymus
Figure 18.7

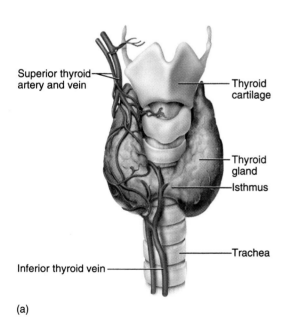

The Thyroid Gland
Figure 18.8

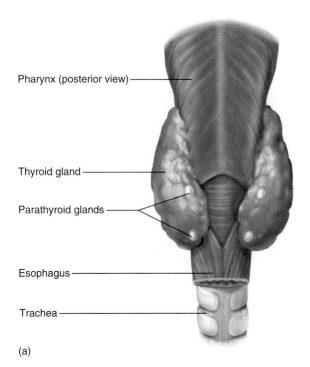

Pharynx (posterior view)

Thyroid gland

Parathyroid glands

Esophagus

Trachea

(a)

The Parathyroid Glands
Figure 18.9

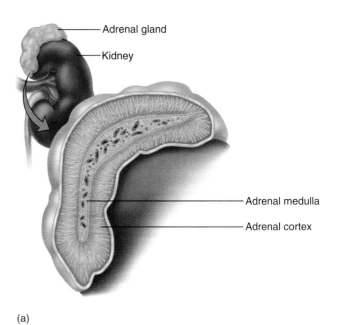

Adrenal gland

Kidney

Adrenal medulla

Adrenal cortex

(a)

The Adrenal Gland
Figure 18.10

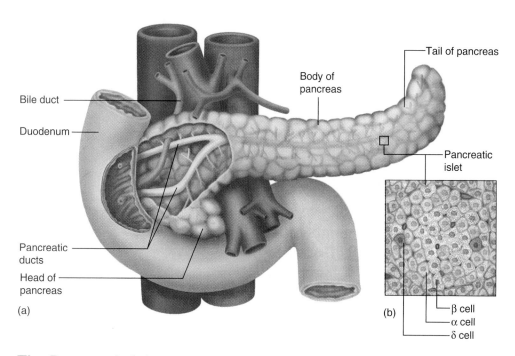

Bile duct

Duodenum

Body of
pancreas

Tail of pancreas

Pancreatic
islet

Pancreatic
ducts

Head of
pancreas

(a)

(b)

β cell

α cell

δ cell

The Pancreatic Islets
Figure 18.11

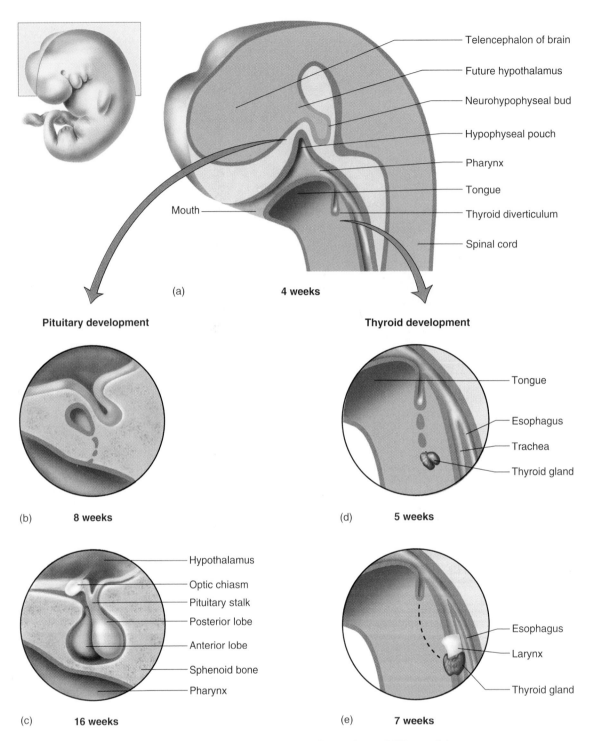

Telencephalon of brain
Future hypothalamus
Neurohypophyseal bud
Hypophyseal pouch
Pharynx
Tongue
Thyroid diverticulum
Spinal cord

Mouth

(a)   4 weeks

Pituitary development

(b)   8 weeks

Hypothalamus
Optic chiasm
Pituitary stalk
Posterior lobe
Anterior lobe
Sphenoid bone
Pharynx

(c)   16 weeks

Thyroid development

Tongue
Esophagus
Trachea
Thyroid gland

(d)   5 weeks

Esophagus
Larynx
Thyroid gland

(e)   7 weeks

Embryonic Development of the Pituitary Gland and Thyroid
Figure 18.13

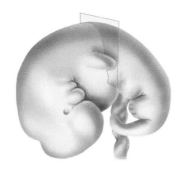

(a)

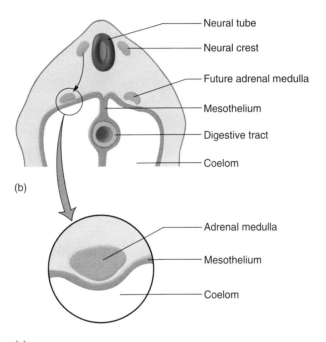

Neural tube

Neural crest

Future adrenal medulla

Mesothelium

Digestive tract

Coelom

(b)

Adrenal medulla

Mesothelium

Coelom

(c)

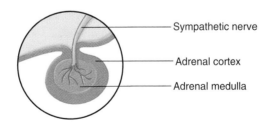

Sympathetic nerve

Adrenal cortex

Adrenal medulla

(d)

**Embryonic Development
of the Adrenal Gland**
Figure 18.14

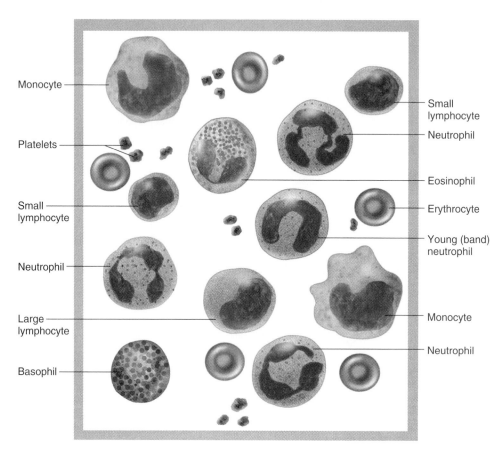

Monocyte

Platelets

Small
lymphocyte

Neutrophil

Large
lymphocyte

Basophil

Small
lymphocyte

Neutrophil

Eosinophil

Erythrocyte

Young (band)
neutrophil

Monocyte

Neutrophil

The Formed Elements of Blood
Figure 19.1

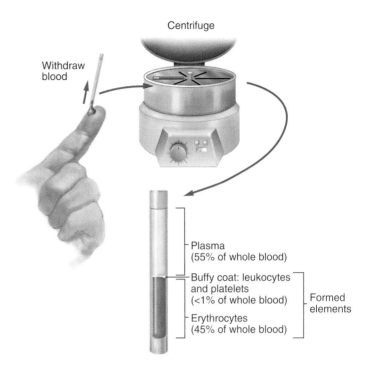

Centrifuge

Withdraw
blood

Plasma
(55% of whole blood)

Buffy coat: leukocytes
and platelets
(<1% of whole blood)

Formed
elements

Erythrocytes
(45% of whole blood)

**Separating the Plasma and Formed Elements
of Blood**
Figure 19.2

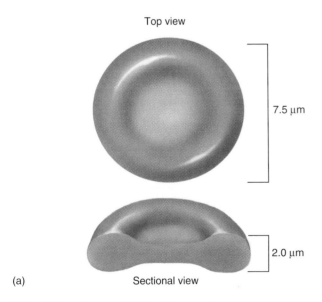

Top view

7.5 μm

2.0 μm

(a)

Sectional view

The Structure of Erythrocytes
Figure 19.3

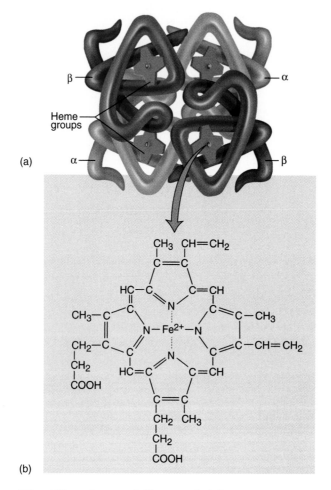

(a)

(b)

The Structure of Hemoglobin
Figure 19.5

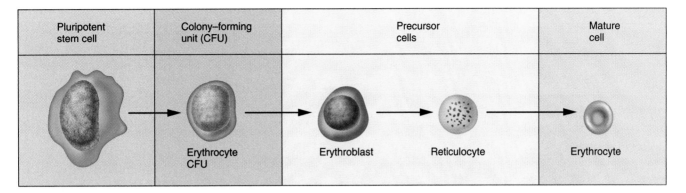

Erythropoiesis
Figure 19.6

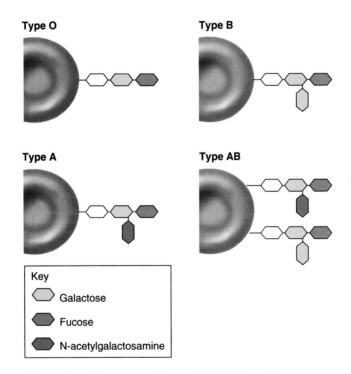

Chemical Basis of the ABO Blood Types
Figure 19.7

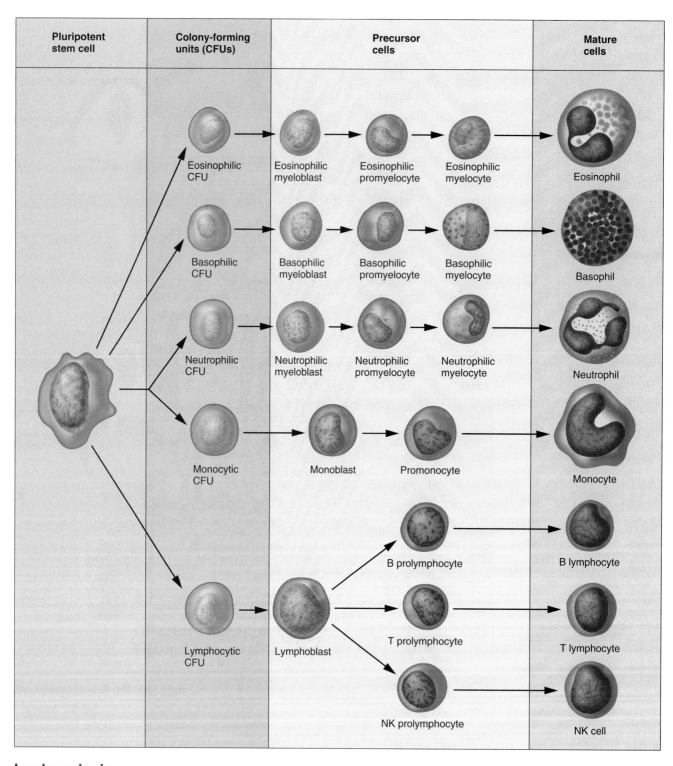

| Pluripotent stem cell | Colony-forming units (CFUs) | Precursor cells | | | Mature cells |

Eosinophilic CFU → **Eosinophilic myeloblast** → **Eosinophilic promyelocyte** → **Eosinophilic myelocyte** → **Eosinophil**

Basophilic CFU → **Basophilic myeloblast** → **Basophilic promyelocyte** → **Basophilic myelocyte** → **Basophil**

Neutrophilic CFU → **Neutrophilic myeloblast** → **Neutrophilic promyelocyte** → **Neutrophilic myelocyte** → **Neutrophil**

Monocytic CFU → **Monoblast** → **Promonocyte** → **Monocyte**

Lymphocytic CFU → **Lymphoblast** → **B prolymphocyte** → **B lymphocyte**

T prolymphocyte → **T lymphocyte**

NK prolymphocyte → **NK cell**

Leukopoiesis
Figure 19.9

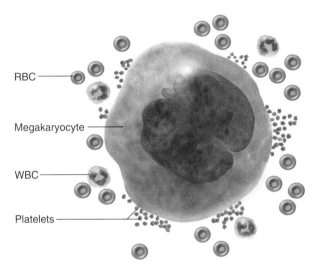

RBC

Megakaryocyte

WBC

Platelets

A Megakaryocyte Producing Platelets
Figure 19.11

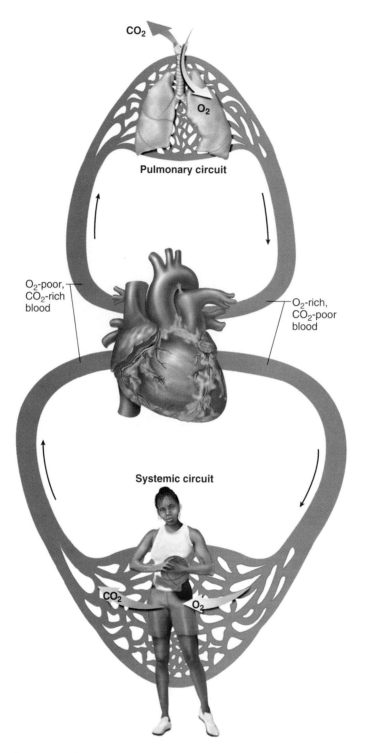

CO₂

O₂

Pulmonary circuit

O₂-poor,
CO₂-rich
blood

O₂-rich,
CO₂-poor
blood

Systemic circuit

CO₂   O₂

**General Schematic of the Cardiovascular
System**
Figure 20.1

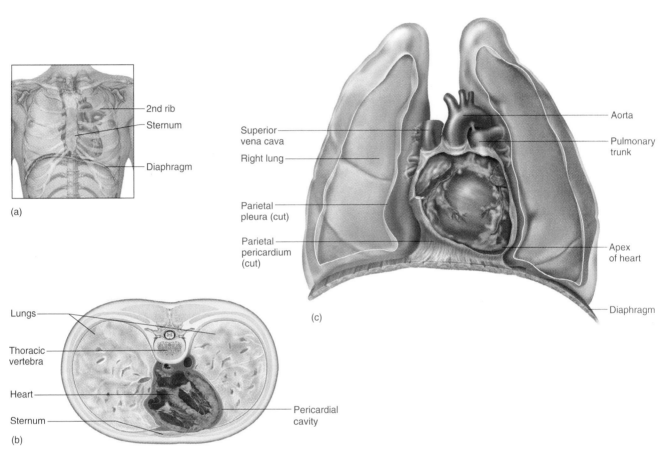

(a)

2nd rib
Sternum
Diaphragm

(b)

Lungs
Thoracic vertebra
Heart
Sternum
Pericardial cavity

(c)

Superior vena cava
Right lung
Parietal pleura (cut)
Parietal pericardium (cut)
Aorta
Pulmonary trunk
Apex of heart
Diaphragm

Position of the Heart in the Thoracic Cavity
Figure 20.2

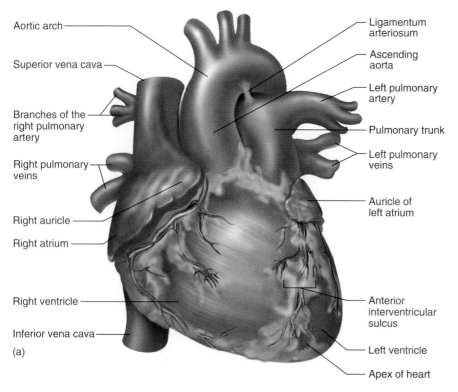

Aortic arch

Superior vena cava

Branches of the right pulmonary artery

Right pulmonary veins

Right auricle

Right atrium

Right ventricle

Inferior vena cava

(a)

Ligamentum arteriosum

Ascending aorta

Left pulmonary artery

Pulmonary trunk

Left pulmonary veins

Auricle of left atrium

Anterior interventricular sulcus

Left ventricle

Apex of heart

External Anatomy of the Heart
Figure 20.3

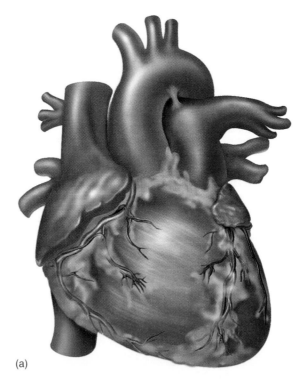

(a)

External Anatomy of the Heart
Figure 20.3

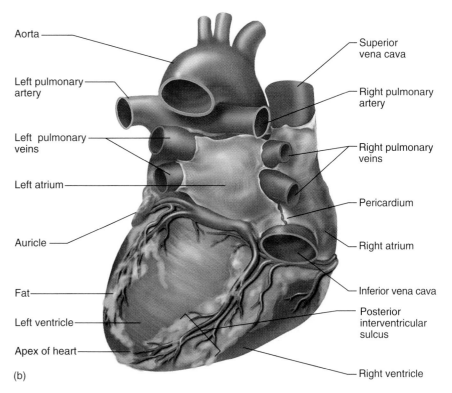

Aorta

Left pulmonary artery

Left pulmonary veins

Left atrium

Auricle

Fat

Left ventricle

Apex of heart

Superior vena cava

Right pulmonary artery

Right pulmonary veins

Pericardium

Right atrium

Inferior vena cava

Posterior interventricular sulcus

Right ventricle

(b)

External Anatomy of the Heart
Figure 20.3

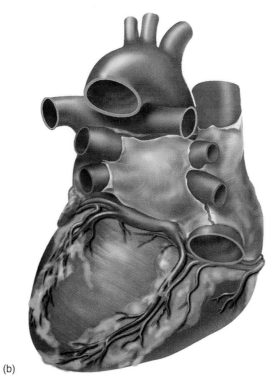

(b)

External Anatomy of the Heart
Figure 20.3

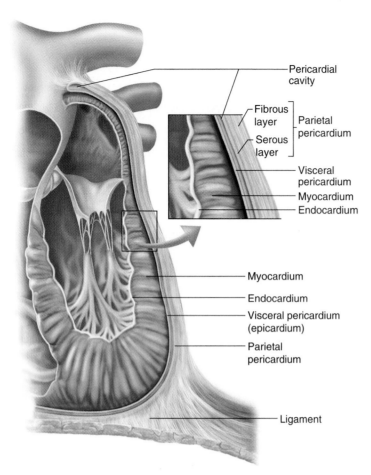

The Pericardium and Heart Wall
Figure 20.4

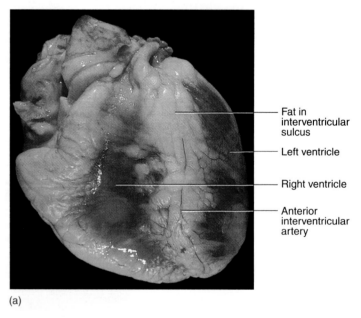

(a)

The Human Heart
Figure 20.5

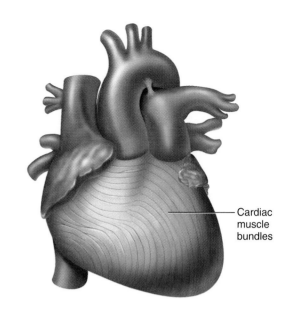

Cardiac muscle bundles

(b)

Twisted Orientation of Myocardial Muscle
Figure 20.6

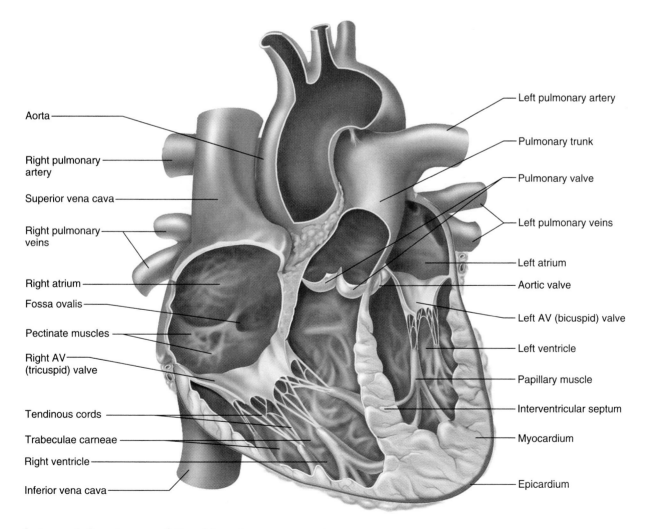

Aorta

Right pulmonary artery

Superior vena cava

Right pulmonary veins

Right atrium

Fossa ovalis

Pectinate muscles

Right AV (tricuspid) valve

Tendinous cords

Trabeculae carneae

Right ventricle

Inferior vena cava

Left pulmonary artery

Pulmonary trunk

Pulmonary valve

Left pulmonary veins

Left atrium

Aortic valve

Left AV (bicuspid) valve

Left ventricle

Papillary muscle

Interventricular septum

Myocardium

Epicardium

Internal Anatomy of the Heart
Figure 20.7

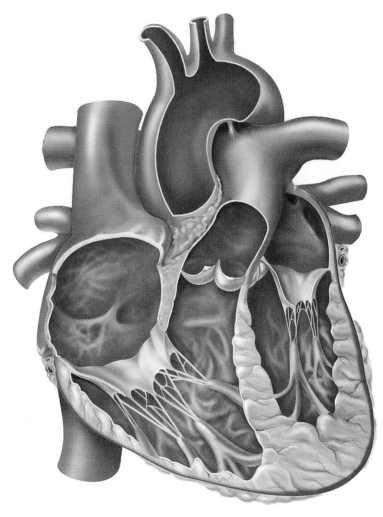

Internal Anatomy of the Heart
Figure 20.7

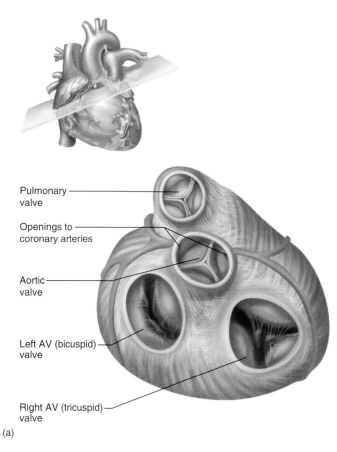

Pulmonary
valve

Openings to
coronary arteries

Aortic
valve

Left AV (bicuspid)
valve

Right AV (tricuspid)
valve

(a)

The Heart Valves
Figure 20.8

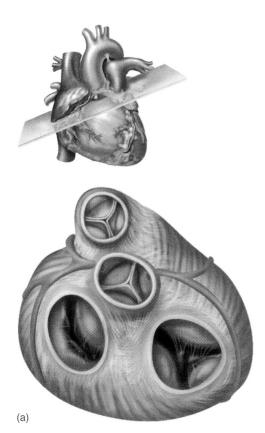

(a)

The Heart Valves
Figure 20.8

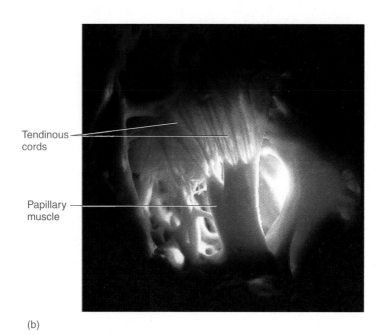

Tendinous
cords

Papillary
muscle

(b)

The Heart Values (*Continued*)
Figure 20.8

b: © The McGraw-Hill Companies, Inc.

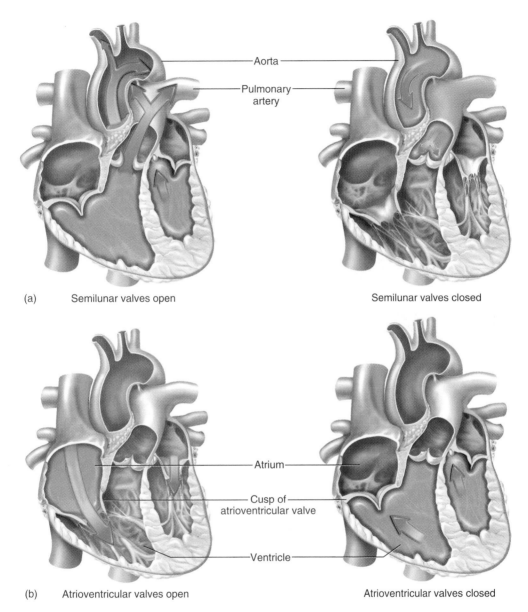

(a)   Semilunar valves open   Semilunar valves closed

Aorta

Pulmonary artery

Atrium

Cusp of atrioventricular valve

Ventricle

(b)   Atrioventricular valves open   Atrioventricular valves closed

Operation of the Heart Valves
Figure 20.9

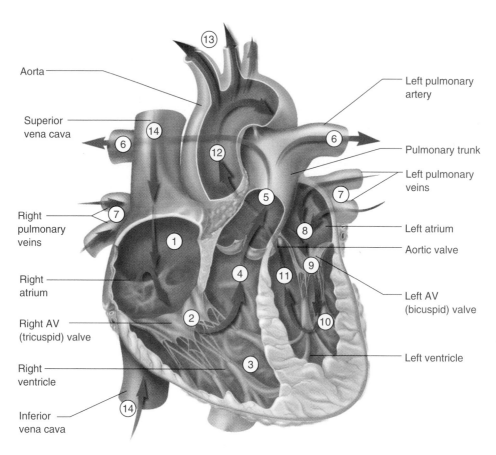

Aorta

Superior
vena cava

Right
pulmonary
veins

Right
atrium

Right AV
(tricuspid) valve

Right
ventricle

Inferior
vena cava

Left pulmonary
artery

Pulmonary trunk

Left pulmonary
veins

Left atrium

Aortic valve

Left AV
(bicuspid) valve

Left ventricle

The Pathway of Blood Flow Through the Heart
Figure 20.10

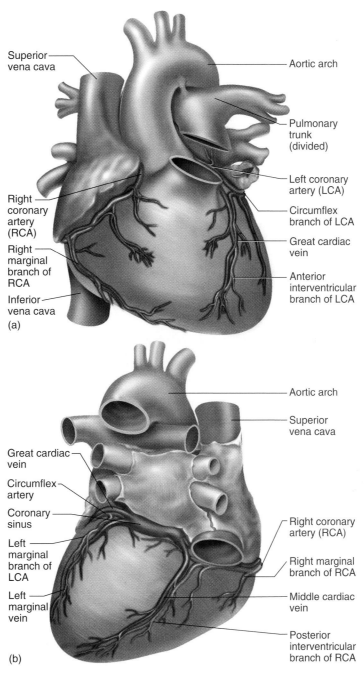

Superior vena cava

Aortic arch

Pulmonary trunk (divided)

Left coronary artery (LCA)

Right coronary artery (RCA)

Circumflex branch of LCA

Great cardiac vein

Right marginal branch of RCA

Anterior interventricular branch of LCA

Inferior vena cava

(a)

Aortic arch

Superior vena cava

Great cardiac vein

Circumflex artery

Coronary sinus

Left marginal branch of LCA

Right coronary artery (RCA)

Right marginal branch of RCA

Left marginal vein

Middle cardiac vein

Posterior interventricular branch of RCA

(b)

The Coronary Blood Vessels
Figure 20.11

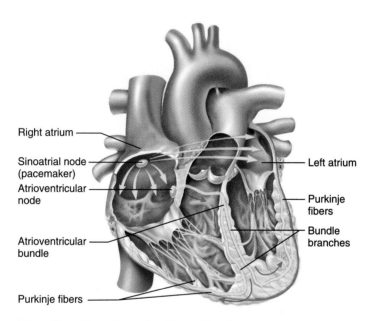

Right atrium

Sinoatrial node
(pacemaker)

Atrioventricular
node

Atrioventricular
bundle

Purkinje fibers

Left atrium

Purkinje
fibers

Bundle
branches

The Cardiac Conduction System
Figure 20.13

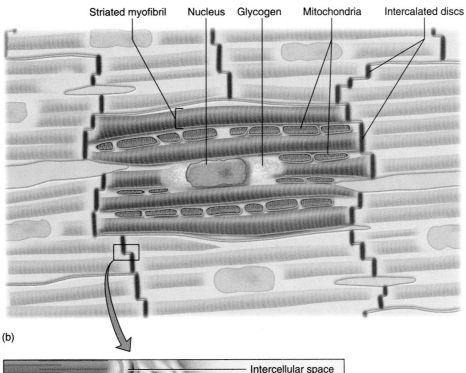

Striated myofibril　Nucleus　Glycogen　Mitochondria　Intercalated discs

(b)

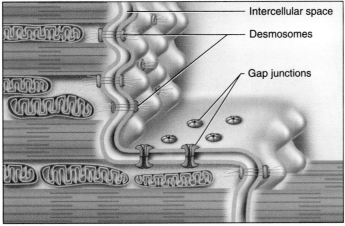

Intercellular space

Desmosomes

Gap junctions

(c)

Cardiac Muscle
Figure 20.14

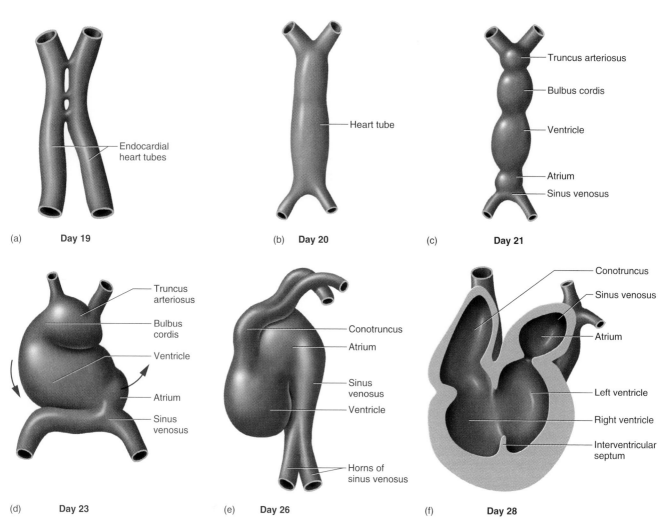

(a) **Day 19**

Endocardial heart tubes

(b) **Day 20**

Heart tube

(c) **Day 21**

Truncus arteriosus

Bulbus cordis

Ventricle

Atrium

Sinus venosus

(d) **Day 23**

Truncus arteriosus

Bulbus cordis

Ventricle

Atrium

Sinus venosus

(e) **Day 26**

Conotruncus

Atrium

Sinus venosus

Ventricle

Horns of sinus venosus

(f) **Day 28**

Conotruncus

Sinus venosus

Atrium

Left ventricle

Right ventricle

Interventricular septum

Embryonic Development of the Heart

Figure 20.15

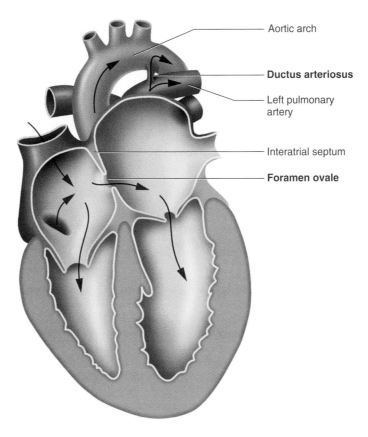

Aortic arch

Ductus arteriosus

Left pulmonary artery

Interatrial septum

Foramen ovale

The Fetal Heart
Figure 20.16

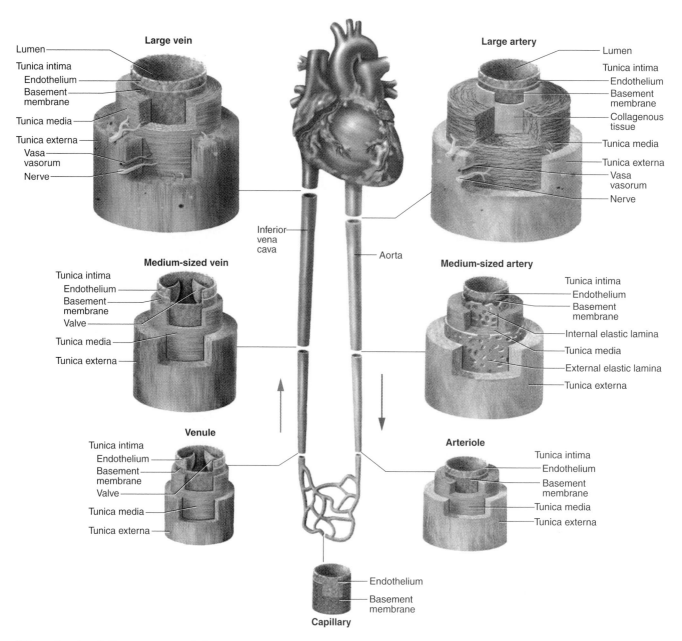

Large vein

Lumen
Tunica intima
Endothelium
Basement membrane
Tunica media
Tunica externa
Vasa vasorum
Nerve

Large artery

Lumen
Tunica intima
Endothelium
Basement membrane
Collagenous tissue
Tunica media
Tunica externa
Vasa vasorum
Nerve

Inferior vena cava

Aorta

Medium-sized vein

Tunica intima
Endothelium
Basement membrane
Valve
Tunica media
Tunica externa

Medium-sized artery

Tunica intima
Endothelium
Basement membrane
Internal elastic lamina
Tunica media
External elastic lamina
Tunica externa

Venule

Tunica intima
Endothelium
Basement membrane
Valve
Tunica media
Tunica externa

Arteriole

Tunica intima
Endothelium
Basement membrane
Tunica media
Tunica externa

Capillary

Endothelium
Basement membrane

Histological Structure of Blood Vessels
Figure 21.1

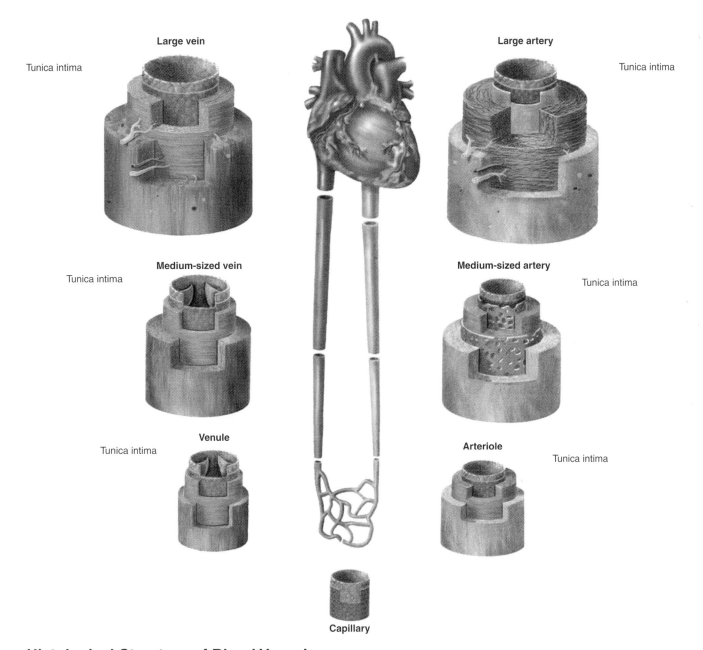

Large vein

Tunica intima

Medium-sized vein

Tunica intima

Venule

Tunica intima

Large artery

Tunica intima

Medium-sized artery

Tunica intima

Arteriole

Tunica intima

Capillary

Histological Structure of Blood Vessels
Figure 21.1

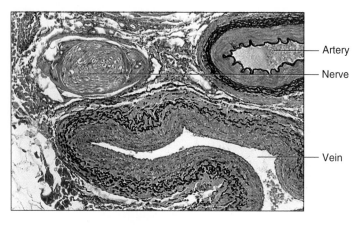

A Neurovascular Bundle
Figure 21.2
© The McGraw-Hill Companies, Inc./Dennis Strete, photographer

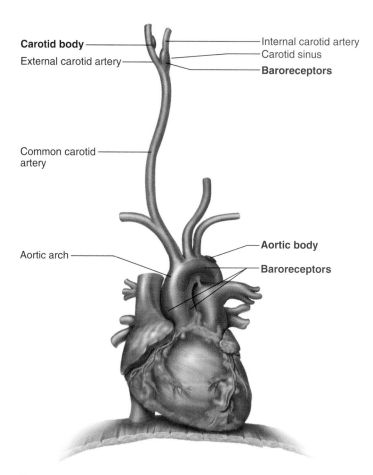

Baroreceptors and Chemoreceptors in the Arteries Superior to the Heart
Figure 21.4

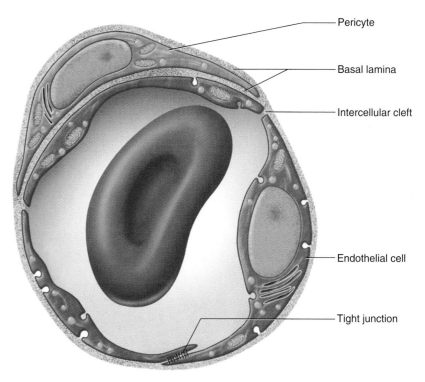

(a)

Structure of a Continuous Capillary
Figure 21.5

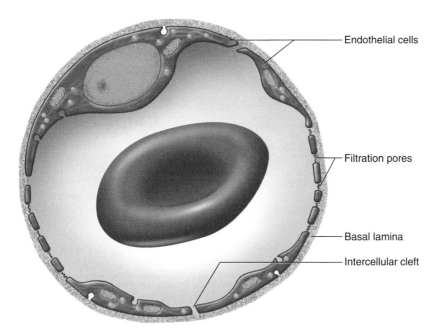

(a)

Structure of a Fenestrated Capillary
Figure 21.6

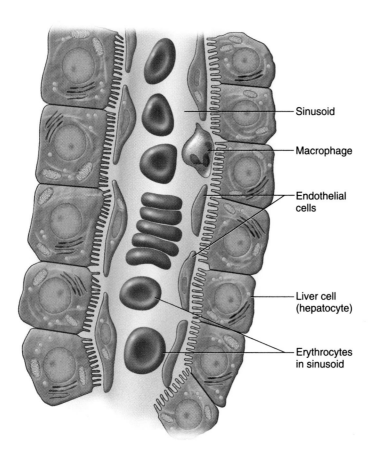

A Sinusoid of the Liver
Figure 21.7

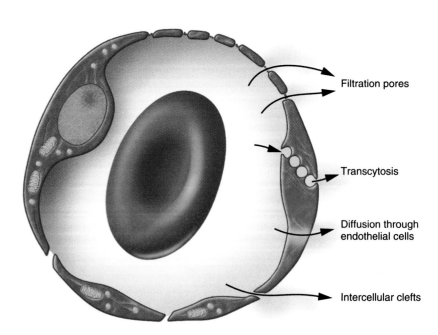

Pathways of Capillary Fluid Exchange
Figure 21.8

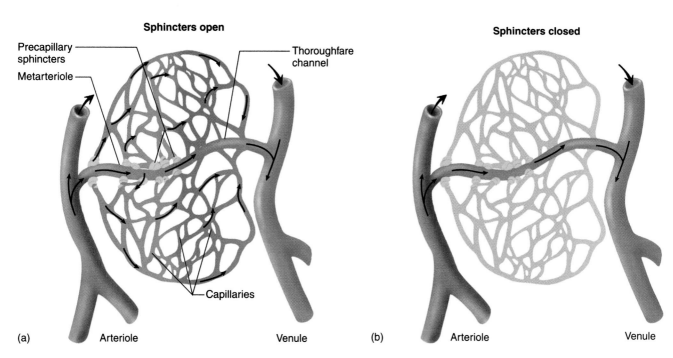

Sphincters open

Precapillary sphincters

Metarteriole

Thoroughfare channel

Capillaries

(a) Arteriole Venule

Sphincters closed

(b) Arteriole Venule

Perfusion of a Capillary Bed
Figure 21.9

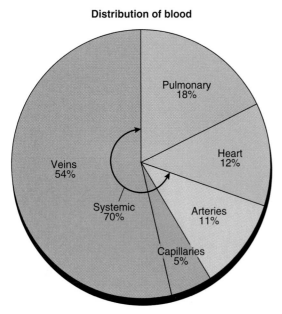

Distribution of blood

Veins
54%

Pulmonary
18%

Heart
12%

Systemic
70%

Arteries
11%

Capillaries
5%

Typical Distribution of the Blood in a Resting Adult
Figure 21.10

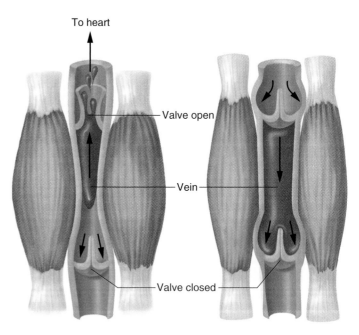

To heart

Valve open

Vein

Valve closed

(a)   Contracted skeletal muscles   (b)   Relaxed skeletal muscles

The Skeletal Muscle Pump
Figure 21.11

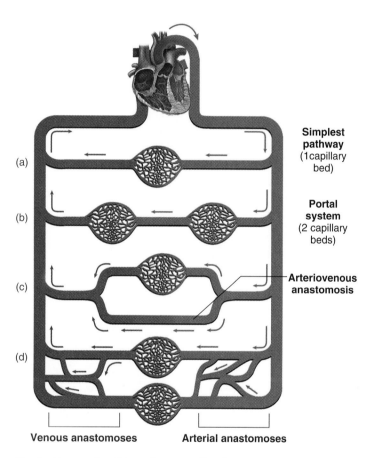

(a)

(b)

(c)

(d)

Simplest pathway (1capillary bed)

Portal system (2 capillary beds)

Arteriovenous anastomosis

Venous anastomoses

Arterial anastomoses

Variations in Circulatory Pathways
Figure 21.12

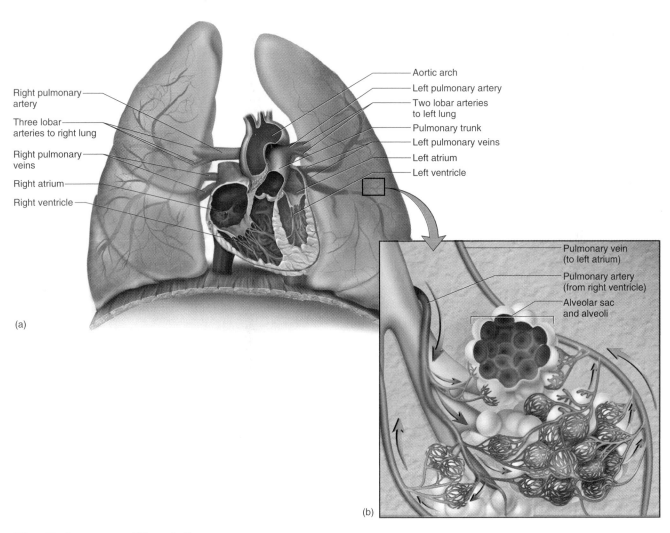

Right pulmonary artery

Three lobar arteries to right lung

Right pulmonary veins

Right atrium

Right ventricle

Aortic arch

Left pulmonary artery

Two lobar arteries to left lung

Pulmonary trunk

Left pulmonary veins

Left atrium

Left ventricle

(a)

Pulmonary vein (to left atrium)

Pulmonary artery (from right ventricle)

Alveolar sac and alveoli

(b)

The Pulmonary Circulation
Figure 21.13

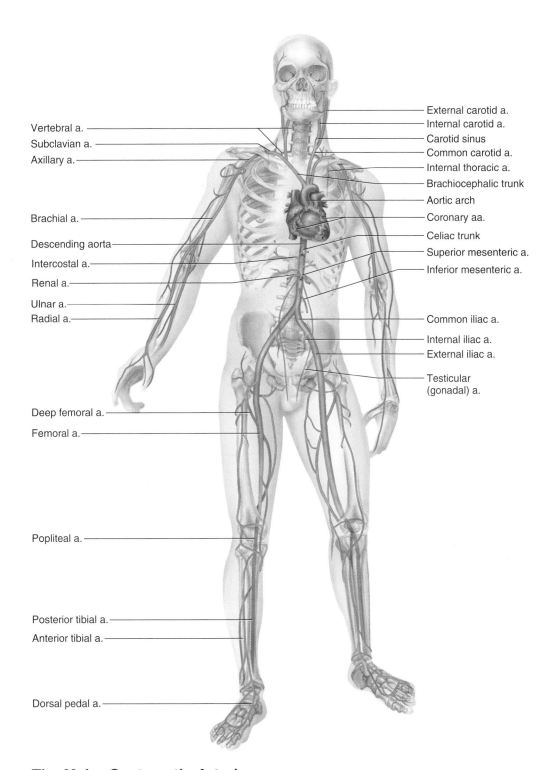

External carotid a.
Internal carotid a.
Carotid sinus
Common carotid a.
Internal thoracic a.
Brachiocephalic trunk
Aortic arch
Coronary aa.
Celiac trunk
Superior mesenteric a.
Inferior mesenteric a.

Common iliac a.
Internal iliac a.
External iliac a.

Testicular
(gonadal) a.

Vertebral a.
Subclavian a.
Axillary a.

Brachial a.
Descending aorta
Intercostal a.
Renal a.
Ulnar a.
Radial a.

Deep femoral a.
Femoral a.

Popliteal a.

Posterior tibial a.
Anterior tibial a.

Dorsal pedal a.

The Major Systematic Arteries
Figure 21.14

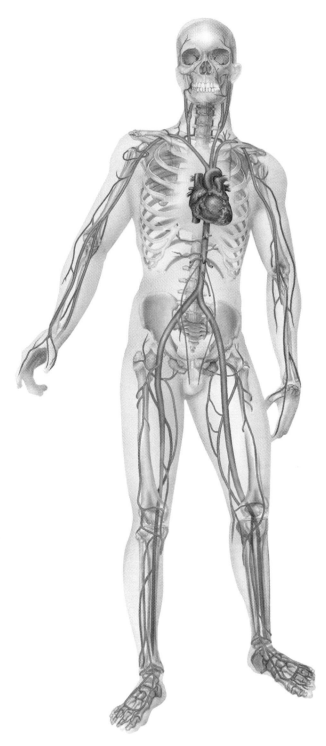

The Major Systematic Arteries
Figure 21.14

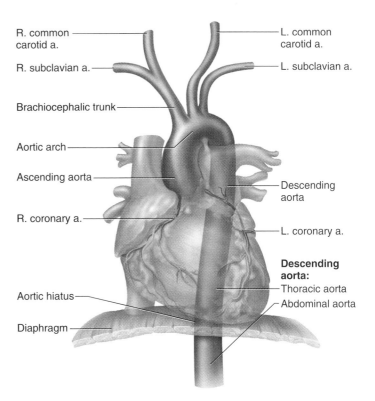

The Thoracic Aorta
Figure 21.15

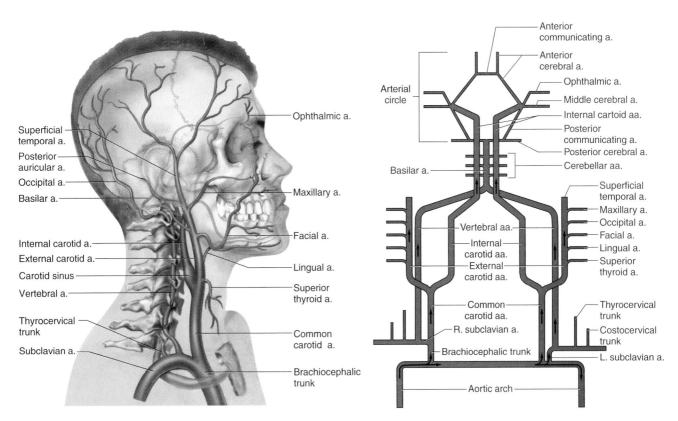

Arteries of the Head and Neck
Figure 21.16

375

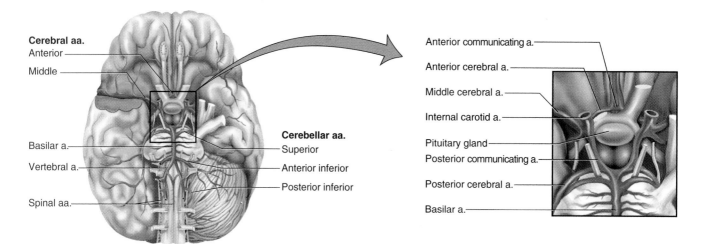

Cerebral aa.
Anterior
Middle

Basilar a.
Vertebral a.
Spinal aa.

Cerebellar aa.
Superior
Anterior inferior
Posterior inferior

Anterior communicating a.
Anterior cerebral a.
Middle cerebral a.
Internal carotid a.
Pituitary gland
Posterior communicating a.
Posterior cerebral a.
Basilar a.

The Cerebral Arterial Circle
Figure 21.17

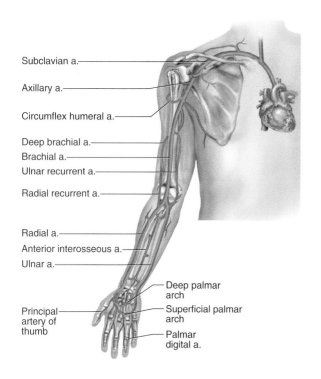

Subclavian a.
Axillary a.
Circumflex humeral a.
Deep brachial a.
Brachial a.
Ulnar recurrent a.
Radial recurrent a.
Radial a.
Anterior interosseous a.
Ulnar a.
Principal artery of thumb
Deep palmar arch
Superficial palmar arch
Palmar digital a.

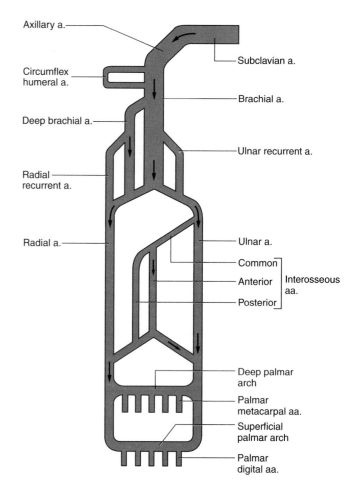

Axillary a.
Circumflex humeral a.
Deep brachial a.
Radial recurrent a.
Radial a.
Subclavian a.
Brachial a.
Ulnar recurrent a.
Ulnar a.
Common
Anterior
Posterior
Interosseous aa.
Deep palmar arch
Palmar metacarpal aa.
Superficial palmar arch
Palmar digital aa.

Arteries of the Upper Limb
Figure 21.18

376

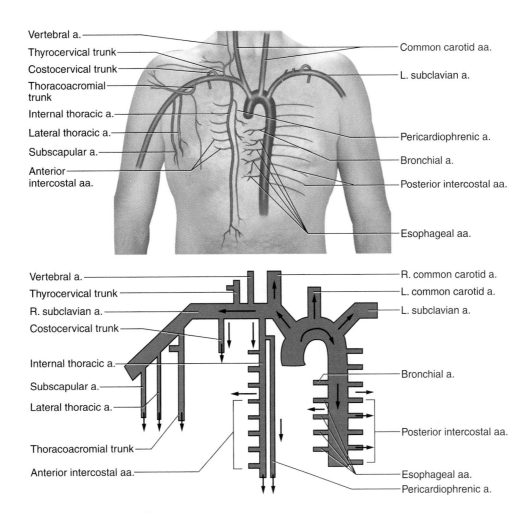

Arteries of the Thorax
Figure 21.19

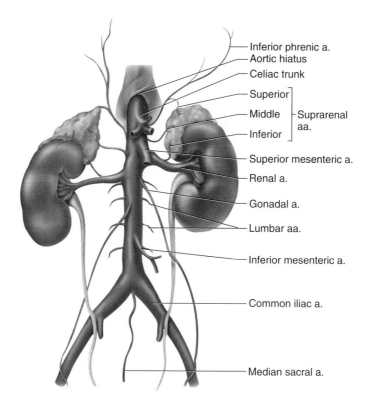

Inferior phrenic a.
Aortic hiatus
Celiac trunk
Superior ⎤
Middle ⎬ Suprarenal aa.
Inferior ⎦
Superior mesenteric a.
Renal a.
Gonadal a.
Lumbar aa.
Inferior mesenteric a.
Common iliac a.
Median sacral a.

The Abdominal Aorta and its Major Branches
Figure 21.20

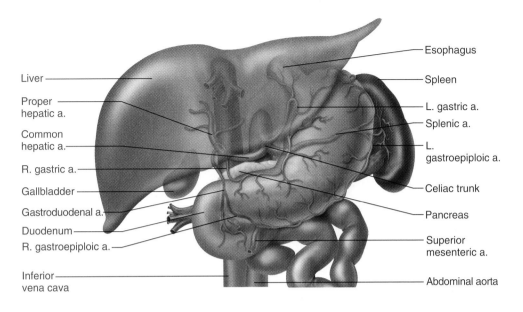

Esophagus

Liver

Proper hepatic a.

Common hepatic a.

R. gastric a.

Gallbladder

Gastroduodenal a.

Duodenum

R. gastroepiploic a.

Inferior vena cava

Spleen

L. gastric a.

Splenic a.

L. gastroepiploic a.

Celiac trunk

Pancreas

Superior mesenteric a.

Abdominal aorta

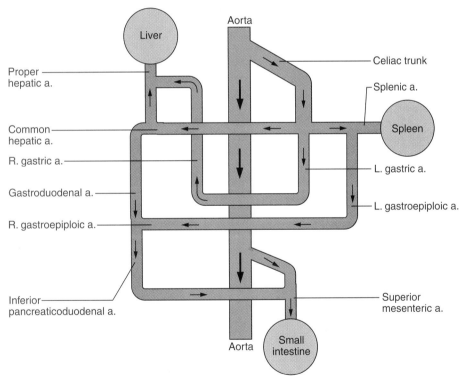

Aorta

Liver

Proper hepatic a.

Common hepatic a.

R. gastric a.

Gastroduodenal a.

R. gastroepiploic a.

Inferior pancreaticoduodenal a.

Aorta

Celiac trunk

Splenic a.

Spleen

L. gastric a.

L. gastroepiploic a.

Superior mesenteric a.

Small intestine

Branches of the Celiac Trunk
Figure 21.21

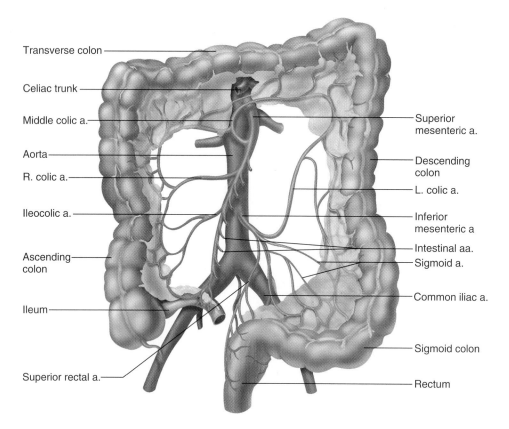

Transverse colon

Celiac trunk

Middle colic a.

Aorta

R. colic a.

Ileocolic a.

Ascending colon

Ileum

Superior rectal a.

Superior mesenteric a.

Descending colon

L. colic a.

Inferior mesenteric a

Intestinal aa.

Sigmoid a.

Common iliac a.

Sigmoid colon

Rectum

The Mesenteric Arteries
Figure 21.22

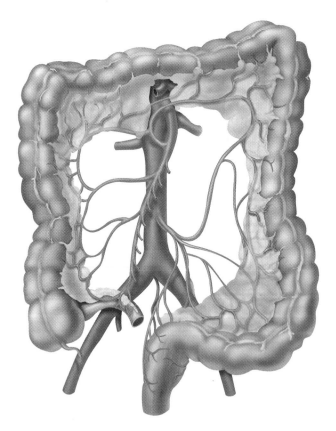

The Mesenteric Arteries
Figure 21.22

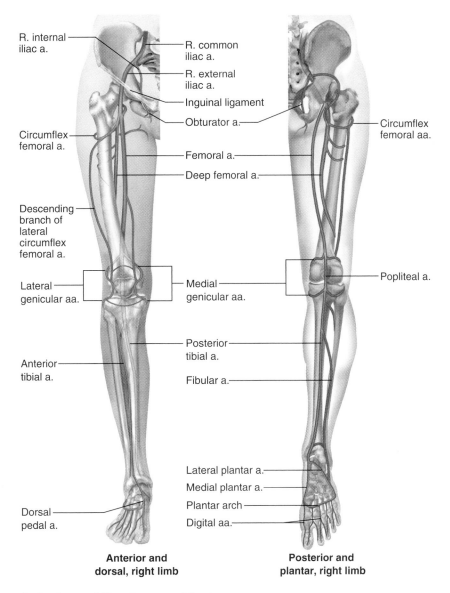

R. internal iliac a.

R. common iliac a.

R. external iliac a.

Inguinal ligament

Obturator a.

Circumflex femoral a.

Circumflex femoral aa.

Femoral a.

Deep femoral a.

Descending branch of lateral circumflex femoral a.

Lateral genicular aa.

Medial genicular aa.

Popliteal a.

Posterior tibial a.

Anterior tibial a.

Fibular a.

Lateral plantar a.

Medial plantar a.

Plantar arch

Digital aa.

Dorsal pedal a.

Anterior and dorsal, right limb

Posterior and plantar, right limb

Arteries of the Lower Limb
Figure 21.23

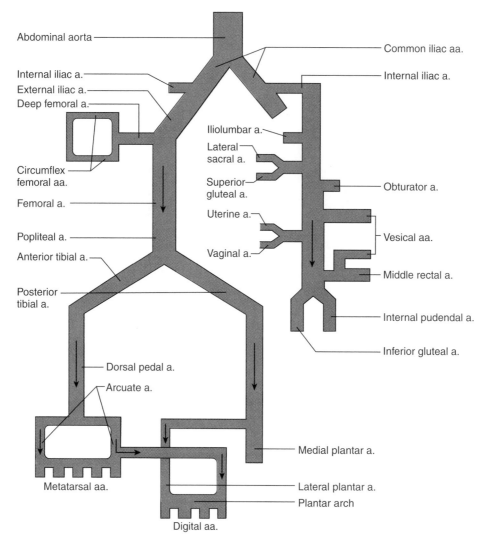

Abdominal aorta

Common iliac aa.

Internal iliac a.

Internal iliac a.

External iliac a.

Deep femoral a.

Iliolumbar a.

Lateral
sacral a.

Obturator a.

Circumflex
femoral aa.

Superior
gluteal a.

Femoral a.

Uterine a.

Vesical aa.

Popliteal a.

Vaginal a.

Middle rectal a.

Anterior tibial a.

Posterior
tibial a.

Internal pudendal a.

Inferior gluteal a.

Dorsal pedal a.

Arcuate a.

Medial plantar a.

Metatarsal aa.

Lateral plantar a.

Plantar arch

Digital aa.

Arterial Flowchart of the Lower Limb
Figure 21.24

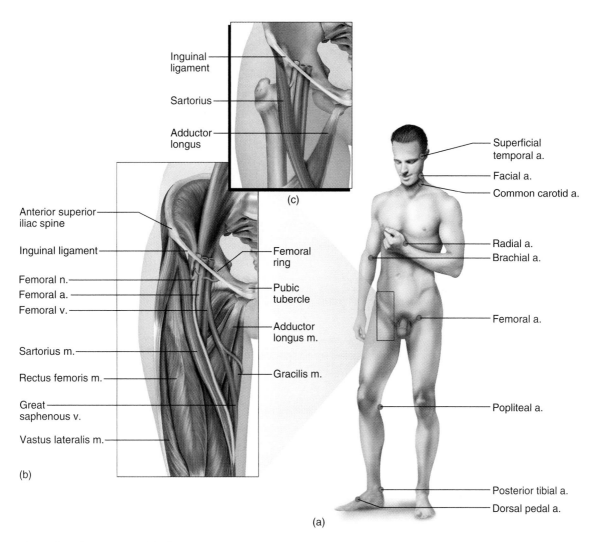

Inguinal ligament

Sartorius

Adductor longus

(c)

Anterior superior iliac spine

Inguinal ligament

Femoral n.

Femoral a.

Femoral v.

Sartorius m.

Rectus femoris m.

Great saphenous v.

Vastus lateralis m.

(b)

Femoral ring

Pubic tubercle

Adductor longus m.

Gracilis m.

Superficial temporal a.

Facial a.

Common carotid a.

Radial a.
Brachial a.

Femoral a.

Popliteal a.

Posterior tibial a.

Dorsal pedal a.

(a)

Arterial Pressure Points
Figure 21.25

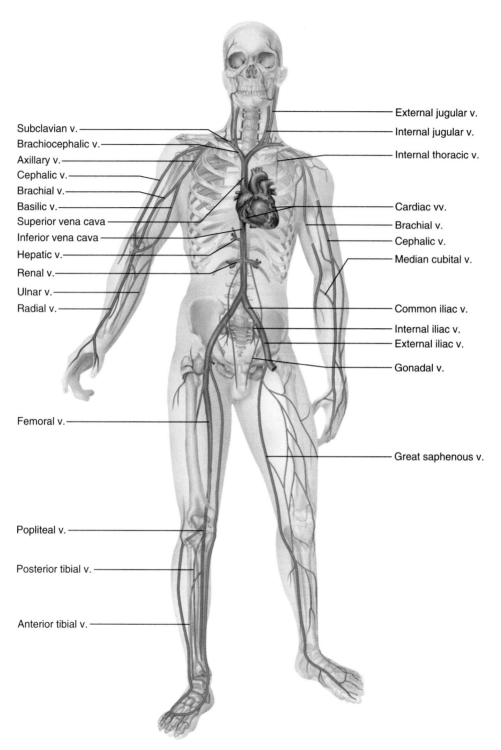

The Major Systematic Veins
Figure 21.26

External jugular v.
Internal jugular v.
Internal thoracic v.
Cardiac vv.
Brachial v.
Cephalic v.
Median cubital v.
Common iliac v.
Internal iliac v.
External iliac v.
Gonadal v.

Subclavian v.
Brachiocephalic v.
Axillary v.
Cephalic v.
Brachial v.
Basilic v.
Superior vena cava
Inferior vena cava
Hepatic v.
Renal v.
Ulnar v.
Radial v.
Femoral v.
Popliteal v.
Posterior tibial v.
Anterior tibial v.
Great saphenous v.

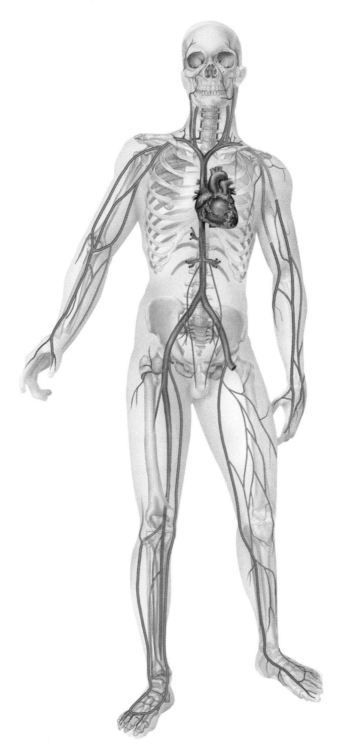

The Major Systematic Veins
Figure 21.26

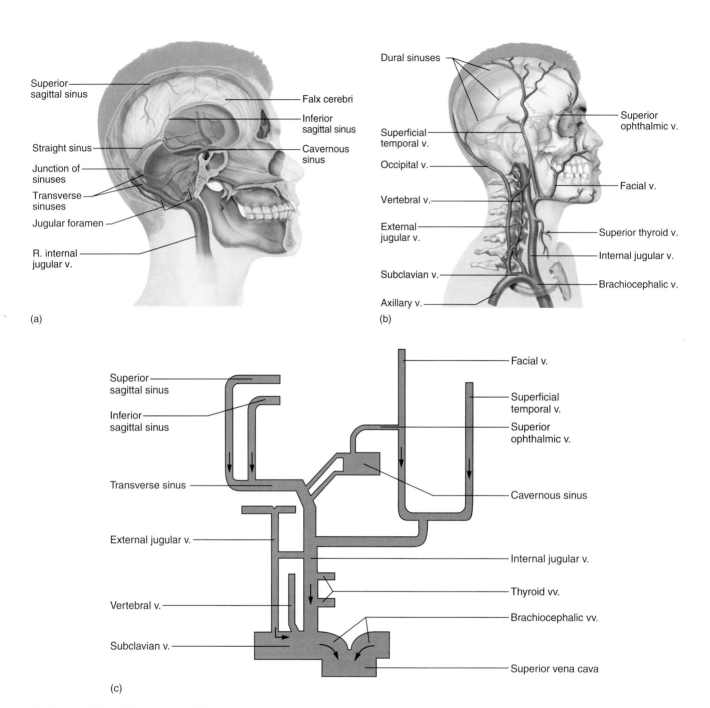

(a)

(b)

(c)

Veins of the Head and Neck

Figure 21.27

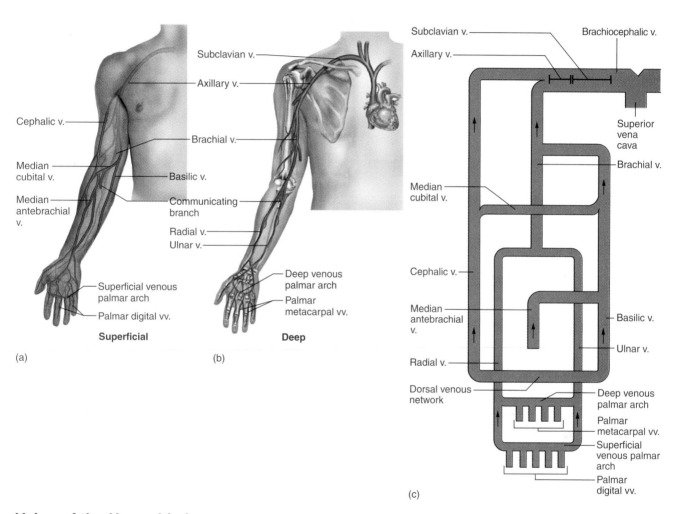

(a)

Superficial

(b)

Deep

(c)

Veins of the Upper Limb
Figure 21.28

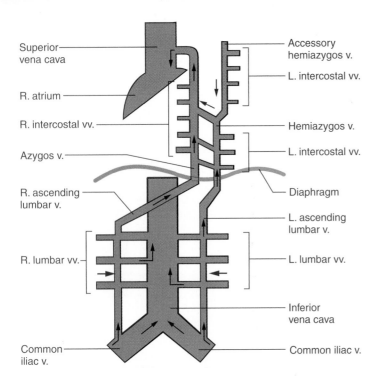

Flowchart of the Azygos System
Figure 21.29

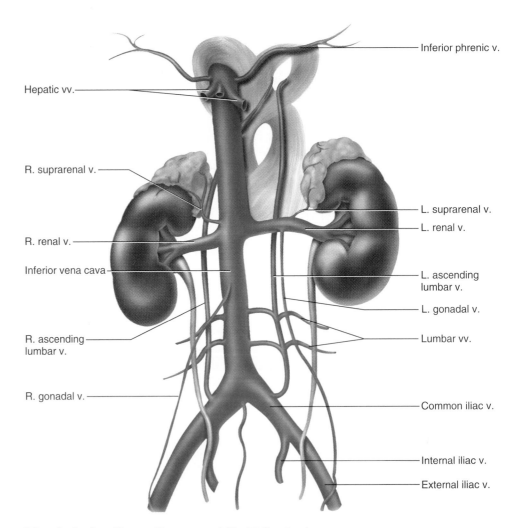

The Inferior Vena Cava and its Tributaries
Figure 21.30

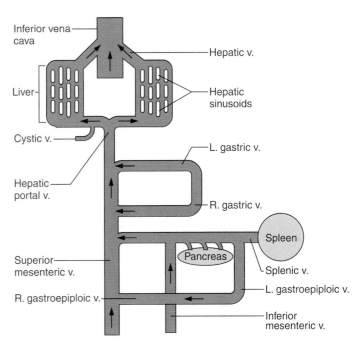

Flowchart of the Hepatic Portal System
Figure 21.31

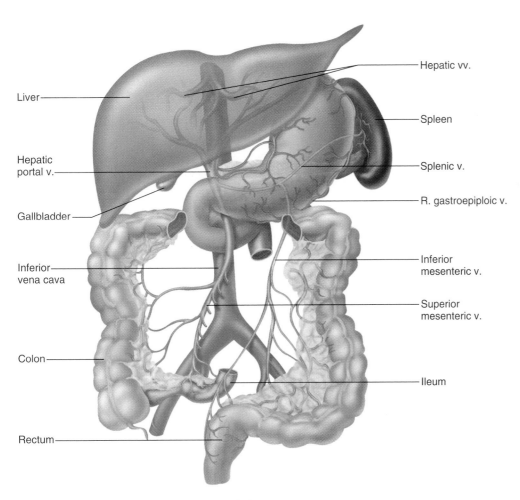

Anatomy of the Hepatic Portal System
Figure 21.32

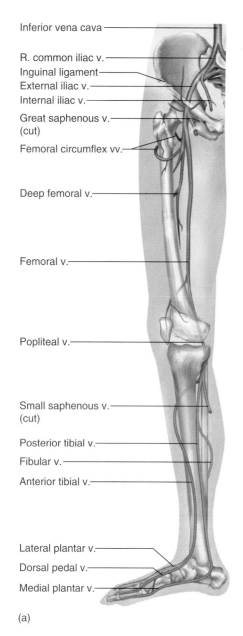

Inferior vena cava

R. common iliac v.

Inguinal ligament

External iliac v.

Internal iliac v.

Great saphenous v.
(cut)

Femoral circumflex vv.

Deep femoral v.

Femoral v.

Popliteal v.

Small saphenous v.
(cut)

Posterior tibial v.

Fibular v.

Anterior tibial v.

Lateral plantar v.

Dorsal pedal v.

Medial plantar v.

(a)

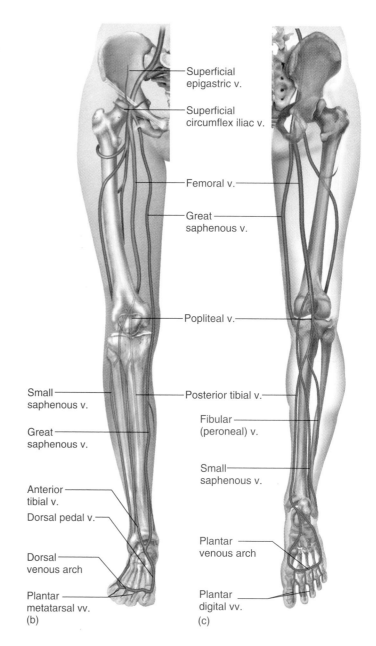

Superficial
epigastric v.

Superficial
circumflex iliac v.

Femoral v.

Great
saphenous v.

Popliteal v.

Small
saphenous v.

Posterior tibial v.

Great
saphenous v.

Fibular
(peroneal) v.

Small
saphenous v.

Anterior
tibial v.

Dorsal pedal v.

Plantar
venous arch

Dorsal
venous arch

Plantar
metatarsal vv.

Plantar
digital vv.

(b)

(c)

Veins of the Lower Limb
Figure 21.33

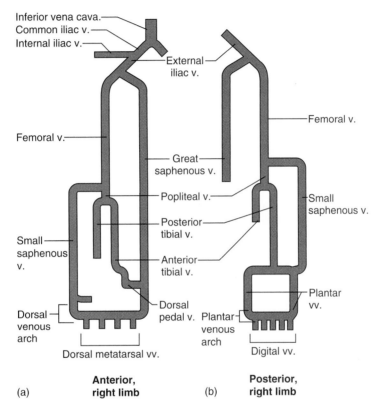

Inferior vena cava.
Common iliac v.
Internal iliac v.
External iliac v.
Femoral v.
Great saphenous v.
Popliteal v.
Posterior tibial v.
Small saphenous v.
Anterior tibial v.
Dorsal venous arch
Dorsal pedal v.
Dorsal metatarsal vv.
Plantar venous arch
Femoral v.
Small saphenous v.
Plantar vv.
Digital vv.

(a) **Anterior, right limb** (b) **Posterior, right limb**

Flowchart of the Lower Limb Veins
Figure 21.34

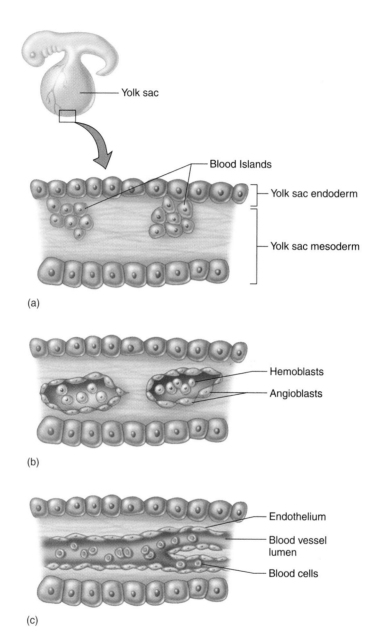

(a)

Yolk sac

Blood Islands

Yolk sac endoderm

Yolk sac mesoderm

(b)

Hemoblasts

Angioblasts

(c)

Endothelium

Blood vessel
lumen

Blood cells

**Development of Blood Vessels and Primitive
Erythrocytes from Embryonic Blood Islands**
Figure 21.35

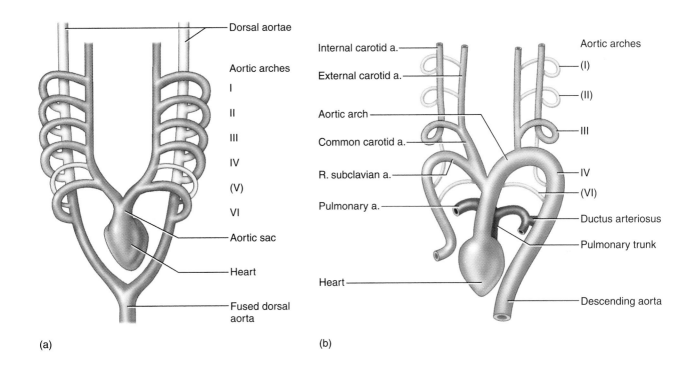

(a)

Labels (left diagram): Dorsal aortae, Aortic arches I, II, III, IV, (V), VI, Aortic sac, Heart, Fused dorsal aorta

(b)

Labels (right diagram): Internal carotid a., External carotid a., Aortic arch, Common carotid a., R. subclavian a., Pulmonary a., Heart, Aortic arches (I), (II), III, IV, (VI), Ductus arteriosus, Pulmonary trunk, Descending aorta

Development of Some Major Arteries from the Embryonic Aortic Arches
Figure 21.36

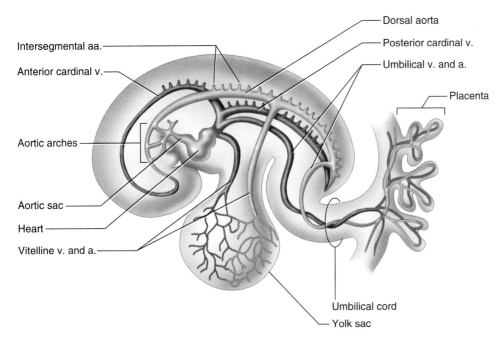

Labels: Intersegmental aa., Anterior cardinal v., Aortic arches, Aortic sac, Heart, Vitelline v. and a., Dorsal aorta, Posterior cardinal v., Umbilical v. and a., Placenta, Umbilical cord, Yolk sac

Major Embryonic Blood Vessels at 26 Days
Figure 21.37

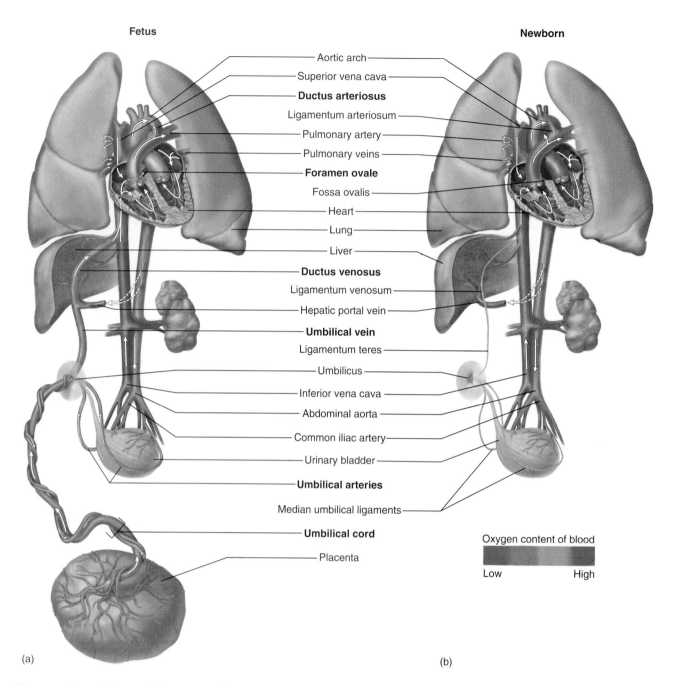

Fetus

Newborn

Aortic arch

Superior vena cava

Ductus arteriosus

Ligamentum arteriosum

Pulmonary artery

Pulmonary veins

Foramen ovale

Fossa ovalis

Heart

Lung

Liver

Ductus venosus

Ligamentum venosum

Hepatic portal vein

Umbilical vein

Ligamentum teres

Umbilicus

Inferior vena cava

Abdominal aorta

Common iliac artery

Urinary bladder

Umbilical arteries

Median umbilical ligaments

Umbilical cord

Placenta

Oxygen content of blood

Low   High

(a)

(b)

Some Circulatory Changes Occurring at Birth
Figure 21.38

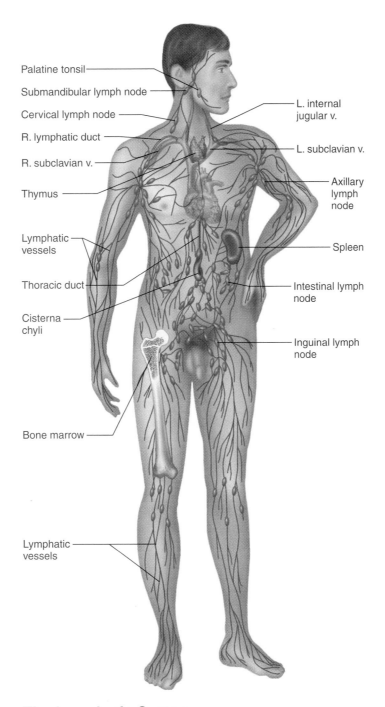

Palatine tonsil
Submandibular lymph node
Cervical lymph node
R. lymphatic duct
R. subclavian v.
Thymus

L. internal jugular v.
L. subclavian v.
Axillary lymph node

Lymphatic vessels
Thoracic duct
Cisterna chyli

Spleen
Intestinal lymph node
Inguinal lymph node

Bone marrow

Lymphatic vessels

The Lymphatic System
Figure 22.1

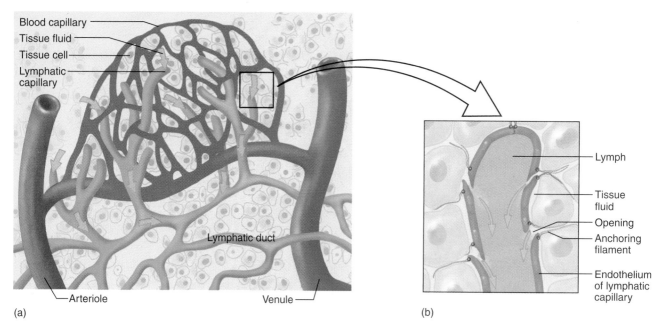

Blood capillary
Tissue fluid
Tissue cell
Lymphatic capillary

Lymphatic duct

Arteriole

Venule

(a)

Lymph
Tissue fluid
Opening
Anchoring filament
Endothelium of lymphatic capillary

(b)

Lymphatic Capillaries
Figure 22.3

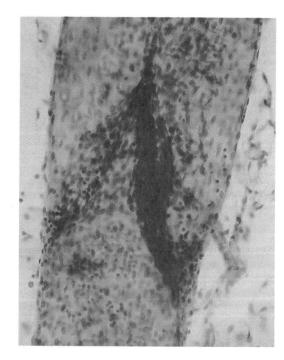

Valve in a Lymphatic Vessel
Figure 22.4

© The McGraw-Hill Companies, Inc./Dennis Strete, photographer

396

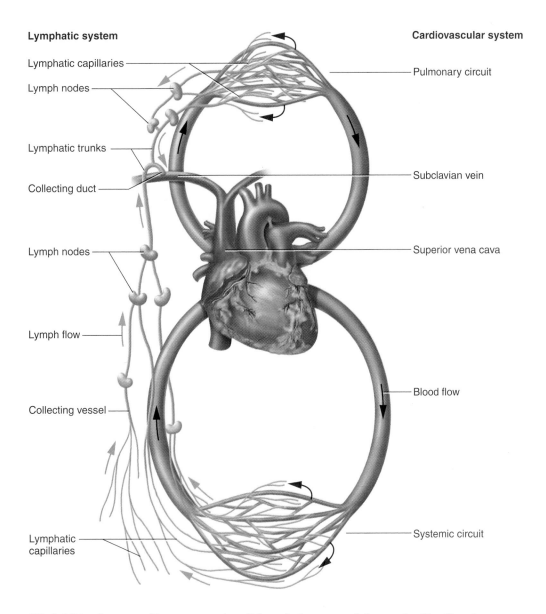

Lymphatic system

Lymphatic capillaries

Lymph nodes

Lymphatic trunks

Collecting duct

Lymph nodes

Lymph flow

Collecting vessel

Lymphatic
capillaries

Cardiovascular system

Pulmonary circuit

Subclavian vein

Superior vena cava

Blood flow

Systemic circuit

Fluid Exchange Between the Circulatory and Lymphatic Systems
Figure 22.5

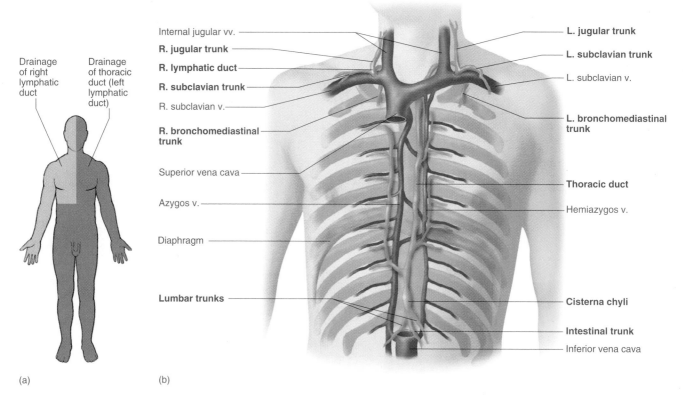

Internal jugular vv.
R. jugular trunk
R. lymphatic duct
R. subclavian trunk
R. subclavian v.
R. bronchomediastinal trunk
Superior vena cava
Azygos v.
Diaphragm
Lumbar trunks

L. jugular trunk
L. subclavian trunk
L. subclavian v.
L. bronchomediastinal trunk
Thoracic duct
Hemiazygos v.
Cisterna chyli
Intestinal trunk
Inferior vena cava

Drainage of right lymphatic duct

Drainage of thoracic duct (left lymphatic duct)

Lymphatics of the Thoracic Region
Figure 22.6

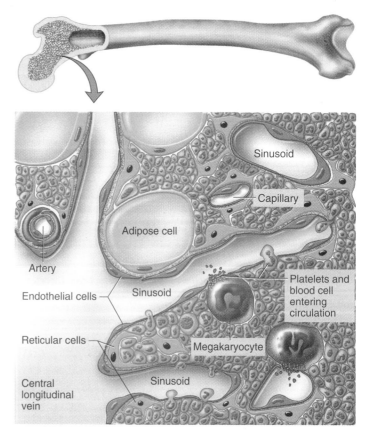

Sinusoid
Capillary
Adipose cell
Platelets and blood cell entering circulation
Artery
Endothelial cells
Sinusoid
Reticular cells
Megakaryocyte
Central longitudinal vein
Sinusoid

Histology of the Red Bone Marrow
Figure 22.9

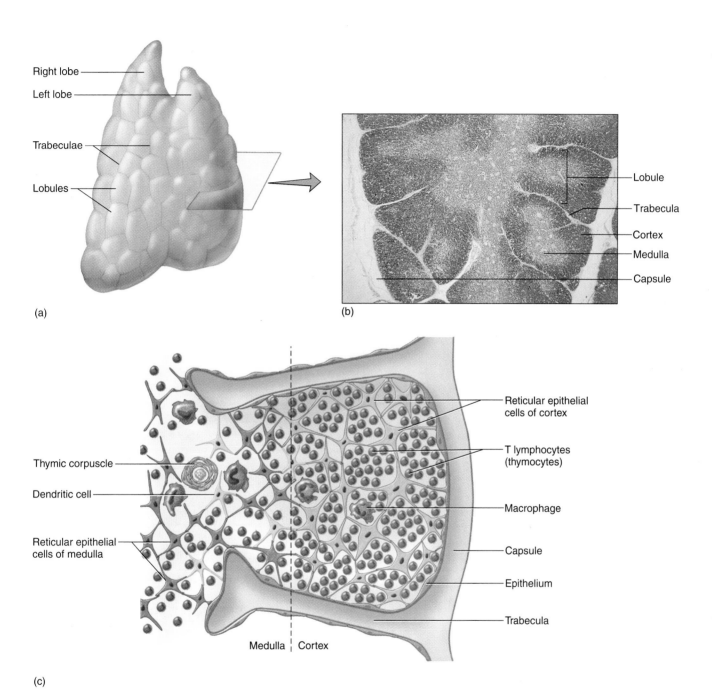

The Thymus

Figure 22.10

b: © The McGraw-Hill Companies, Inc./Dennis Strete, photographer

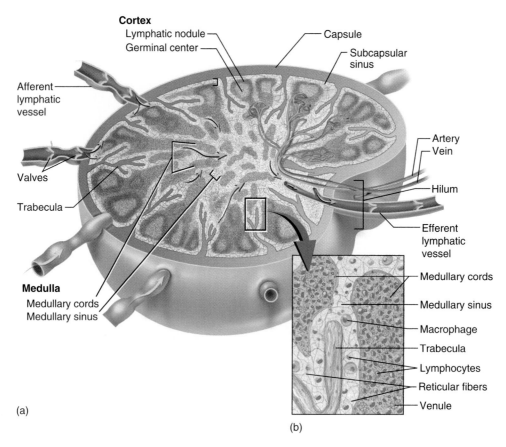

Cortex
Lymphatic nodule
Germinal center
Capsule
Subcapsular sinus
Afferent lymphatic vessel
Valves
Trabecula
Medulla
Medullary cords
Medullary sinus
Artery
Vein
Hilum
Efferent lymphatic vessel
Medullary cords
Medullary sinus
Macrophage
Trabecula
Lymphocytes
Reticular fibers
Venule

(a)

(b)

Anatomy of a Lymph Node
Figure 22.12

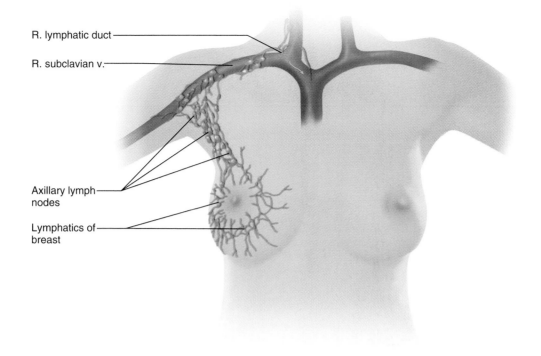

R. lymphatic duct

R. subclavian v.

Axillary lymph nodes

Lymphatics of breast

(a)

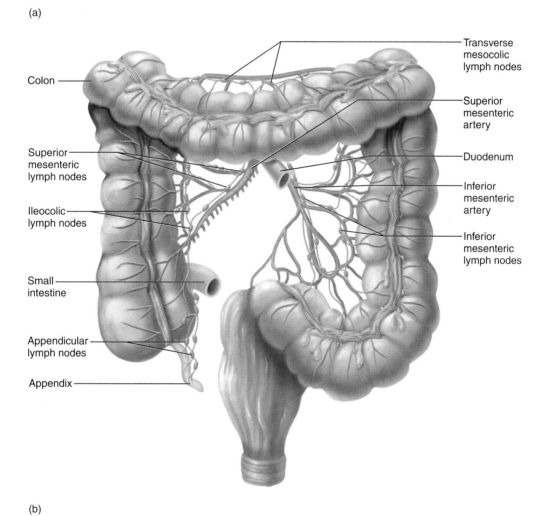

Colon

Transverse mesocolic lymph nodes

Superior mesenteric artery

Superior mesenteric lymph nodes

Ileocolic lymph nodes

Duodenum

Inferior mesenteric artery

Inferior mesenteric lymph nodes

Small intestine

Appendicular lymph nodes

Appendix

(b)

Some Areas of Lymph Node Concentration
Figure 22.13

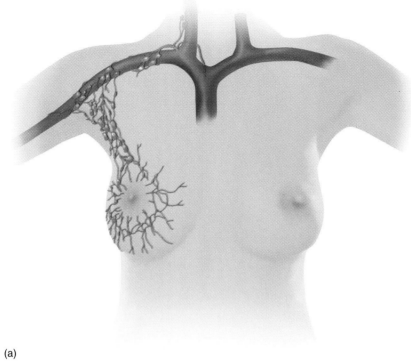

(a)

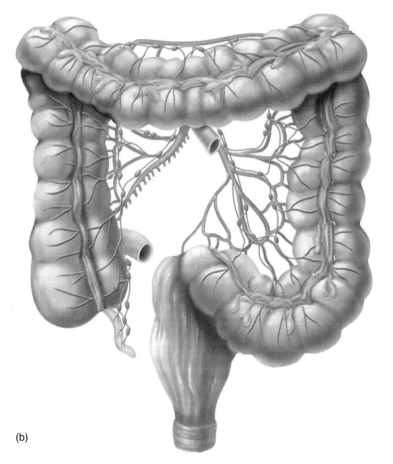

(b)

Some Areas of Lymph Node Concentration
Figure 22.13

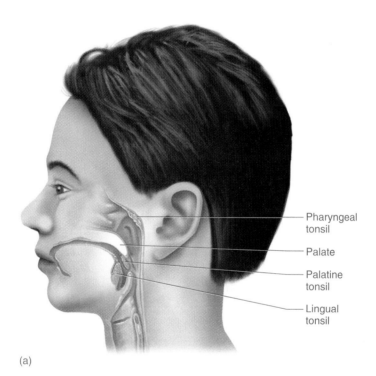

Pharyngeal
tonsil

Palate

Palatine
tonsil

Lingual
tonsil

(a)

The Tonsils
Figure 22.14

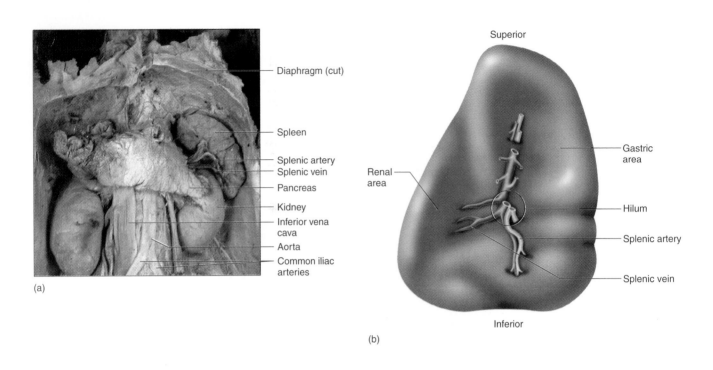

Diaphragm (cut)

Spleen

Splenic artery
Splenic vein

Pancreas

Kidney

Inferior vena
cava

Aorta

Common iliac
arteries

(a)

Superior

Renal
area

Gastric
area

Hilum

Splenic artery

Splenic vein

Inferior

(b)

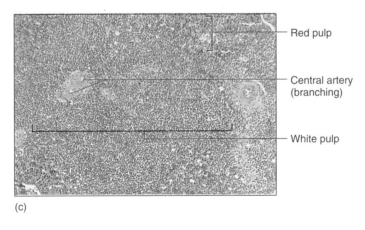

Red pulp

Central artery
(branching)

White pulp

(c)

The Spleen
Figure 22.15

a,c: © The McGraw-Hill Companies, Inc./Dennis Strete, photographer

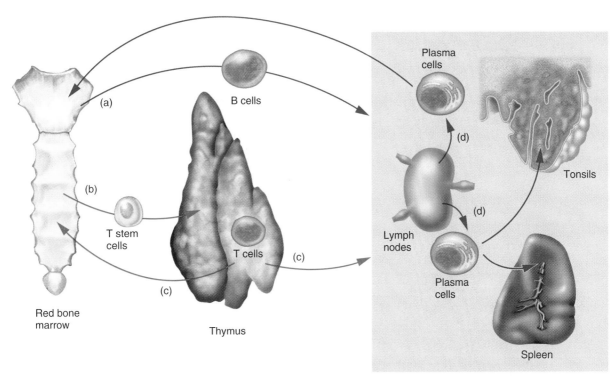

(a)

B cells

(b)

T stem
cells

T cells

(c)

(c)

Red bone
marrow

Thymus

Plasma
cells

(d)

Lymph
nodes

(d)

Plasma
cells

Tonsils

Spleen

Other lymphatic tissues
and organs

The Life History and Migrations of B and T Cells
Figure 22.16

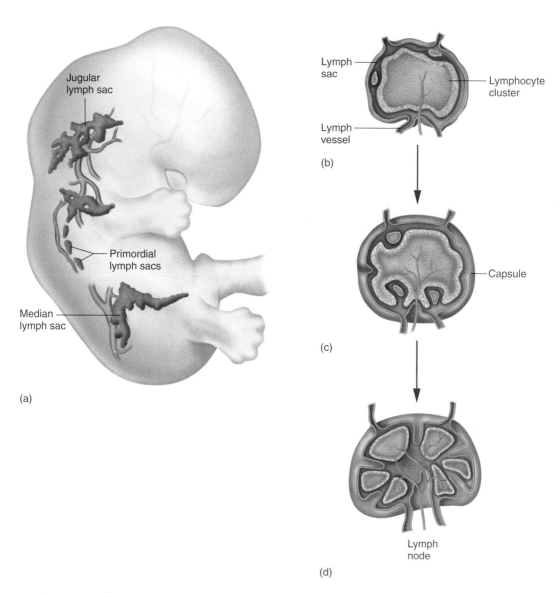

(a)

Jugular
lymph sac

Primordial
lymph sacs

Median
lymph sac

(b)

Lymph
sac

Lymph
vessel

Lymphocyte
cluster

(c)

Capsule

(d)

Lymph
node

Embryonic Development of the Lymphatic Vessels and Lymph Nodes
Figure 22.18

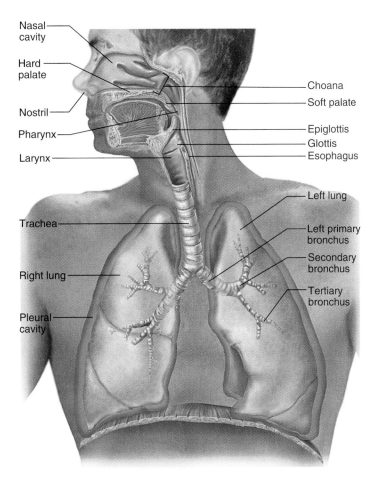

The Respiratory System
Figure 23.1

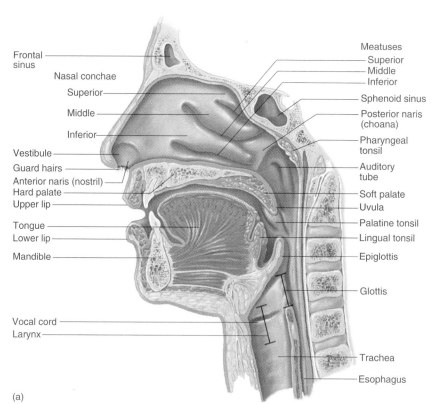

Frontal sinus

Nasal conchae
Superior
Middle
Inferior

Vestibule
Guard hairs
Anterior naris (nostril)
Hard palate
Upper lip

Tongue
Lower lip

Mandible

Vocal cord
Larynx

Meatuses
Superior
Middle
Inferior

Sphenoid sinus
Posterior naris (choana)

Pharyngeal tonsil

Auditory tube

Soft palate
Uvula
Palatine tonsil
Lingual tonsil
Epiglottis

Glottis

Trachea
Esophagus

(a)

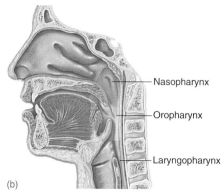

Nasopharynx

Oropharynx

Laryngopharynx

(b)

Anatomy of the Upper Limb Respiratory Tract
Figure 23.2

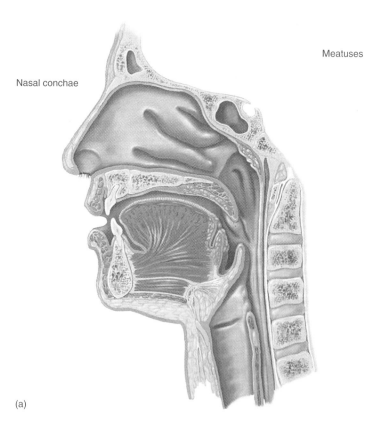

Nasal conchae

Meatuses

(a)

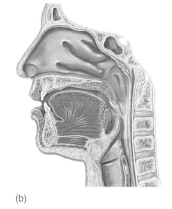

(b)

Anatomy of the Upper Limb Respiratory Tract
Figure 23.2

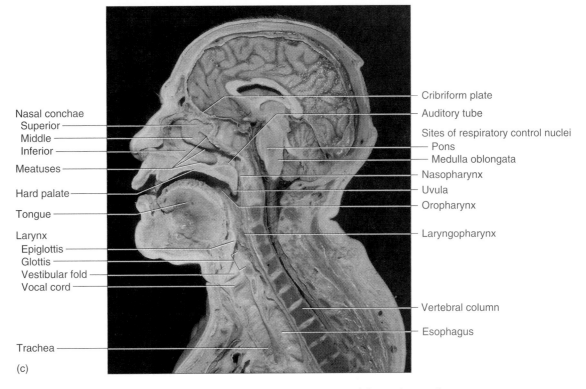

Anatomy of the Upper Limb Respiratory Tract (*Continued*)
Figure 23.2

c: © The McGraw-Hill Companies, Inc./Rebecca Gray, photographer/Don Kincaid, dissections

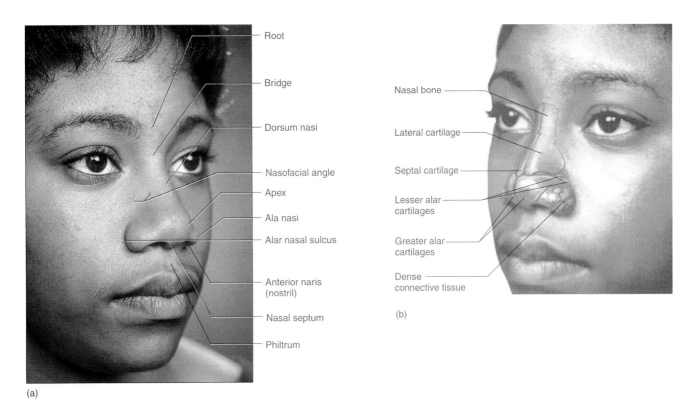

Anatomy of the Nasal Region
Figure 23.3

a,b: © The McGraw-Hill Companies, Inc./Joe DeGrandis, photographer

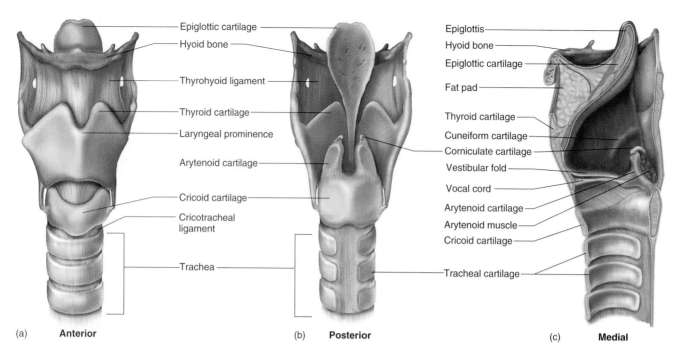

Anatomy of the Larynx
Figure 23.4

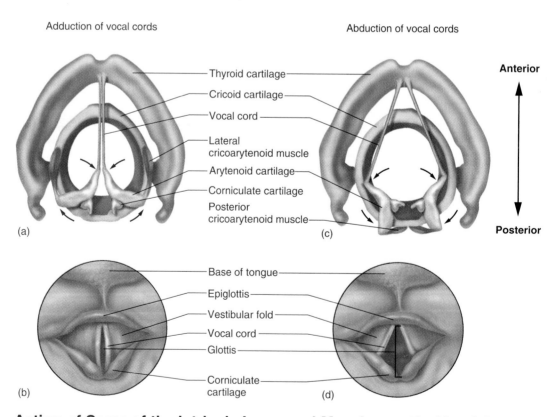

Action of Some of the Intrinsic Laryngeal Muscles on the Vocal Cords
Figure 23.6

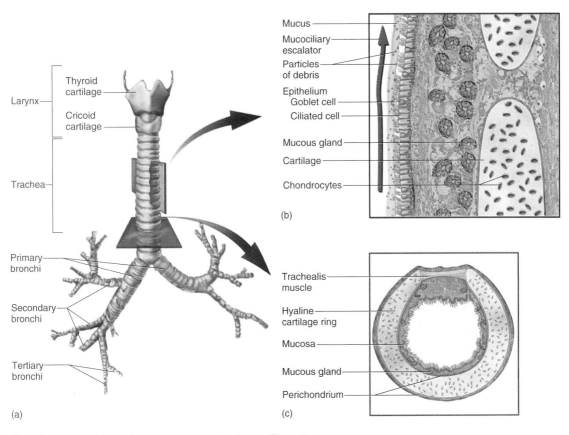

Labels for figure (a):

Larynx
— Thyroid cartilage
— Cricoid cartilage

Trachea

Primary bronchi

Secondary bronchi

Tertiary bronchi

(a)

Labels for figure (b):

Mucus
Mucociliary escalator
Particles of debris
Epithelium
  Goblet cell
  Ciliated cell
Mucous gland
Cartilage
Chondrocytes

(b)

Labels for figure (c):

Trachealis muscle
Hyaline cartilage ring
Mucosa
Mucous gland
Perichondrium

(c)

Anatomy of the Lower Respiratory Tract
Figure 23.7

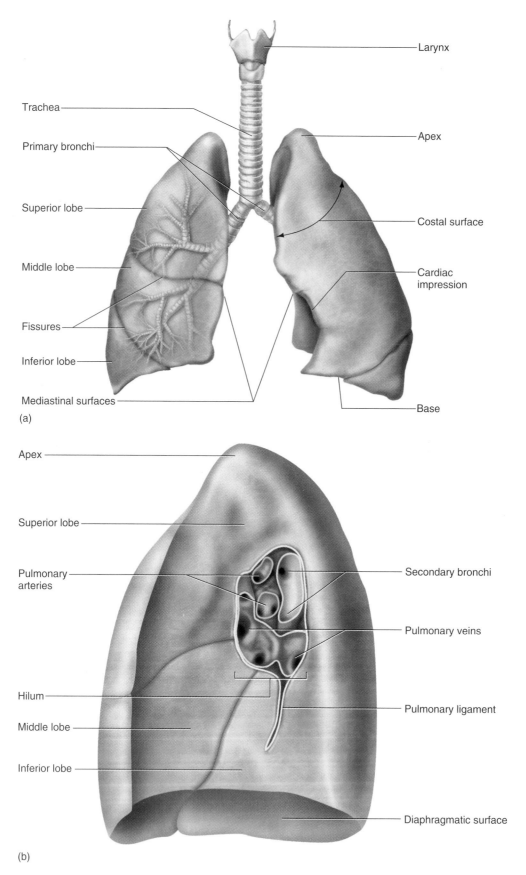

Gross Anatomy of the Lungs
Figure 23.9

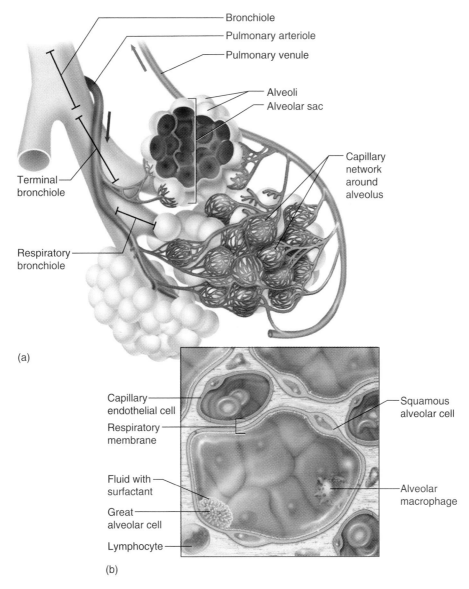

(a)

(b)

Bronchiole
Pulmonary arteriole
Pulmonary venule
Alveoli
Alveolar sac
Capillary network around alveolus
Terminal bronchiole
Respiratory bronchiole

Capillary endothelial cell
Respiratory membrane
Squamous alveolar cell
Fluid with surfactant
Great alveolar cell
Lymphocyte
Alveolar macrophage

Pulmonary Alveoli
Figure 23.11

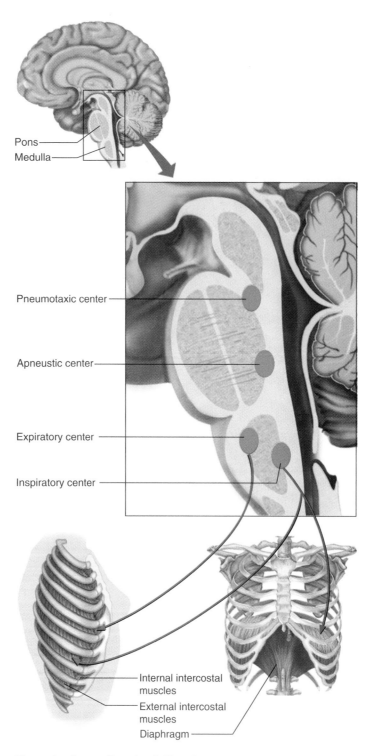

Pons

Medulla

Pneumotaxic center

Apneustic center

Expiratory center

Inspiratory center

Internal intercostal muscles

External intercostal muscles

Diaphragm

Respiratory Control Centers
Figure 23.13

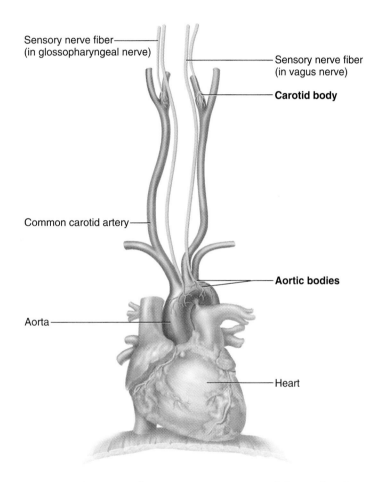

Sensory nerve fiber
(in glossopharyngeal nerve)

Sensory nerve fiber
(in vagus nerve)

Carotid body

Common carotid artery

Aortic bodies

Aorta

Heart

The Peripheral Chemoreceptors of Respiration
Figure 23.14

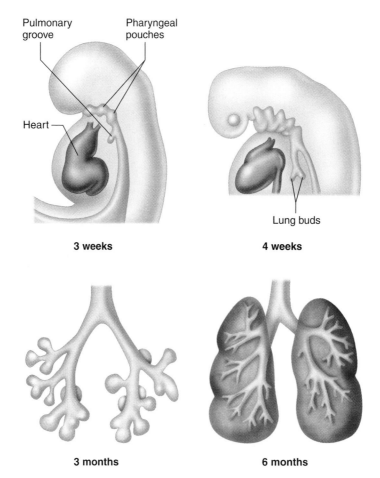

Pulmonary groove

Pharyngeal pouches

Heart

Lung buds

3 weeks

4 weeks

3 months

6 months

Embryonic Development of the Respiratory System
Figure 23.15

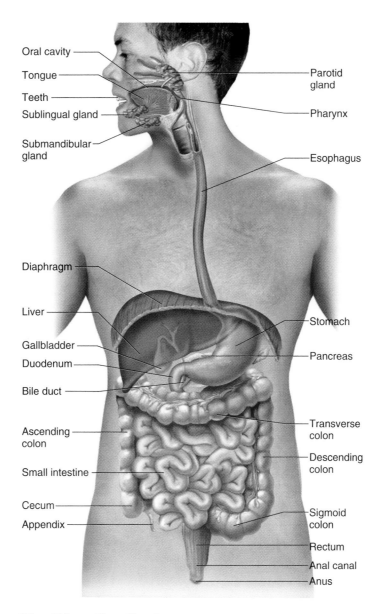

Oral cavity

Tongue

Teeth

Sublingual gland

Submandibular gland

Parotid gland

Pharynx

Esophagus

Diaphragm

Liver

Gallbladder

Duodenum

Bile duct

Stomach

Pancreas

Ascending colon

Small intestine

Cecum

Appendix

Transverse colon

Descending colon

Sigmoid colon

Rectum

Anal canal

Anus

The Digestive System
Figure 24.1

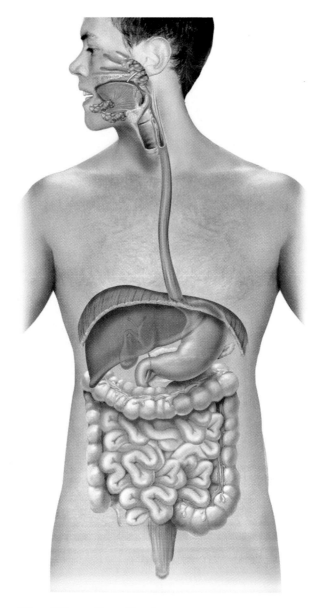

The Digestive System
Figure 24.1

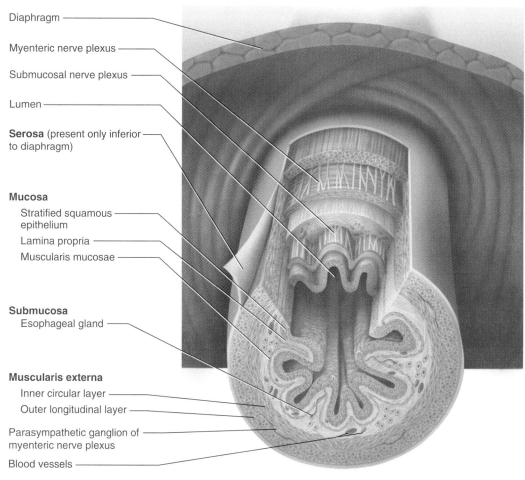

Diaphragm

Myenteric nerve plexus

Submucosal nerve plexus

Lumen

Serosa (present only inferior to diaphragm)

Mucosa
  Stratified squamous epithelium
  Lamina propria
  Muscularis mucosae

Submucosa
  Esophageal gland

Muscularis externa
  Inner circular layer
  Outer longitudinal layer

Parasympathetic ganglion of myenteric nerve plexus

Blood vessels

Tissue Layers of the Digestive Tract
Figure 24.2

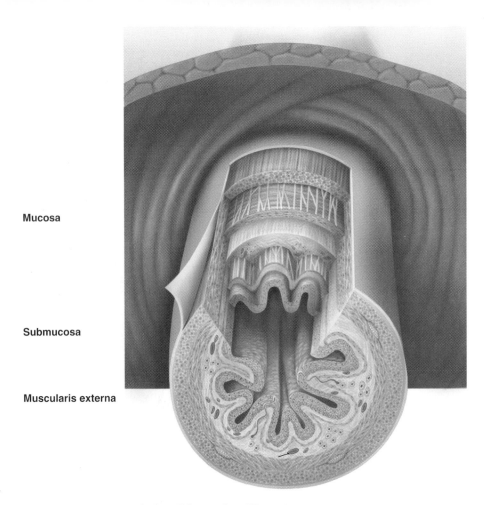

Mucosa

Submucosa

Muscularis externa

Tissue Layers of the Digestive Tract
Figure 24.2

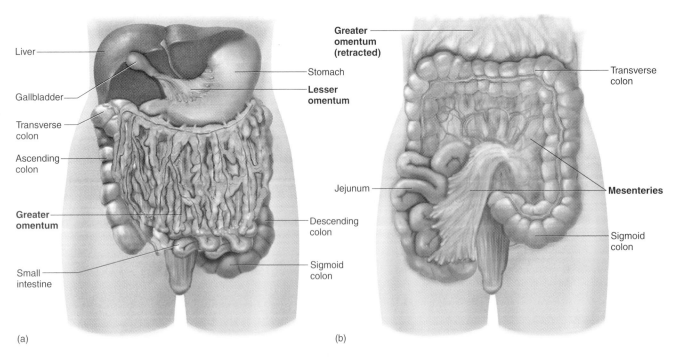

Liver

Gallbladder

Transverse
colon

Ascending
colon

**Greater
omentum**

Small
intestine

(a)

Stomach

**Lesser
omentum**

Descending
colon

Sigmoid
colon

**Greater
omentum
(retracted)**

Transverse
colon

Mesenteries

Jejunum

Sigmoid
colon

(b)

Serous Membranes Associated with the Digestive Tract
Figure 24.3

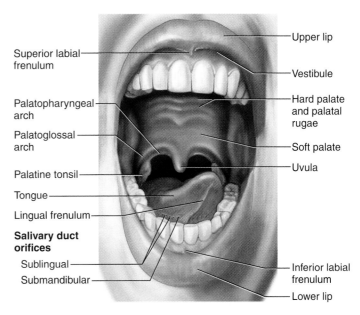

Upper lip
Superior labial frenulum
Vestibule
Palatopharyngeal arch
Hard palate and palatal rugae
Palatoglossal arch
Soft palate
Palatine tonsil
Uvula
Tongue
Lingual frenulum
Salivary duct orifices
Sublingual
Submandibular
Inferior labial frenulum
Lower lip

The Oral Cavity
Figure 24.4

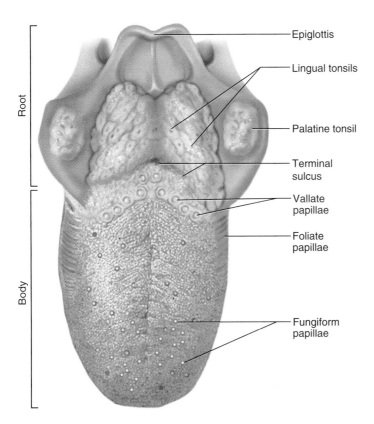

Root
Body

Epiglottis
Lingual tonsils
Palatine tonsil
Terminal sulcus
Vallate papillae
Foliate papillae
Fungiform papillae

The Tongue
Figure 24.5

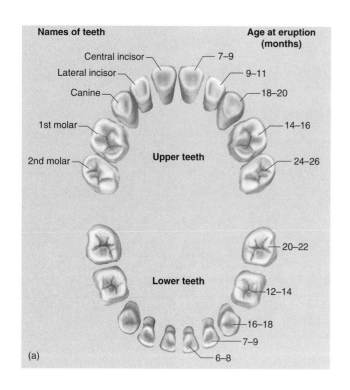

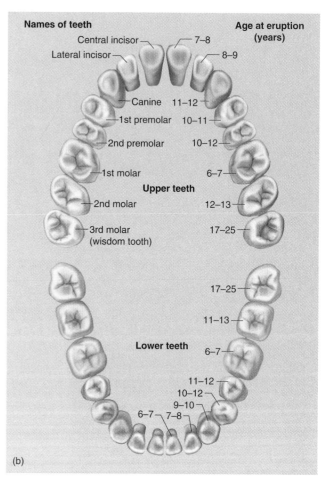

The Dentition and Ages at Which the Teeth Erupt
Figure 24.6

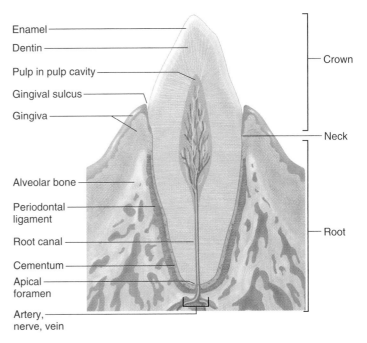

Enamel
Dentin
Pulp in pulp cavity
Gingival sulcus
Gingiva

Alveolar bone
Periodontal ligament
Root canal
Cementum
Apical foramen
Artery, nerve, vein

Crown
Neck
Root

Median Section of a Canine Tooth and Its Alveolus
Figure 24.7

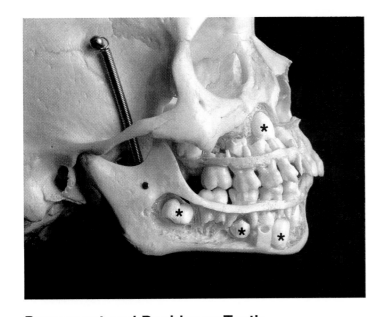

Permanent and Deciduous Teeth in a Child's Skull
Figure 24.8

© The McGraw-Hill Companies, Inc./Rebecca Gray, photographer/Don Kincaid, dissections

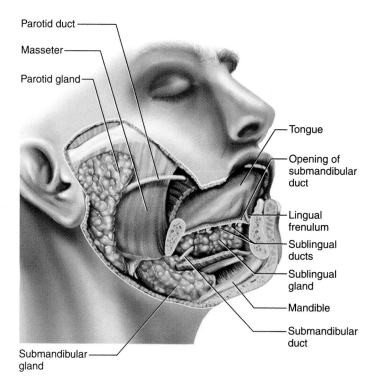

Parotid duct

Masseter

Parotid gland

Tongue

Opening of
submandibular
duct

Lingual
frenulum

Sublingual
ducts

Sublingual
gland

Mandible

Submandibular
duct

Submandibular
gland

The Extrinsic Salivary Glands
Figure 24.9

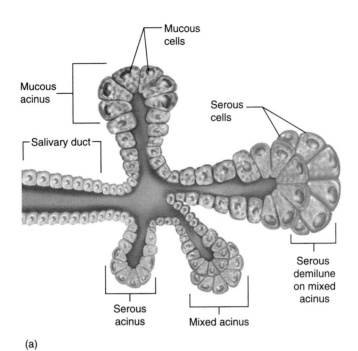

(a)

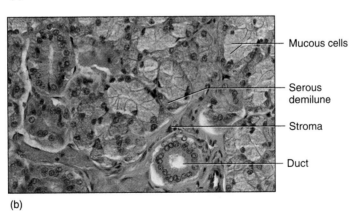

Mucous cells

Serous demilune

Stroma

Duct

(b)

Anatomy of the Salivary Glands
Figure 24.10

b: © The McGraw-Hill Companies, Inc./Dennis Strete, photographer

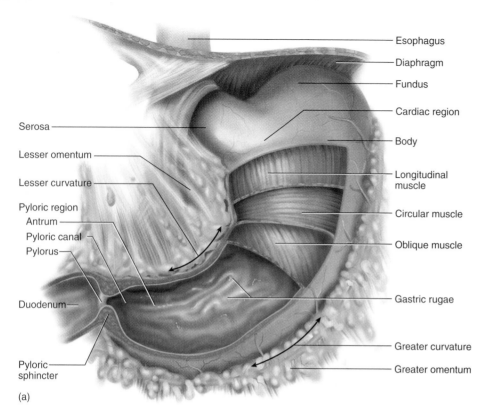

Esophagus
Diaphragm
Fundus
Cardiac region
Body
Longitudinal muscle
Circular muscle
Oblique muscle

Serosa
Lesser omentum
Lesser curvature
Pyloric region
Antrum
Pyloric canal
Pylorus
Duodenum
Pyloric sphincter

Gastric rugae

Greater curvature
Greater omentum

(a)

The Stomach
Figure 24.11

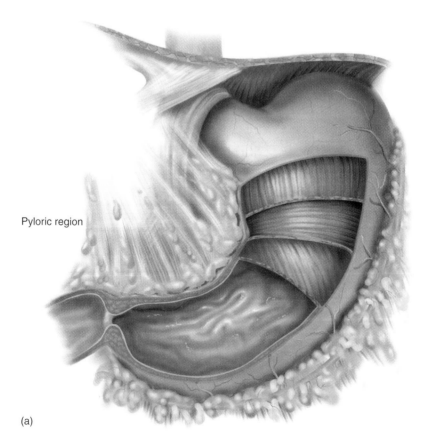

Pyloric region

(a)

The Stomach
Figure 24.11

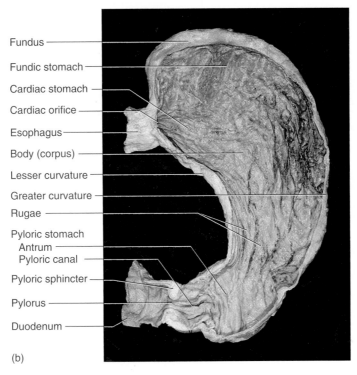

Fundus

Fundic stomach

Cardiac stomach

Cardiac orifice

Esophagus

Body (corpus)

Lesser curvature

Greater curvature

Rugae

Pyloric stomach
  Antrum
  Pyloric canal

Pyloric sphincter

Pylorus

Duodenum

(b)

The Stomach (*Continued*)
Figure 24.11

b: © The McGraw-Hill Companies, Inc./Rebecca Gray,
photographer/Don Kincaid, dissections

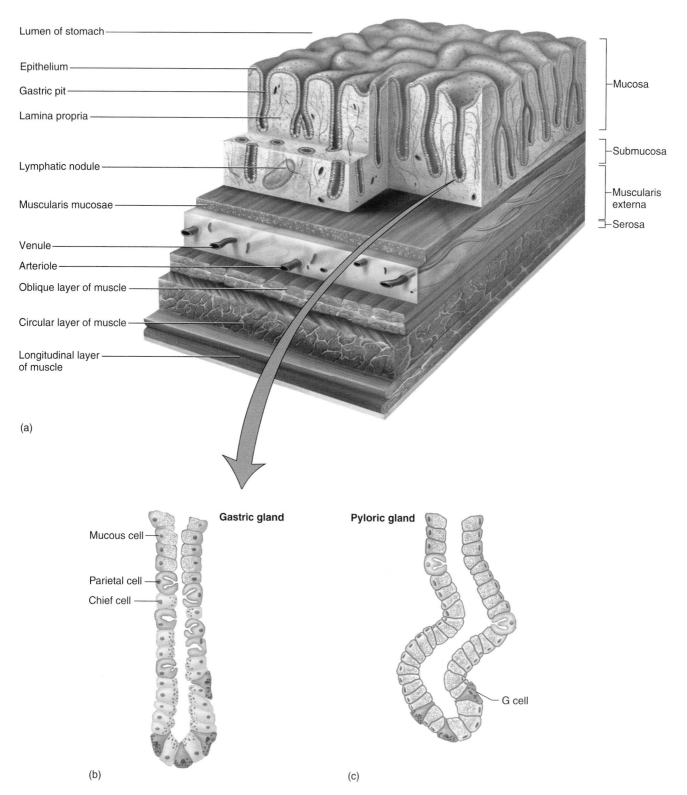

Lumen of stomach

Epithelium

Gastric pit

Lamina propria

Lymphatic nodule

Muscularis mucosae

Venule

Arteriole

Oblique layer of muscle

Circular layer of muscle

Longitudinal layer
of muscle

Mucosa

Submucosa

Muscularis
externa

Serosa

(a)

Gastric gland

Mucous cell

Parietal cell

Chief cell

Pyloric gland

G cell

(b)

(c)

Microscopic Anatomy of the Stomach Wall
Figure 24.12

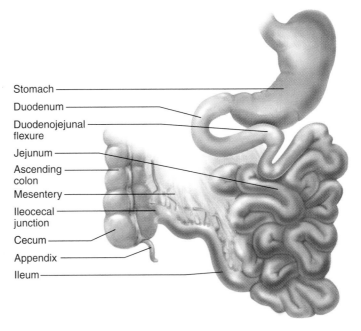

Stomach
Duodenum
Duodenojejunal
flexure
Jejunum
Ascending
colon
Mesentery
Ileocecal
junction
Cecum
Appendix
Ileum

Gross Anatomy of the Small Intestine
Figure 24.14

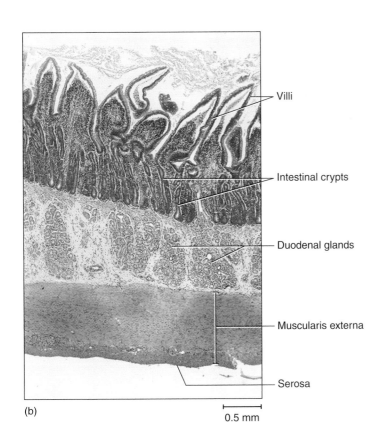

Villi

Intestinal crypts

Duodenal glands

Muscularis externa

Serosa

(b)

0.5 mm

Intestinal Villi
Figure 24.15

b: © The McGraw-Hill Companies, Inc./Dennis Strete,
photographer

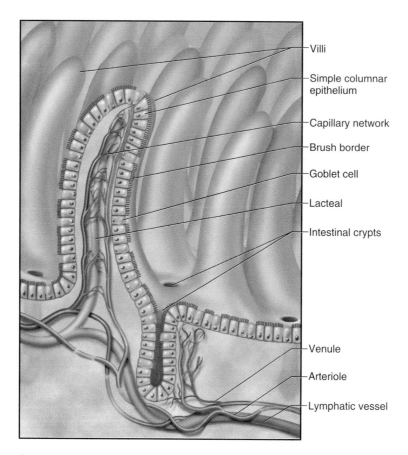

Structure of a Villus
Figure 24.16

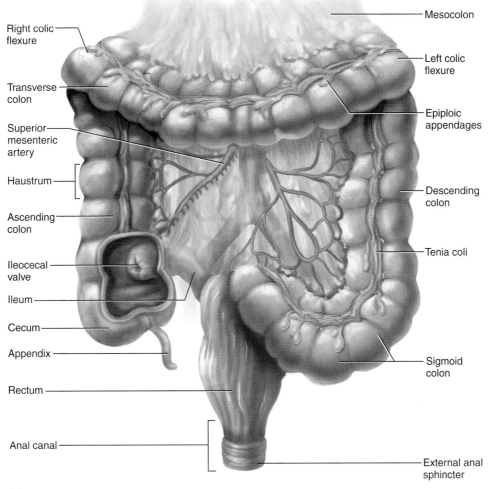

Right colic flexure

Transverse colon

Superior mesenteric artery

Haustrum

Ascending colon

Ileocecal valve

Ileum

Cecum

Appendix

Rectum

Anal canal

Mesocolon

Left colic flexure

Epiploic appendages

Descending colon

Tenia coli

Sigmoid colon

External anal sphincter

(a)

The Large Intestine
Figure 24.17

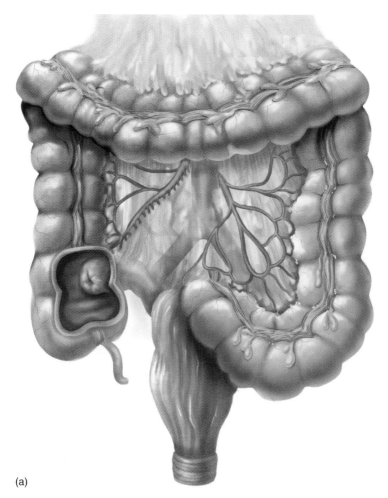

(a)

The Large Intestine
Figure 24.17

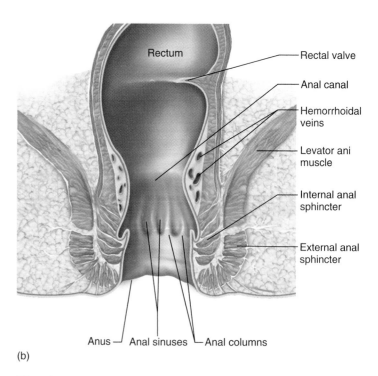

(b)

The Large Intestine
Figure 24.17

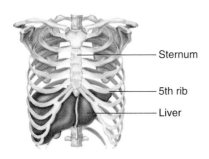

Sternum

5th rib

Liver

(a)

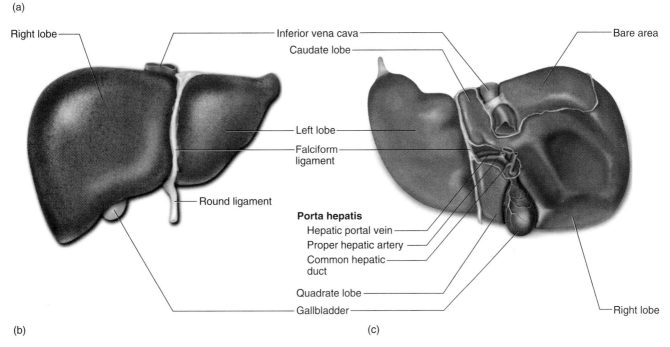

Right lobe

Inferior vena cava

Caudate lobe

Bare area

Left lobe

Falciform ligament

Round ligament

Porta hepatis

Hepatic portal vein

Proper hepatic artery

Common hepatic duct

Quadrate lobe

Gallbladder

Right lobe

(b)

(c)

Gross Anatomy of the Liver
Figure 24.18

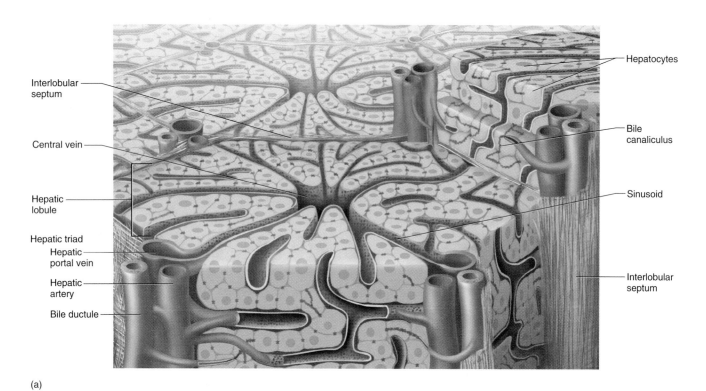

Interlobular septum

Central vein

Hepatic lobule

Hepatic triad
Hepatic portal vein
Hepatic artery
Bile ductule

Hepatocytes

Bile canaliculus

Sinusoid

Interlobular septum

(a)

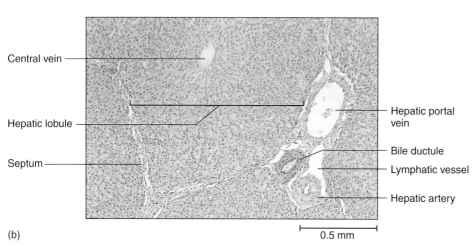

Central vein

Hepatic lobule

Septum

Hepatic portal vein

Bile ductule
Lymphatic vessel

Hepatic artery

0.5 mm

(b)

Microscopic Anatomy of the Liver
Figure 24.19

b: © The McGraw-Hill Companies, Inc./Dennis Strete, photographer

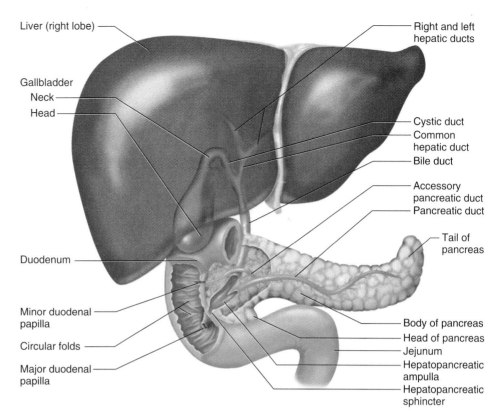

Liver (right lobe)

Gallbladder
Neck
Head

Duodenum

Minor duodenal
papilla

Circular folds

Major duodenal
papilla

Right and left
hepatic ducts

Cystic duct
Common
hepatic duct
Bile duct

Accessory
pancreatic duct
Pancreatic duct

Tail of
pancreas

Body of pancreas
Head of pancreas
Jejunum
Hepatopancreatic
ampulla
Hepatopancreatic
sphincter

**Gross Anatomy of the Gallbladder, Pancreas,
and Bile Passages**
Figure 24.20

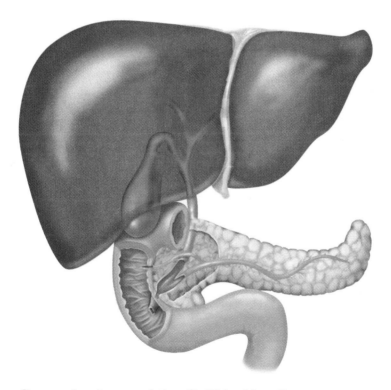

**Gross Anatomy of the Gallbladder, Pancreas,
and Bile Passages**
Figure 24.20

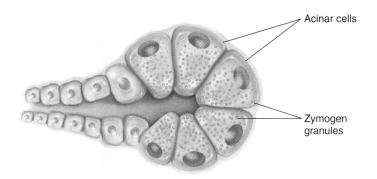

(a)

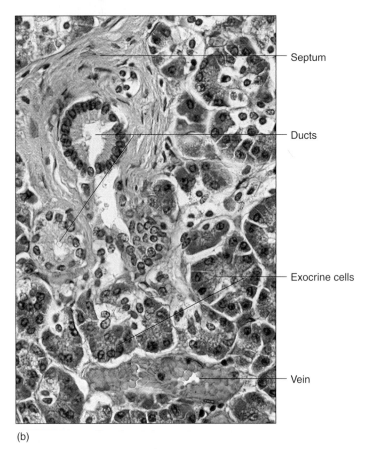

(b)

Histology of the Pancreas
Figure 24.21

b: © The McGraw-Hill Companies, Inc./Dennis Strete, photographer

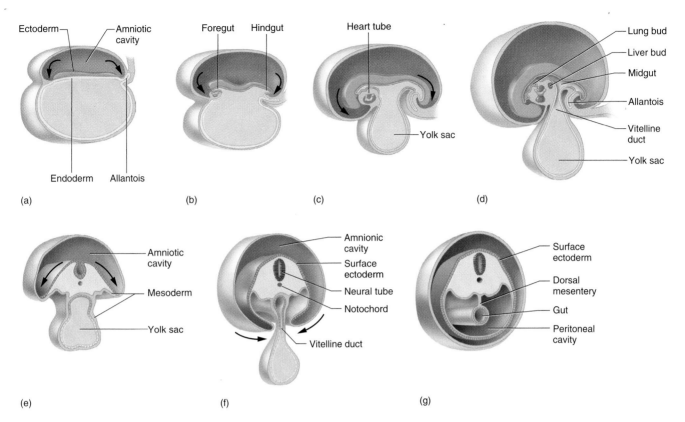

(a)
Ectoderm — Amniotic cavity
Endoderm — Allantois

(b)
Foregut — Hindgut

(c)
Heart tube
Yolk sac

(d)
Lung bud
Liver bud
Midgut
Allantois
Vitelline duct
Yolk sac

(e)
Amniotic cavity
Mesoderm
Yolk sac

(f)
Amnionic cavity
Surface ectoderm
Neural tube
Notochord
Vitelline duct

(g)
Surface ectoderm
Dorsal mesentery
Gut
Peritoneal cavity

Embryonic Development of the Digestive Tract
Figure 24.22

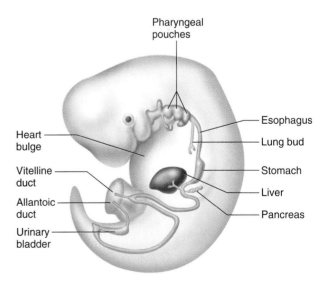

Pharyngeal pouches
Heart bulge
Vitelline duct
Allantoic duct
Urinary bladder
Esophagus
Lung bud
Stomach
Liver
Pancreas

Lateral View of the 5-week Embryo
Figure 24.23

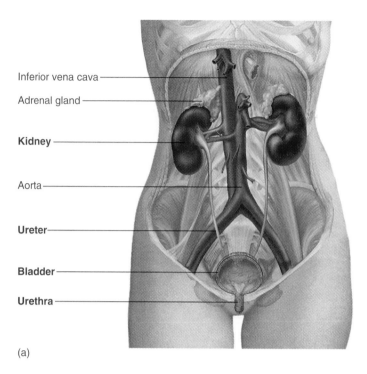

(a)

Inferior vena cava
Adrenal gland
Kidney
Aorta
Ureter
Bladder
Urethra

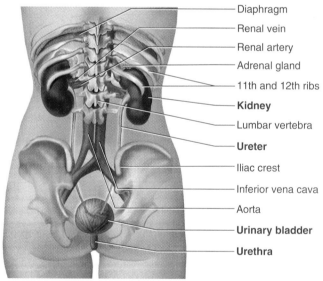

(b)

Diaphragm
Renal vein
Renal artery
Adrenal gland
11th and 12th ribs
Kidney
Lumbar vertebra
Ureter
Iliac crest
Inferior vena cava
Aorta
Urinary bladder
Urethra

The Urinary System
Figure 25.1

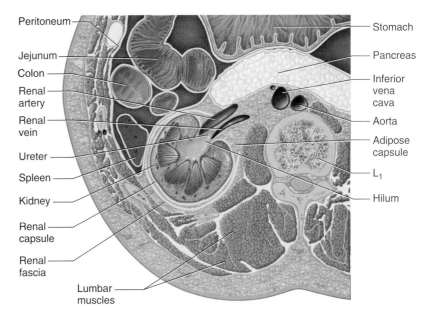

Peritoneum
Jejunum
Colon
Renal artery
Renal vein
Ureter
Spleen
Kidney
Renal capsule
Renal fascia
Lumbar muscles

Stomach
Pancreas
Inferior vena cava
Aorta
Adipose capsule
L_1
Hilum

Location of the Kidney
Figure 25.2

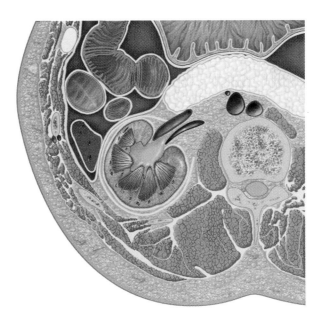

Location of the Kidney
Figure 25.2

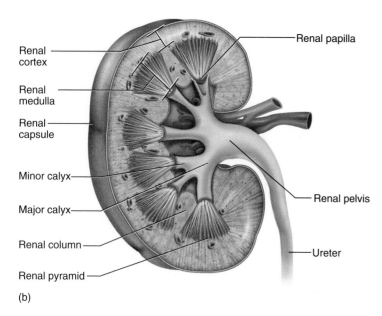

(b)

Gross Anatomy of the Kidney
Figure 25.3

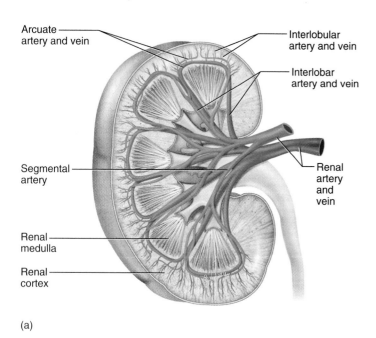

(a)

Renal Circulation
Figure 25.4

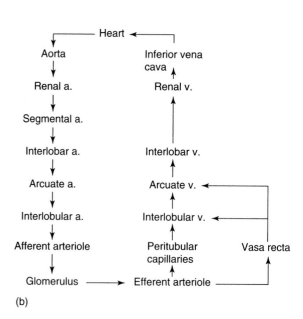

(b)

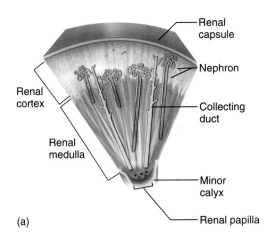

(a)

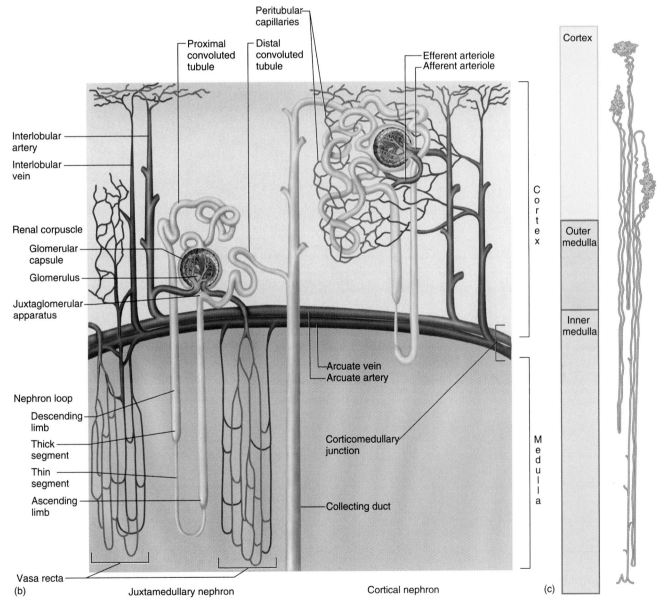

(b)

(c)

Structure of the Nephron
Figure 25.5

(a)

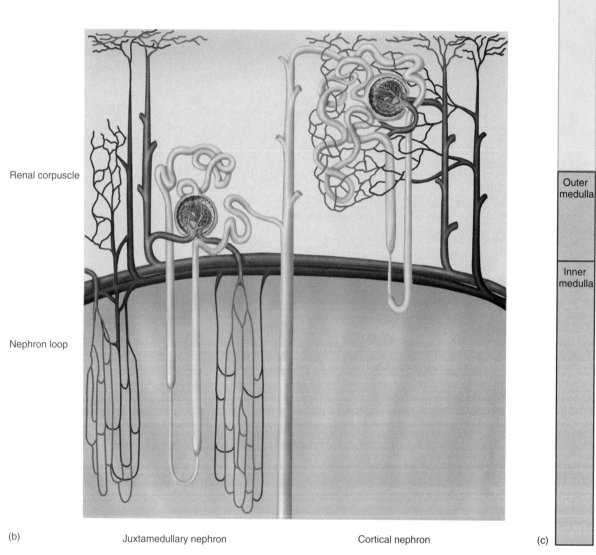

Renal corpuscle

Nephron loop

(b)

Juxtamedullary nephron   Cortical nephron   (c)

Cortex

Outer medulla

Inner medulla

Structure of the Nephron
Figure 25.5

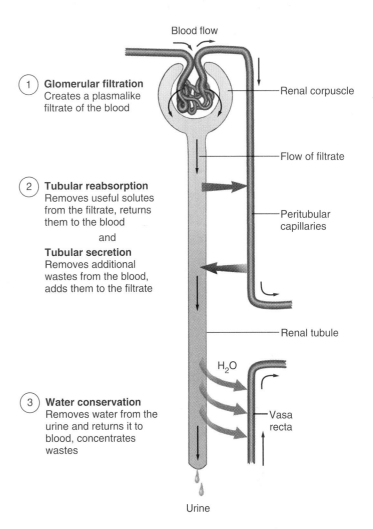

1. **Glomerular filtration**
Creates a plasmalike filtrate of the blood

2. **Tubular reabsorption**
Removes useful solutes from the filtrate, returns them to the blood

and

Tubular secretion
Removes additional wastes from the blood, adds them to the filtrate

3. **Water conservation**
Removes water from the urine and returns it to blood, concentrates wastes

Blood flow

Renal corpuscle

Flow of filtrate

Peritubular capillaries

Renal tubule

H_2O

Vasa recta

Urine

Basic Steps in the Formation of Urine
Figure 25.6

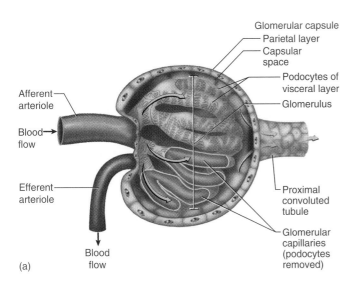

Glomerular capsule
Parietal layer
Capsular space

Podocytes of visceral layer

Glomerulus

Proximal convoluted tubule

Glomerular capillaries (podocytes removed)

Afferent arteriole

Blood flow

Efferent arteriole

Blood flow

(a)

The Renal Corpuscle
Figure 25.7

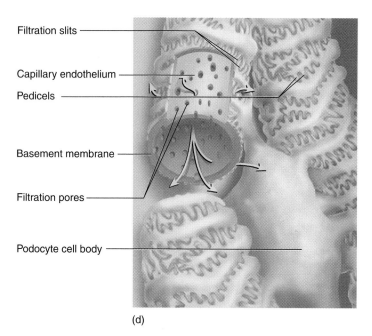

Filtration slits

Capillary endothelium

Pedicels

Basement membrane

Filtration pores

Podocyte cell body

(d)

Structure of the Glomerulus
Figure 25.8

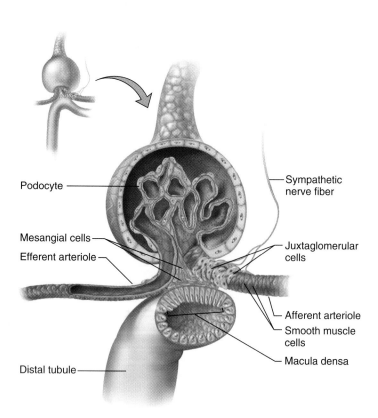

Podocyte

Mesangial cells

Efferent arteriole

Distal tubule

Sympathetic nerve fiber

Juxtaglomerular cells

Afferent arteriole

Smooth muscle cells

Macula densa

The Juxtaglomerular Apparatus
Figure 25.9

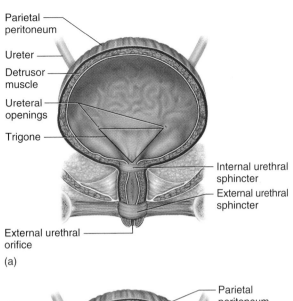

Parietal peritoneum

Ureter

Detrusor muscle

Ureteral openings

Trigone

Internal urethral sphincter

External urethral sphincter

External urethral orifice

(a)

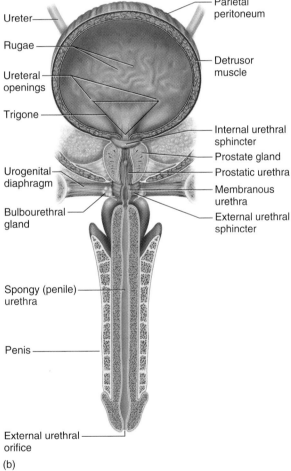

Ureter

Rugae

Ureteral openings

Trigone

Parietal peritoneum

Detrusor muscle

Internal urethral sphincter

Prostate gland

Prostatic urethra

Urogenital diaphragm

Membranous urethra

Bulbourethral gland

External urethral sphincter

Spongy (penile) urethra

Penis

External urethral orifice

(b)

Anatomy of the Urinary Bladder and Urethra
Figure 25.10

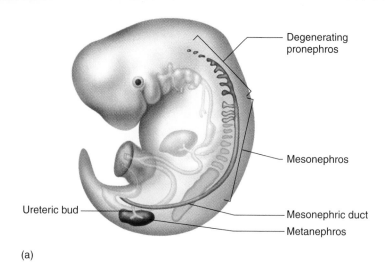

(a)

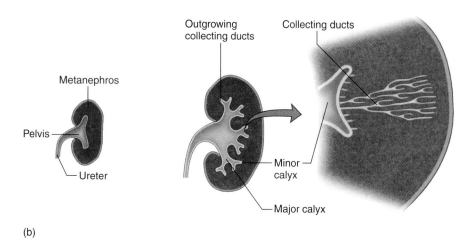

(b)

Embryonic Development of the Urinary Tract
Figure 25.11

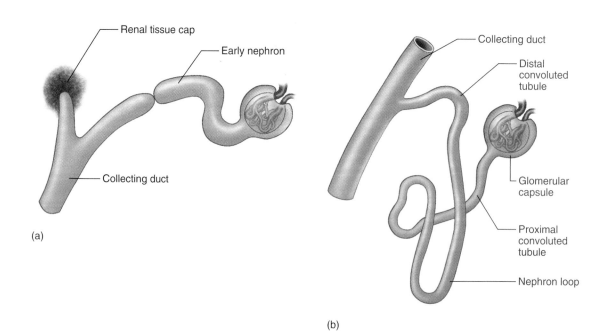

(a)

(b)

Embryonic Development of the Nephron
Figure 25.12

446

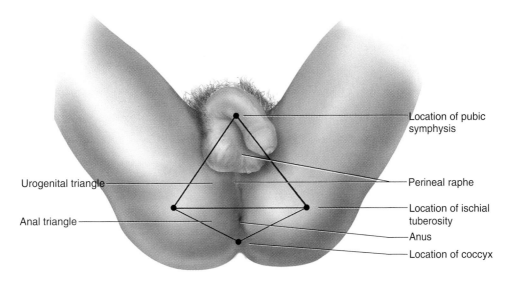

Location of pubic
symphysis

Urogenital triangle

Anal triangle

Perineal raphe

Location of ischial
tuberosity

Anus

Location of coccyx

The Male Perineum
Figure 26.1

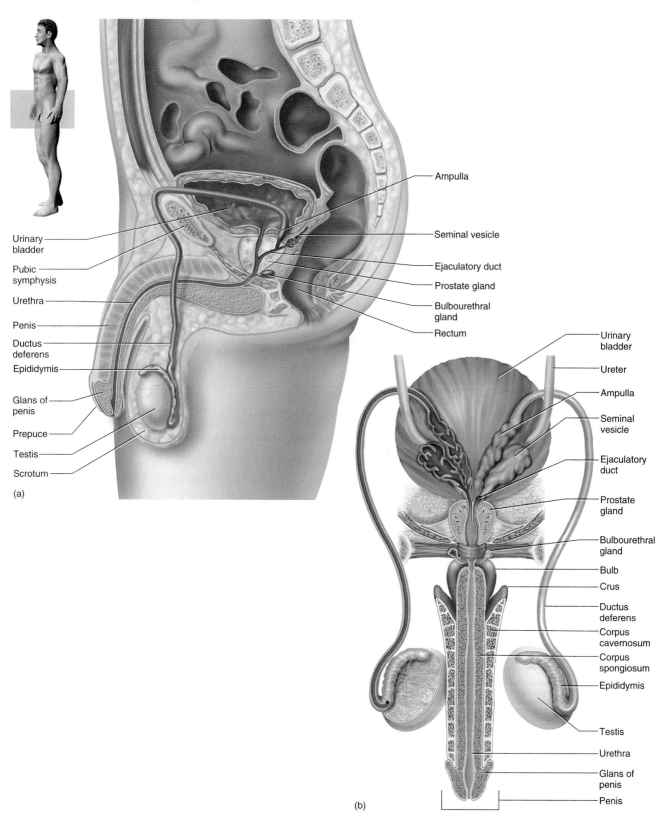

Urinary
bladder

Pubic
symphysis

Urethra

Penis

Ductus
deferens

Epididymis

Glans of
penis

Prepuce

Testis

Scrotum

(a)

Ampulla

Seminal vesicle

Ejaculatory duct

Prostate gland

Bulbourethral
gland

Rectum

Urinary
bladder

Ureter

Ampulla

Seminal
vesicle

Ejaculatory
duct

Prostate
gland

Bulbourethral
gland

Bulb

Crus

Ductus
deferens

Corpus
cavernosum

Corpus
spongiosum

Epididymis

Testis

Urethra

Glans of
penis

Penis

(b)

The Male Reproductive System
Figure 26.2

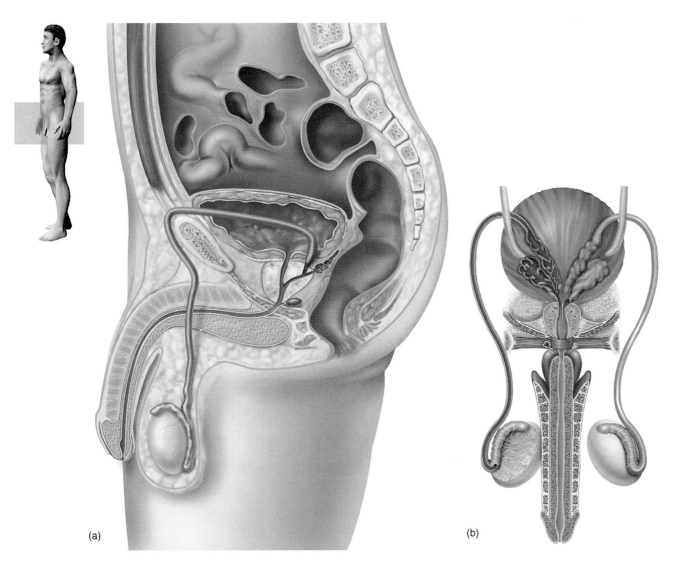

(a)

(b)

The Male Reproductive System
Figure 26.2

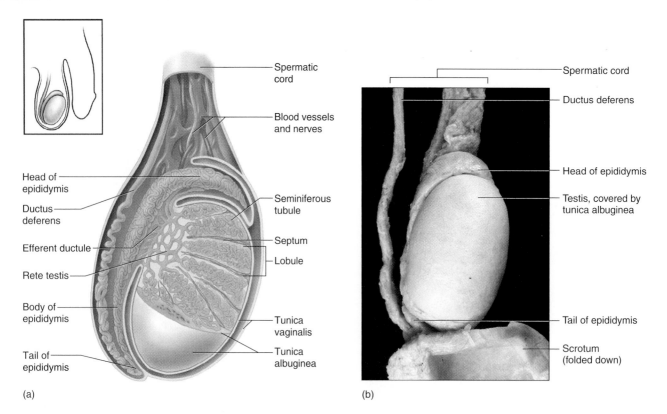

(a)

(b)

The Testis and Associated Structures
Figure 26.3

b: © The McGraw-Hill Companies, Inc./Dennis Strete, photographer

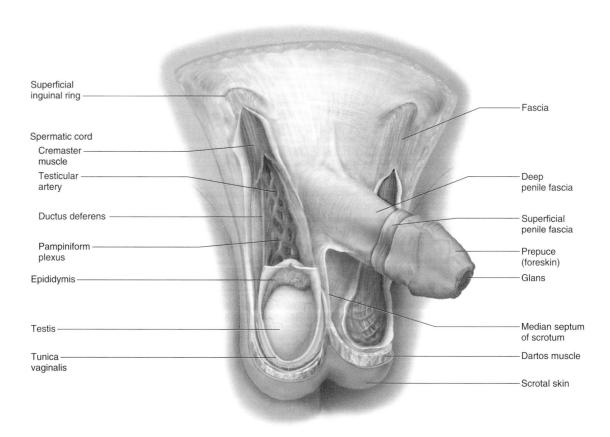

Anatomy of the Male Inguinal Region and External Genitalia
Figure 26.5

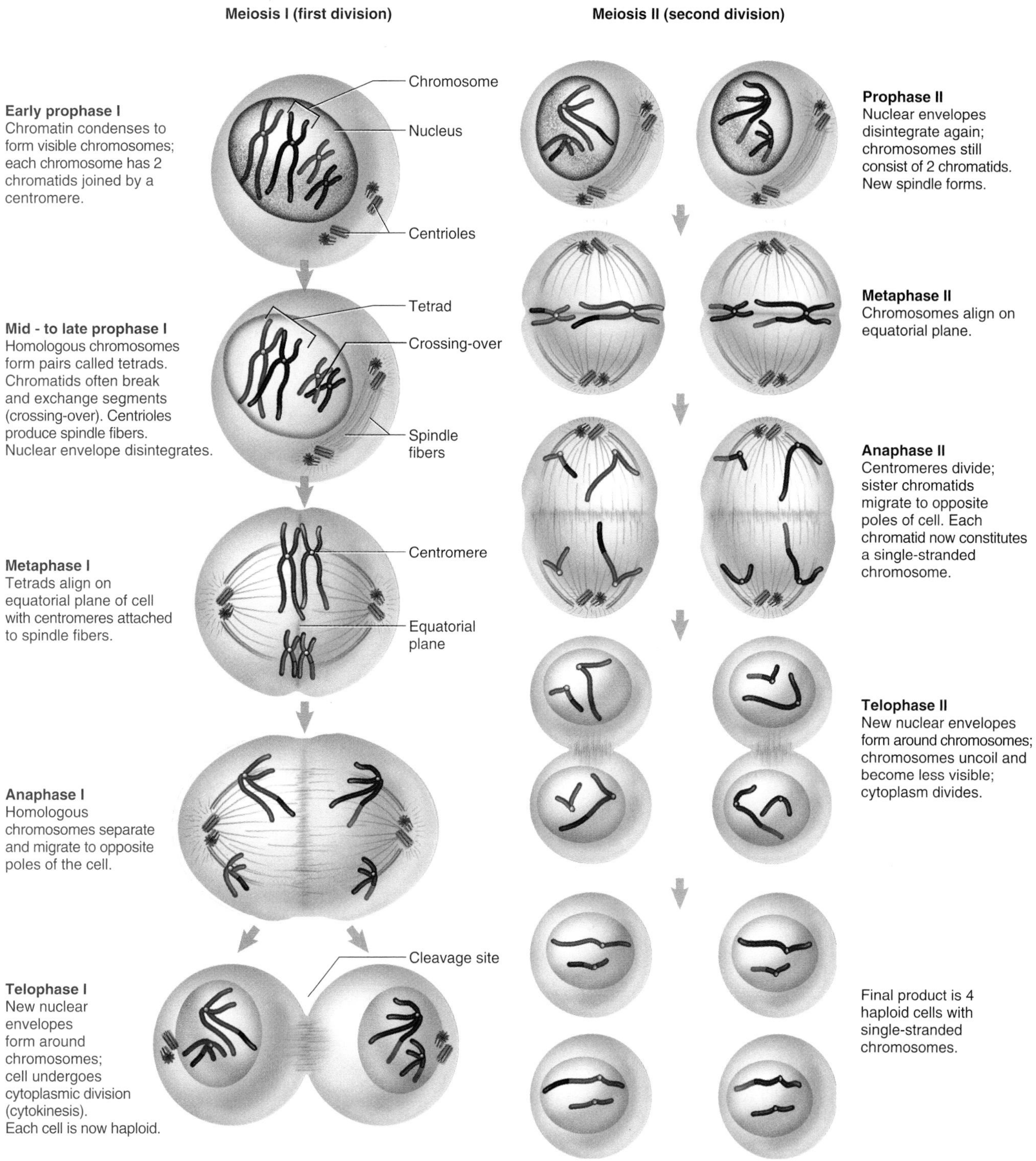

Meiosis I (first division)

Early prophase I
Chromatin condenses to form visible chromosomes; each chromosome has 2 chromatids joined by a centromere.

Mid - to late prophase I
Homologous chromosomes form pairs called tetrads. Chromatids often break and exchange segments (crossing-over). Centrioles produce spindle fibers. Nuclear envelope disintegrates.

Metaphase I
Tetrads align on equatorial plane of cell with centromeres attached to spindle fibers.

Anaphase I
Homologous chromosomes separate and migrate to opposite poles of the cell.

Telophase I
New nuclear envelopes form around chromosomes; cell undergoes cytoplasmic division (cytokinesis). Each cell is now haploid.

Chromosome
Nucleus
Centrioles
Tetrad
Crossing-over
Spindle fibers
Centromere
Equatorial plane
Cleavage site

Meiosis II (second division)

Prophase II
Nuclear envelopes disintegrate again; chromosomes still consist of 2 chromatids. New spindle forms.

Metaphase II
Chromosomes align on equatorial plane.

Anaphase II
Centromeres divide; sister chromatids migrate to opposite poles of cell. Each chromatid now constitutes a single-stranded chromosome.

Telophase II
New nuclear envelopes form around chromosomes; chromosomes uncoil and become less visible; cytoplasm divides.

Final product is 4 haploid cells with single-stranded chromosomes.

Meiosis
Figure 26.6

Cross section of
seminiferous tubules

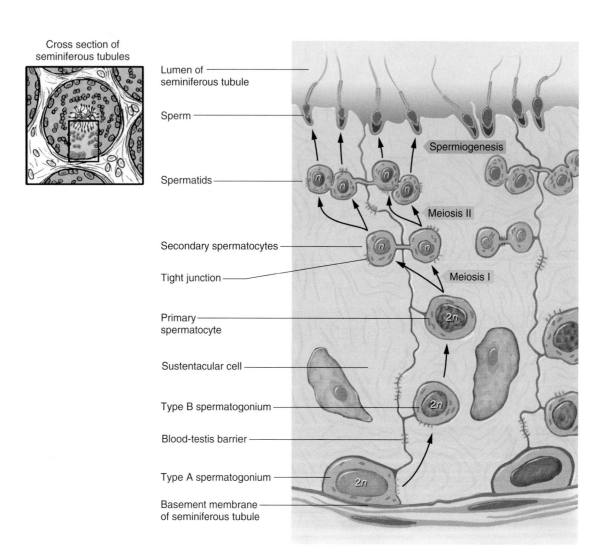

Lumen of
seminiferous tubule

Sperm

Spermiogenesis

Spermatids

Meiosis II

Secondary spermatocytes

Tight junction

Meiosis I

Primary
spermatocyte

Sustentacular cell

Type B spermatogonium

Blood-testis barrier

Type A spermatogonium

Basement membrane
of seminiferous tubule

Spermatogenesis
Figure 26.7

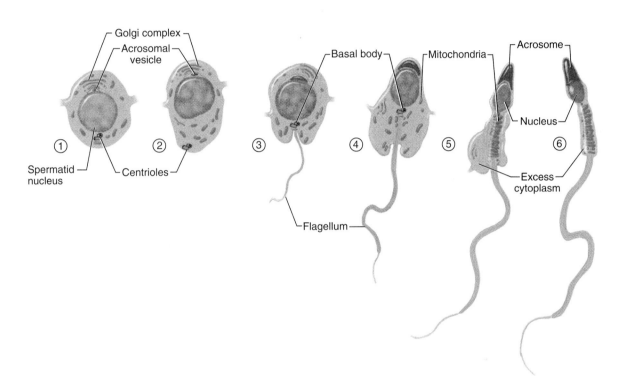

Spermiogenesis
Figure 26.8

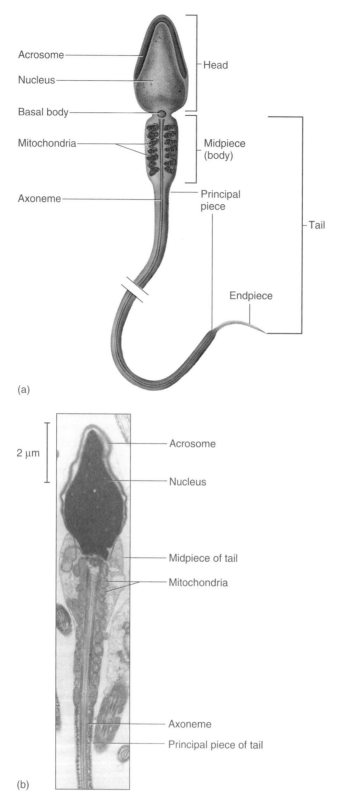

(a)

Acrosome

Nucleus

Basal body

Mitochondria

Axoneme

Head

Midpiece
(body)

Principal
piece

Tail

Endpiece

2 μm

Acrosome

Nucleus

Midpiece of tail

Mitochondria

Axoneme

Principal piece of tail

(b)

The Mature Spermatozoon
Figure 26.9

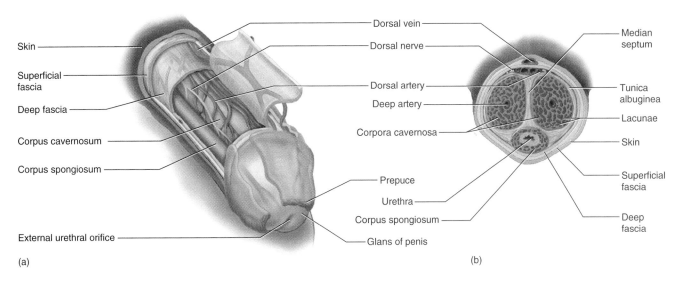

Skin

Superficial fascia

Deep fascia

Corpus cavernosum

Corpus spongiosum

External urethral orifice

Dorsal vein

Dorsal nerve

Dorsal artery

Deep artery

Corpora cavernosa

Prepuce

Urethra

Corpus spongiosum

Glans of penis

Median septum

Tunica albuginea

Lacunae

Skin

Superficial fascia

Deep fascia

(a)

(b)

Anatomy of the Penis

Figure 26.10

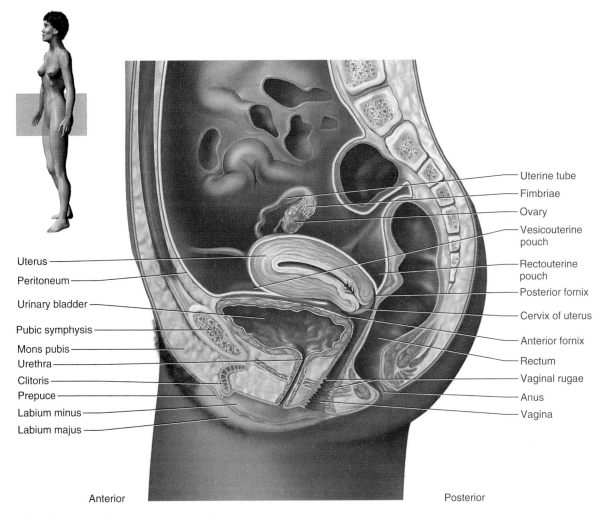

Uterus

Peritoneum

Urinary bladder

Pubic symphysis

Mons pubis

Urethra

Clitoris

Prepuce

Labium minus

Labium majus

Uterine tube

Fimbriae

Ovary

Vesicouterine pouch

Rectouterine pouch

Posterior fornix

Cervix of uterus

Anterior fornix

Rectum

Vaginal rugae

Anus

Vagina

Anterior

Posterior

The Female Reproductive System

Figure 26.11

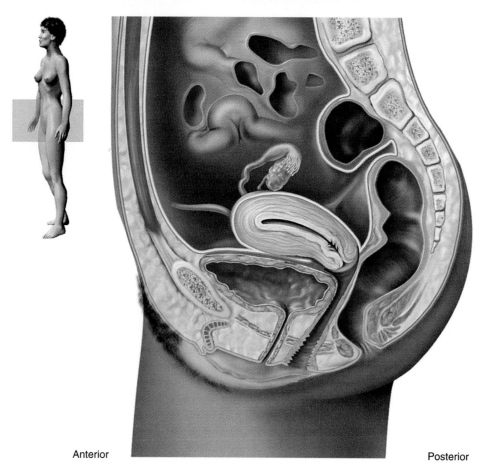

Anterior   Posterior

The Female Reproductive System
Figure 26.11

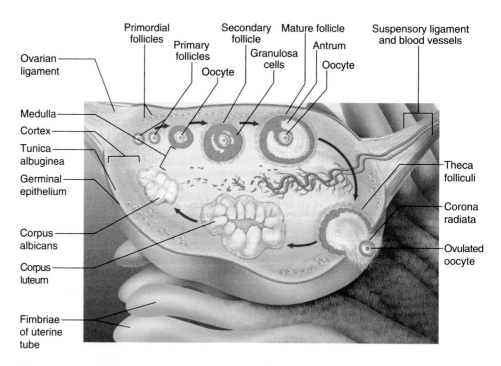

Ovarian
ligament

Primordial
follicles
Primary
follicles
Oocyte

Secondary
follicle
Granulosa
cells

Mature follicle
Antrum
Oocyte

Suspensory ligament
and blood vessels

Medulla
Cortex
Tunica
albuginea
Germinal
epithelium

Corpus
albicans
Corpus
luteum

Theca
folliculi

Corona
radiata

Ovulated
oocyte

Fimbriae
of uterine
tube

**Structure of the Ovary and the Developmental Sequence
of the Ovarian Follicles**
Figure 26.12

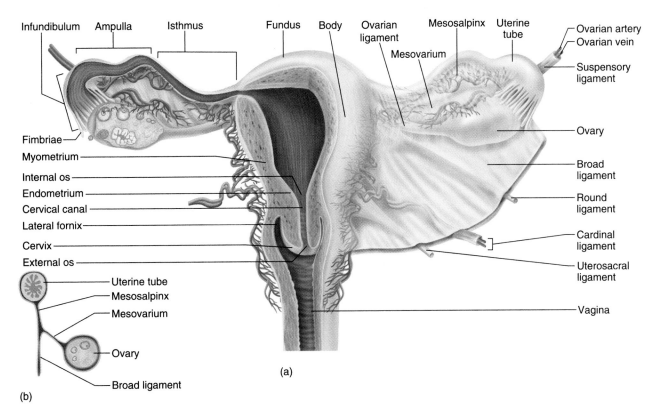

Infundibulum Ampulla Isthmus Fundus Body Ovarian ligament Mesosalpinx Uterine tube Ovarian artery Ovarian vein

Mesovarium

Suspensory ligament

Fimbriae

Myometrium

Internal os

Endometrium

Cervical canal

Lateral fornix

Cervix

External os

Ovary

Broad ligament

Round ligament

Cardinal ligament

Uterosacral ligament

Uterine tube

Mesosalpinx

Mesovarium

Ovary

Broad ligament

Vagina

(a)

(b)

The Female Reproductive Tract and Supportive Ligaments
Figure 26.13

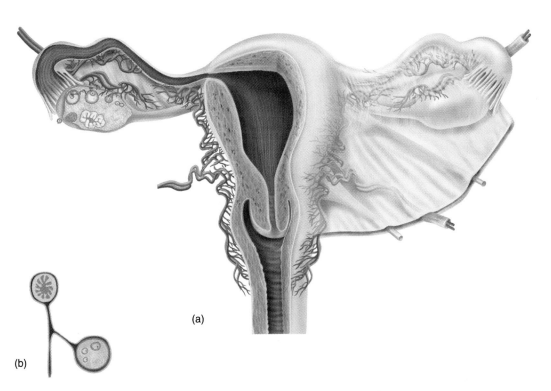

(a)

(b)

The Female Reproductive Tract and Supportive Ligaments
Figure 26.13

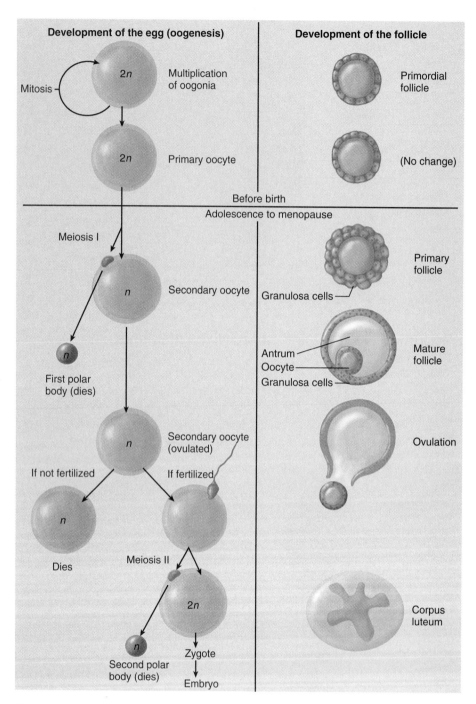

Oogenesis *(left)* and Corresponding Development of the Follicle *(right)*
Figure 26.14

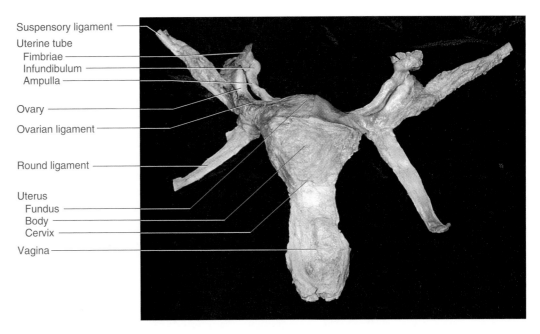

Dissection of the Female Reproductive Tract
Figure 26.17

© The McGraw-Hill Companies, Inc./Rebecca Gray, photographer/Don Kincaid, dissections

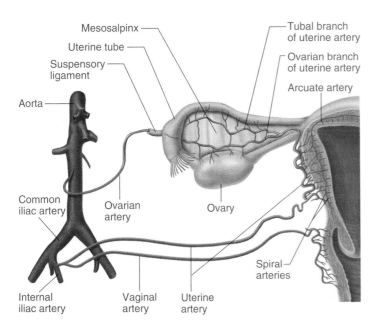

Blood Supply to the Female Reproductive Tract
Figure 26.20

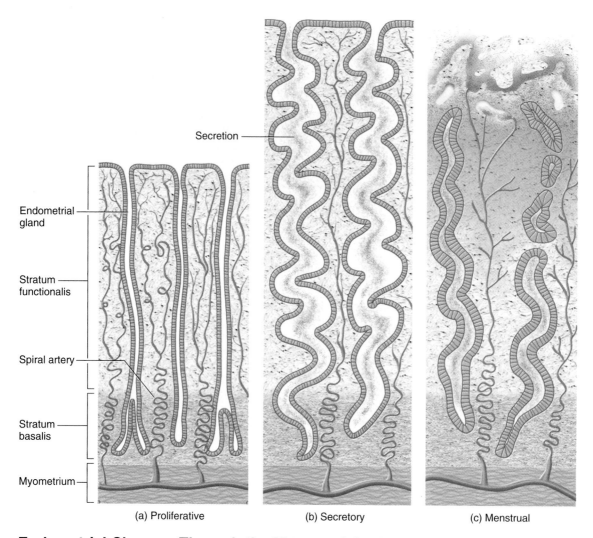

Secretion

Endometrial
gland

Stratum
functionalis

Spiral artery

Stratum
basalis

Myometrium

(a) Proliferative   (b) Secretory   (c) Menstrual

Endometrial Changes Through the Menstrual Cycle
Figure 26.21

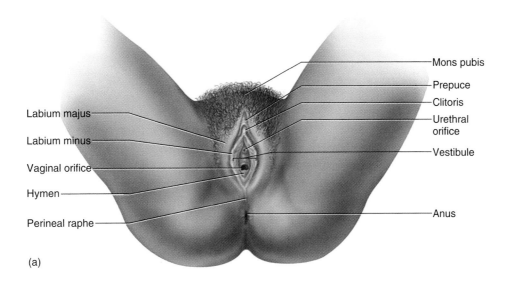

(a)

Mons pubis

Prepuce

Clitoris

Urethral orifice

Vestibule

Labium majus

Labium minus

Vaginal orifice

Hymen

Perineal raphe

Anus

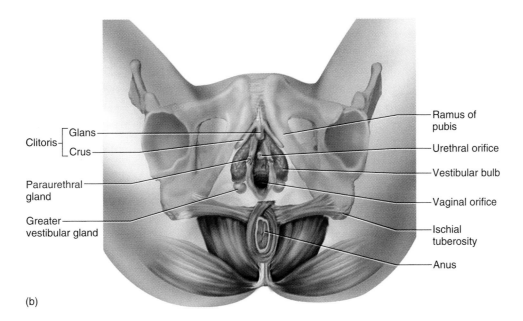

(b)

Clitoris — Glans
         — Crus

Paraurethral gland

Greater vestibular gland

Ramus of pubis

Urethral orifice

Vestibular bulb

Vaginal orifice

Ischial tuberosity

Anus

The Female Perineum
Figure 26.22

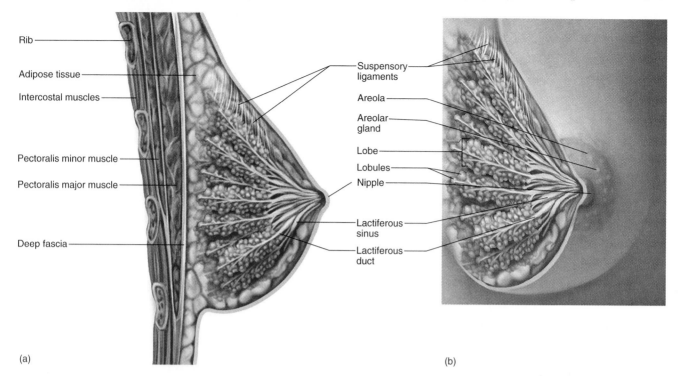

Anatomy of the Lactating Breast
Figure 26.23

(a)

(b)

Labels for (a):
Rib
Adipose tissue
Intercostal muscles
Pectoralis minor muscle
Pectoralis major muscle
Deep fascia

Labels for (b):
Suspensory ligaments
Areola
Areolar gland
Lobe
Lobules
Nipple
Lactiferous sinus
Lactiferous duct

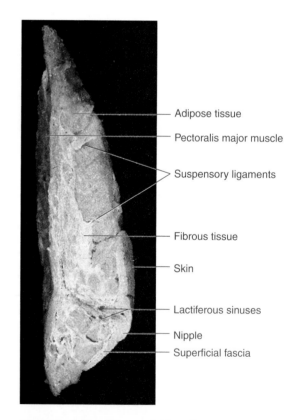

Sagittal Section of the Breast of a Cadaver
Figure 26.24

Labels:
Adipose tissue
Pectoralis major muscle
Suspensory ligaments
Fibrous tissue
Skin
Lactiferous sinuses
Nipple
Superficial fascia

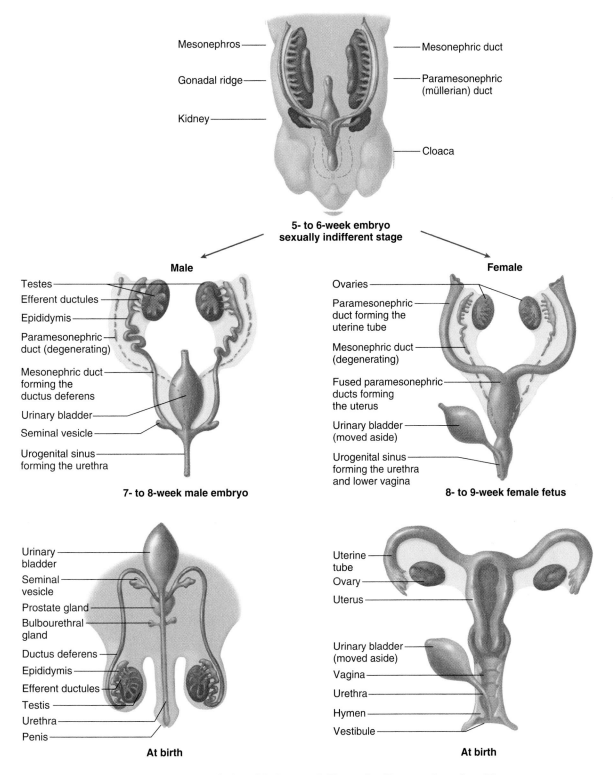

Mesonephros ——

Gonadal ridge ——

Kidney ——

—— **Mesonephric duct**

—— **Paramesonephric (müllerian) duct**

—— **Cloaca**

**5- to 6-week embryo
sexually indifferent stage**

Male

Testes ——
Efferent ductules ——
Epididymis ——
**Paramesonephric
duct (degenerating)** ——
**Mesonephric duct
forming the
ductus deferens** ——
Urinary bladder ——
Seminal vesicle ——
**Urogenital sinus
forming the urethra** ——

7- to 8-week male embryo

Female

Ovaries ——
**Paramesonephric
duct forming the
uterine tube** ——
**Mesonephric duct
(degenerating)** ——
**Fused paramesonephric
ducts forming
the uterus** ——
**Urinary bladder
(moved aside)** ——
**Urogenital sinus
forming the urethra
and lower vagina** ——

8- to 9-week female fetus

**Urinary
bladder** ——
**Seminal
vesicle** ——
Prostate gland ——
**Bulbourethral
gland** ——
Ductus deferens ——
Epididymis ——
Efferent ductules ——
Testis ——
Urethra ——
Penis ——

At birth

**Uterine
tube** ——
Ovary ——
Uterus ——
**Urinary bladder
(moved aside)** ——
Vagina ——
Urethra ——
Hymen ——
Vestibule ——

At birth

Embryonic Development of the Male and Female Reproductive Tracts
Figure 26.26

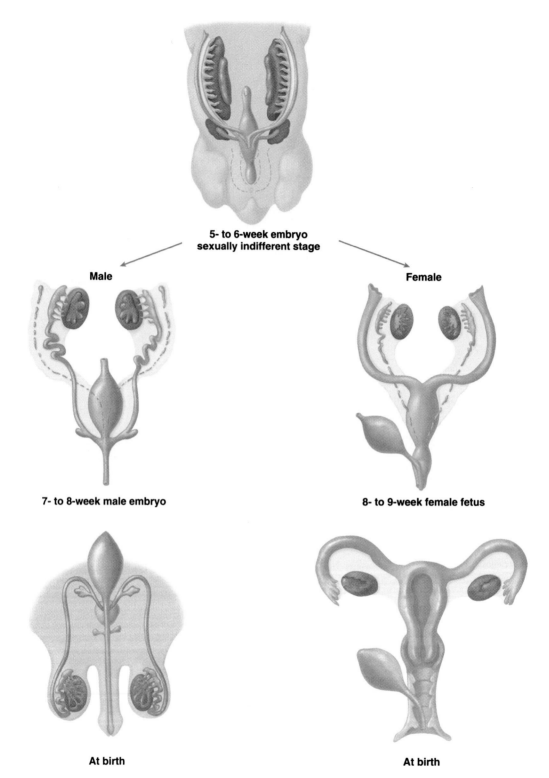

**5- to 6-week embryo
sexually indifferent stage**

Male

Female

7- to 8-week male embryo

8- to 9-week female fetus

At birth

At birth

**Embryonic Development of the Male and Female
Reproductive Tracts**
Figure 26.26

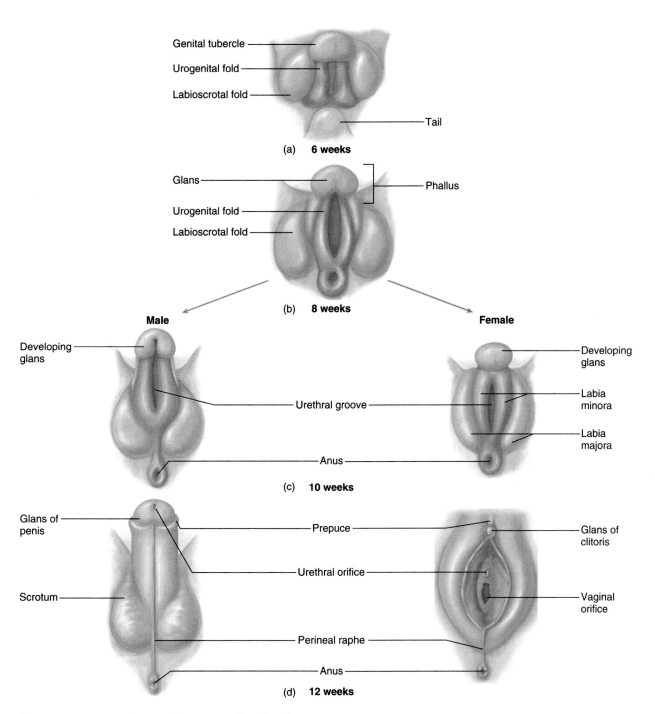

Development of the External Genitalia
Figure 26.27

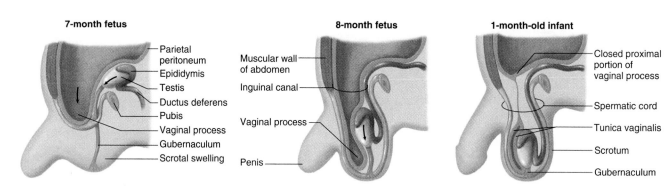

7-month fetus

- Parietal peritoneum
- Epididymis
- Testis
- Ductus deferens
- Pubis
- Vaginal process
- Gubernaculum
- Scrotal swelling

8-month fetus

- Muscular wall of abdomen
- Inguinal canal
- Vaginal process
- Penis

1-month-old infant

- Closed proximal portion of vaginal process
- Spermatic cord
- Tunica vaginalis
- Scrotum
- Gubernaculum

Descent of the Testis
Figure 26.28